BARRON'S

The Leader in Test Preparation

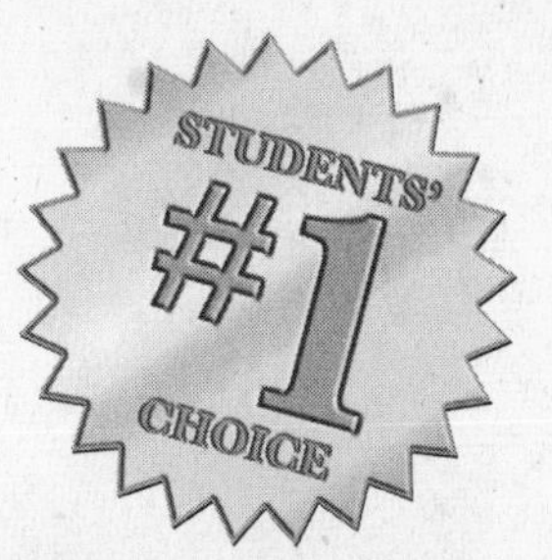

SAT® SUBJECT TEST

PHYSICS

9TH EDITION

Herman Gewirtz
Formerly, Bronx High School of Science, Bronx, NY

and

Jonathan S. Wolf, M.A., Ed.M.
Scarsdale High School, Scarsdale, NY

All inquiries should be addressed to:
Barron's Educational Series, Inc.
250 Wireless Boulevard
Hauppauge, New York 11788
www.barronseduc.com

ISBN-13: 978-0-7641-3663-4
ISBN-10: 0-7641-3663-1

Library of Congress Control Number: 2006029959

Library of Congress Cataloging-in-Publication Data

Gewirtz, Herman.
Barron's SAT subject test : Physics.—9th ed. / Herman Gewirtz and Jonathan S. Wolf.
p. cm.
Includes index.
ISBN-13: 978-0-7641-3663-4
ISBN-10: 0-7641-3663-1
1. Physics—Examinations, questions, etc. 2. SAT (Educational test)—Study guides. 3. Universities and colleges—United States—Entrance requirements—Study guides. 4. Universities and colleges—United States—Examinations—Study guides. I. Wolf, Jonathan S. II. Barron's Educational Series, inc. III. Title.

QC32.G445 2007
530.076—dc22 2006029959

PRINTED IN THE UNITED STATES OF AMERICA
9 8

Contents

Preface

This ninth edition of Barron's *SAT Subject Test Physics* reflects changes that have taken place in both the examination and the curriculum since the release of the eighth edition.

The book begins with an introduction, which sets the stage for your review. In it, you will learn about the structure of the SAT subject test in physics, methods for solving physics problems, and some general information about the scoring of the test. This introduction is followed by a diagnostic test (with full answers and explanations) that you can use as a preliminary assessment. After taking this diagnostic test, refer to the appropriate review chapters for additional information. The remainder of the book contains an extensive review of the material covered in a typical high school physics class. Challenging questions are identified with the icon . You may need to use a calculator for some of these questions. *However, you may NOT use a calculator for the practice tests or the actual SAT subject test.*

Additional practice tests, with answers and explanations, are provided. These practice tests are comparable to the actual test. The appendices at the back of the book assist your review by discussing background math skills as well as providing useful charts, formulas, and tables. There is also a glossary of physics-related terms.

It is both an honor and a pleasure to assist in the preparation of this latest edition, and I would like to thank several people. Linda Turner, Senior Editor at Barron's, has always been helpful to me with her guidance for all of my projects. My colleagues Bob Draper, Pat Jablonowski, and Joe Vaughan have been generous with their constant encouragement and professional advice in regard to both content and pedagogy. Finally, I thank my wife, Karen, as well as my daughters, Marissa and Ilana, for their understanding, love, and support.

Jonathan S. Wolf
August 2006

Introduction

Preliminary

Before we discuss examinations, you should make sure that you have a copy of the College Board pamphlet: *SAT Subject Tests Preparation Booklet.* If you don't have them, you should be able to obtain them from your high school advisor or guidance office. If not, write to

College Board SAT Program
P.O. Box 6200
Princeton, NJ 08541-6200

There is also a College Board Internet site at *www.collegeboard.com.*

In the pamphlet, start reading the material at the beginning, such as Planning to Take the Tests, How to Register, How to Prepare for the Tests, and The Day Before the Tests.

Then turn to the section on the Physics Subject Test, which describes the test and includes sample questions and answers. Because the instructions for the test may have changed, be sure to read them carefully. This will save you time on the actual test.

The Examination Contents

The College Board does not publish copies of former examinations each year. The physics examination is made up annually by a group of experts who are guided by a knowledge of what is commonly taught throughout the country. You will, therefore, be well prepared for the test if you know and understand what is taught in a good secondary school course in physics and if you get some practice in the types of questions used. It is the aim of this book to help you in both areas.

The questions on the physics test are based on the large subject-matter areas of mechanics; optics and waves; electricity and magnetism; heat, kinetic theory, and thermodynamics; and modern physics. Some of the questions ask for mere recall of knowledge. Other questions are designed to see if you really understand concepts and principles, if you can reason quantitatively, and if you can apply scientific concepts and principles to familiar and unfamiliar situations. Some questions involve more than one physical relationship. All questions are in the multiple-choice format, requiring you to choose the best answer from among the five choices given. The practice tests at the end of this book reflect the subject-matter contents and the question types you will encounter on the actual test.

The percentages of topics covered on a given exam can vary from year to year. Approximate average percentages are as follows:

Mechanics—35%
Electricity and magnetism—20%
Waves (includes sound and optics)—20%
Heat and thermodynamics—10%
Modern physics—10%
Miscellaneous—5%

Additionally, some questions will ask for a simple recall of information, while others may require the use of several concepts in a multistep solution.

How to Solve Physics Problems

Have you ever tried to solve a physics problem only to quit in frustration because you didn't know how to get started? Many students find themselves in this situation. In physics, success and achievement require more than just being able to memorize and use formulas, and there is also more to mastering physics than just learning how to solve problems. First, you must understand what the problem is asking you to do. Then, you must access from your memory all the information you feel is related to the ideas being discussed in the problem. Next, you must determine which information is relevant to the problem. Finally, you must decide on a solution path that will hopefully lead you to the correct answer.

A **problem** can be defined as a situation in which you want to achieve a goal but are unsure how to go about it. There are many elements in a problem presented to you. Some of this information is explicit and some may be implicit. Additionally, the goal of the problem may be implicit. For example, to answer a "yes or no" problem, you must first ascertain which quantity or quantities need to be determined before you can answer in the affirmative or negative. Effective studying and reviewing means that you must develop an instinct for certain familiar problem-solving types and techniques.

The first element of a problem consists of the **givens**. This is information that is explicitly provided in the problem statement. As you read a problem, the words that are associated with concepts begin to access information from your memory. What kind of information is retrieved and the form that recollection takes, depend on what you already know about the subject, the type of problem, your experience, and your expertise.

"Difficult" problems are difficult because sometimes the information given is implicit or not well defined. If you are used to solving certain types of problems, you come to expect that a problem will look and read in a certain way. Even so, identifying the givens is the first step after accepting the problem in your mind; that is, after you initially read the problem, you begin to form a mental representation of it. This representation may start with a redescription of the problem (perhaps as a mental image or model of the situation), followed by a translation of the problem into symbols that link concepts already known with the given information.

The representation may be similar to a map, in which associated knowledge is linked by "operators" that tell your brain how to deal with particular concepts using formulas or learned "rules of thumb" for problem solving. **Problem solving** consists of the mental and behavioral operations necessary to reach the goal of a solution. When you begin working on a problem, you rely on stored knowledge about the particular problem, the subject area, and your past problem-solving experience.

The second element in a problem consists of the **obstacles**. These are the factors that prevent you from immediately achieving your goal. In a multiple-choice question, the correct answer is accompanied by a series of "distractors" that may be closely linked to the answer. If the question is not a simple recall of information, you may have to use some of the methods mentioned above. Since you may not use a calculator on the exam and you are not provided with a formula sheet, you must rely on your own memory and problem-solving skills. Deciding on the proper method (or path) for solving the problem most efficiently is the main focus of problem solving as a process.

How to Take the Tests

The examinations at the end of this book are not copies of former tests, but practice with them should be valuable to you. However, be sure to read the instructions given with your actual test. Note the amount of time allotted, and recognize that you may not be able to answer all questions in the available time. No calculators are permitted.

The questions are not of equal difficulty, but each question usually gets the same credit. Do not waste time on questions that seem difficult or time consuming. Go back to them at the end of the test, if you have time. Time can often be saved in numerical examples by making approximate calculations. Don't worry if you can't answer all the questions. Probably no one taking the test can, and completion is not needed for a perfect score. Should you guess? If you know anything about the subject matter of the question and can eliminate some of the choices, it is advisable to guess. A completely random guess, however, may cause you to lose a fraction of a point for an incorrect answer.

Pay attention to units. By doing so, you can catch many mistakes. All problems in this book are in (metric) SI units, which are explained more fully in Chapter 2.

When taking the practice tests in this book, do not stop at the end of 1 hour. Follow the above advice, and try to answer all the questions eventually. Practice keeping track of time without wasting it.

Each of the practice tests includes 75 questions. Some of the questions were deliberately made rather difficult to challenge you. On the actual examination the questions are carefully evaluated, and individual performance is compared with group performance before a score is given.

Also, as cautioned above, note whether the **directions** on your test are the same as those used on the practice tests in this book. *When you take the test, be sure to read the instructions given.*

Your final score on an actual SAT Subject Test in Physics ranges from 200 to 800. These scores are based on a scaling formula developed by ETS at the time of the exam. It is not possible to determine your actual score on a practice exam, but you can determine your raw score and then get an approximate sense of where you might fall on a scaled curve.

To calculate your raw score, count the number of questions correctly answered and then subtract one quarter of the number of incorrect questions:

Raw score = # correct – (1/4) × (# incorrect) = ____________

The second term in the equation is a "guessing penalty." While you should always try to answer as many questions correctly as you can, random guessing will not significantly increase your score. It is always better to eliminate as many choices as possible before even attempting an "educated guess."

Typically, raw scores between 65 and 75 will earn you a scaled score of approximately 800. A raw score around 45 will typically compute to a scaled score of approximately 700. **These scaled-score ranges are only approximations. You should not use them as absolute indicators or predictors of what your own score might be on the actual exam!**

CHAPTER 1

Diagnostic Test

The first of the four practice tests for the SAT Subject Test Physics is to be used as a diagnostic. This means that you should take this exam to assess your level of understanding. On the basis of how well you perform, you may want to organize a review schedule for some of the specific content material in the chapters that follow. If you are already comfortable with the content, view this diagnostic test as just another practice exam. Do not use a calculator.

Here is some helpful information:

1. All questions provide five choices, designated as (A)–(E). Many questions come in sets and are based on the same information.
2. Some questions call for the interpretation of graphs.
3. Sometimes there are several questions on the same topic.
4. In some questions four choices are correct, and you are asked to select the exception. These questions contain a word in capital letters (NOT, EXCEPT).
5. Sometimes the answer choices are given first and are then followed by the question.
6. In some questions with more than one correct answer, you are asked to select a combination of statements that provide the best answer. On the actual test, the individual statements will probably be designated by the Roman numerals I, II, and III, as in this book.
7. It is advisable to consider all the lettered choices before you select your answer.

GOOD LUCK!

Answer Sheet: Diagnostic Test

1. (A) (B) (C) (D) (E)
2. (A) (B) (C) (D) (E)
3. (A) (B) (C) (D) (E)
4. (A) (B) (C) (D) (E)
5. (A) (B) (C) (D) (E)
6. (A) (B) (C) (D) (E)
7. (A) (B) (C) (D) (E)
8. (A) (B) (C) (D) (E)
9. (A) (B) (C) (D) (E)
10. (A) (B) (C) (D) (E)
11. (A) (B) (C) (D) (E)
12. (A) (B) (C) (D) (E)
13. (A) (B) (C) (D) (E)
14. (A) (B) (C) (D) (E)
15. (A) (B) (C) (D) (E)
16. (A) (B) (C) (D) (E)
17. (A) (B) (C) (D) (E)
18. (A) (B) (C) (D) (E)
19. (A) (B) (C) (D) (E)
20. (A) (B) (C) (D) (E)
21. (A) (B) (C) (D) (E)
22. (A) (B) (C) (D) (E)
23. (A) (B) (C) (D) (E)
24. (A) (B) (C) (D) (E)
25. (A) (B) (C) (D) (E)
26. (A) (B) (C) (D) (E)
27. (A) (B) (C) (D) (E)
28. (A) (B) (C) (D) (E)
29. (A) (B) (C) (D) (E)
30. (A) (B) (C) (D) (E)
31. (A) (B) (C) (D) (E)
32. (A) (B) (C) (D) (E)
33. (A) (B) (C) (D) (E)
34. (A) (B) (C) (D) (E)
35. (A) (B) (C) (D) (E)
36. (A) (B) (C) (D) (E)
37. (A) (B) (C) (D) (E)
38. (A) (B) (C) (D) (E)
39. (A) (B) (C) (D) (E)
40. (A) (B) (C) (D) (E)
41. (A) (B) (C) (D) (E)
42. (A) (B) (C) (D) (E)
43. (A) (B) (C) (D) (E)
44. (A) (B) (C) (D) (E)
45. (A) (B) (C) (D) (E)
46. (A) (B) (C) (D) (E)
47. (A) (B) (C) (D) (E)
48. (A) (B) (C) (D) (E)
49. (A) (B) (C) (D) (E)
50. (A) (B) (C) (D) (E)
51. (A) (B) (C) (D) (E)
52. (A) (B) (C) (D) (E)
53. (A) (B) (C) (D) (E)
54. (A) (B) (C) (D) (E)
55. (A) (B) (C) (D) (E)
56. (A) (B) (C) (D) (E)
57. (A) (B) (C) (D) (E)
58. (A) (B) (C) (D) (E)
59. (A) (B) (C) (D) (E)
60. (A) (B) (C) (D) (E)
61. (A) (B) (C) (D) (E)
62. (A) (B) (C) (D) (E)
63. (A) (B) (C) (D) (E)
64. (A) (B) (C) (D) (E)
65. (A) (B) (C) (D) (E)
66. (A) (B) (C) (D) (E)
67. (A) (B) (C) (D) (E)
68. (A) (B) (C) (D) (E)
69. (A) (B) (C) (D) (E)
70. (A) (B) (C) (D) (E)
71. (A) (B) (C) (D) (E)
72. (A) (B) (C) (D) (E)
73. (A) (B) (C) (D) (E)
74. (A) (B) (C) (D) (E)
75. (A) (B) (C) (D) (E)

DIAGNOSTIC TEST

Part A

Instructions: Read the section in the Introduction entitled How to Take the Tests.

Directions: For each group of questions below, there is a set of five lettered choices, followed by numbered questions. For each question select the one choice in the set that best answers the question and fill in the corresponding oval on the answer sheet. You may use a lettered choice once, more than once, or not at all in each set. Do not use a calculator.

Questions 1–3 refer to the graph below, which represents the speed of an object moving along a straight line. The time of observation is represented by *t*.

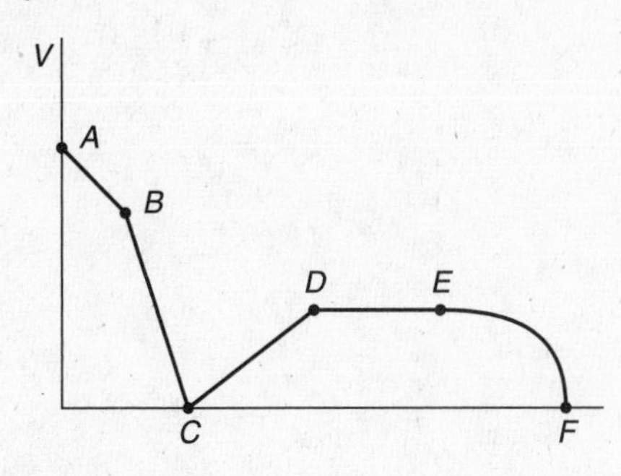

(A) *AB*
(B) *BC*
(C) *CD*
(D) *DE*
(E) *EF*

1. Which choice represents the interval during which the object moves with constant speed?

2. Which choice represents the interval during which the object's speed is increasing?

3. Which choice represents the interval during which the acceleration is changing?

Questions 4–8 refer to the following concepts:

(A) Energy
(B) Power
(C) Momentum
(D) Acceleration
(E) Torque

To which choice is each of the following units most closely related?

4. m/s^2

5. $m \cdot N$

6. W

7. $kg \cdot m/s$

8. J

GO ON TO THE NEXT PAGE ➤

Questions 9–13 may express a relationship to speed or velocity given in these choices:

(A) is proportional to its velocity
(B) is proportional to the square of its speed
(C) is proportional to the square root of its speed
(D) is inversely proportional to its velocity
(E) is not described by any of the above

For each question select the choice that best completes it:

9. The kinetic energy of a given body

10. The acceleration toward the center of an object moving with constant speed around a given circle

11. The momentum of a given body

12. The displacement per second of an object in equilibrium

13. The displacement of an object starting from rest and moving with constant acceleration

Questions 14 and 15 deal with the characteristics of a simple circuit in the following situation:

The following five lengths of thin wire, all of which have the same diameter and length, are connected in a circuit to a battery:

(A) 3 m of nichrome wire
(B) 3 m of copper wire
(C) 3 m of lead wire
(D) 3 m of steel wire
(E) 3 m of iron wire

14. In which length of wire is the current greatest?

15. Which length of wire generates the greatest power?

GO ON TO THE NEXT PAGE ►

Part B

Directions: Each of the questions or incomplete statements is followed by five suggested answers or completions. Select the choice that best answers the question or completes the statement and fill in the corresponding circle on the answer sheet. Do not use a calculator.

16. The resultant of a 3-newton and a 4-newton force that act on an object in opposite directions to each other is, in newtons,

(A) 0
(B) 1
(C) 5
(D) 7
(E) 12

17. Two forces, one of 6 newtons and the other of 8 newtons, act on a point at right angles to each other. The resultant of these forces is, in newtons,

(A) 0
(B) 2
(C) 5
(D) 10
(E) 14

Questions 18–20

As shown in the diagram below, two weights, one of 10 newtons and the other of 6 newtons, are tied to the ends of a flexible string. The string is placed over a pulley that is attached to the ceiling. Frictional losses and the weight of the pulley may be neglected as the weights and the string are allowed to move.

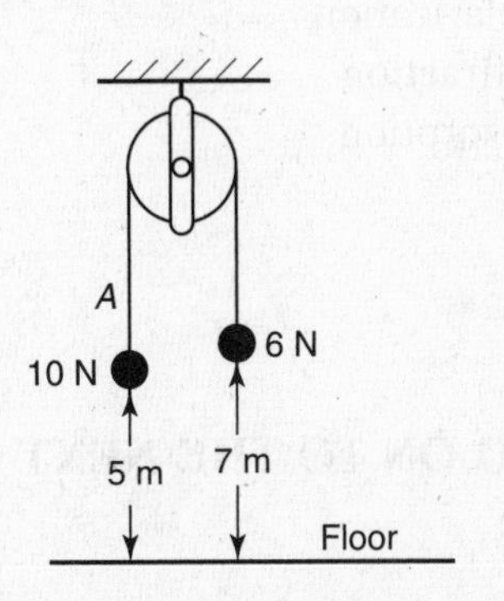

18. At the instant shown in the diagram, the potential energy of the 10-newton object with respect to the floor is, in joules,

(A) 0
(B) 2
(C) 20
(D) 50
(E) 70

19. At the instant shown, the acceleration of the moving 10-newton object is

(A) 0
(B) less than **g**
(C) **g**
(D) 5 **g**
(E) 10 **g**

20. At the instant shown, the tension in rope *A* is

(A) less than 3 N
(B) 3 N
(C) 6 N
(D) more than 6 N but less than 10 N
(E) 10 N

---

21. How many meters will a 2.00-kilogram ball starting from rest fall freely in 1.00 second?

(A) 4.90
(B) 2.00
(C) 9.81
(D) 19.6
(E) 32

GO ON TO THE NEXT PAGE ➤

22. An object is observed to have zero acceleration. Which of the following statements is correct?

(A) The object may be in motion.
(B) The object must be at rest.
(C) The object must have zero net force acting on it.
(D) Both (A) and (C) are correct.
(E) None of the above is correct.

23. A beam of parallel rays is reflected from a smooth plane surface. After reflection the rays will be

(A) converging
(B) diverging
(C) parallel
(D) diffused
(E) focused

Questions 24–28 are based on the following diagram showing the response of 2 kilograms of a gas at 140°C. Heat is released at a rate of 6 kilojoules per minute.

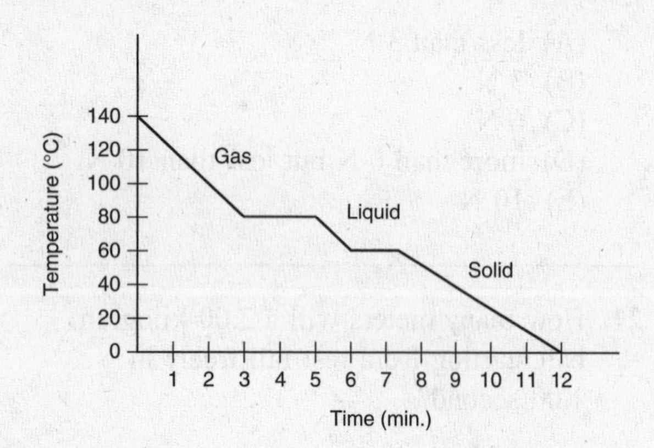

24. The boiling point of this substance is

(A) 0°C
(B) 60°C
(C) 80°C
(D) 140°C
(E) 200°C

25. The melting point of this substance is

(A) 0°C
(B) 60°C
(C) 80°C
(D) 140°C
(E) 200°C

26. The heat of vaporization for this substance is

(A) 80 kJ/kg
(B) 12 kJ/kg
(C) 6 kJ/kg
(D) 2 kJ/kg
(E) 24 kJ/kg

27. The heat of fusion for this substance is

(A) 3 kJ/kg
(B) 6 kJ/kg
(C) 12 kJ/kg
(D) 20 kJ/kg
(E) 24 kJ/kg

28. Compared to the specific heat of the substance in the gas state, the specific heat of the substance in the solid state is

(A) more
(B) less
(C) the same
(D) sometimes more, sometimes less
(E) Not enough information is provided to decide.

---

29. Light passes through two parallel slits and falls on a screen. The pattern produced is due to interference and

(A) reflection
(B) refraction
(C) polarization
(D) diffraction
(E) absorption

GO ON TO THE NEXT PAGE ➤

30. A change in temperature of 50°C is equivalent to a temperature change of

(A) 323 K
(B) 273 K
(C) –273 K
(D) 50 K
(E) 100 K

Questions 31–33 are based on the velocity versus time graph shown below for an object undergoing one-dimensional motion:

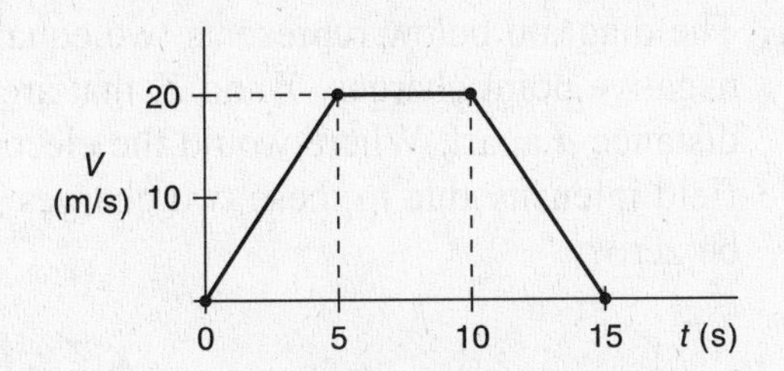

31. Which of the following sketches represents the corresponding displacement versus time graph for this motion?

(A)
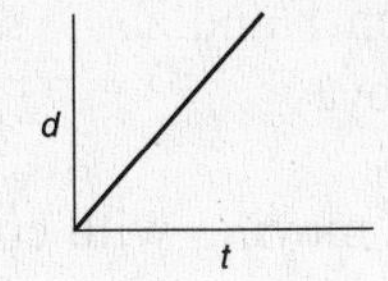

(B)
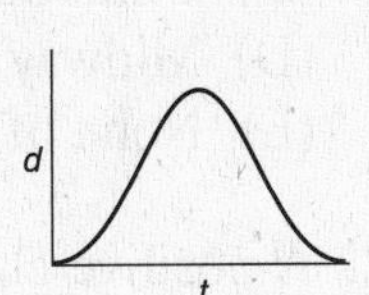

(C)
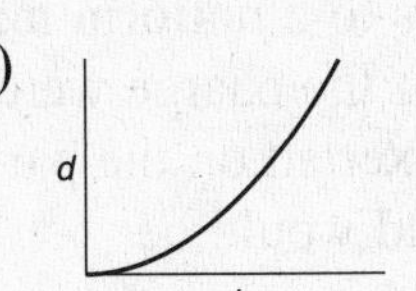

(D)
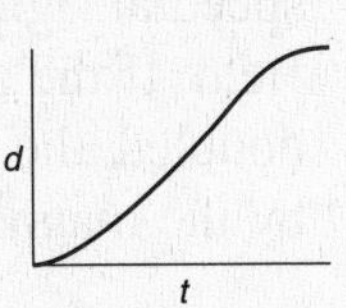

(E)
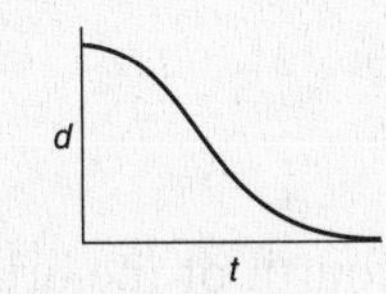

32. What is the total distance traveled by this object during the entire 15-second interval?

(A) 300 m
(B) 250 m
(C) 200 m
(D) 150 m
(E) 100 m

33. The object is initially traveling east. During which interval of time is a net force acting west on the object?

(A) 0–5 s
(B) 5–10 s
(C) 10–15 s
(D) 0–15 s
(E) None of the above

34. The amplitude of a sound wave is detected as its

(A) wavelength
(B) frequency
(C) pitch
(D) resonance
(E) loudness

35. If the velocity of light in a medium depends on its frequency, the medium is said to be

(A) coherent
(B) refractive
(C) dispersive
(D) diffractive
(E) resonant

Questions 36–38 are based on the parallel circuit shown below:

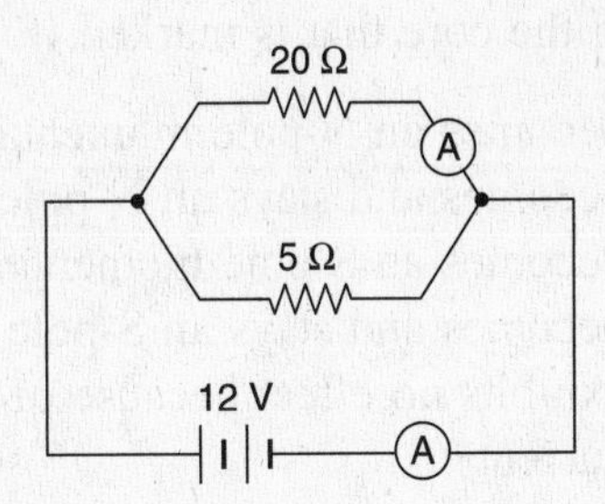

36. What is the value of the equivalent resistance of this circuit?

(A) 25 Ω
(B) 20 Ω
(C) 10 Ω
(D) 5 Ω
(E) 4 Ω

GO ON TO THE NEXT PAGE ➤

37. What is the value of the current flowing in the circuit?

(A) 48 A
(B) 3 A
(C) 12 A
(D) 5 A
(E) 16 A

38. What is the value of the current flowing through the 20-Ω resistor?

(A) 4 A
(B) 12 A
(C) 240 A
(D) 0.6 A
(E) 2.4 A

Questions 39 and 40

X is a coil of copper wire with many turns wound on a soft iron core. Another coil wound on an iron core is near it, as shown below.

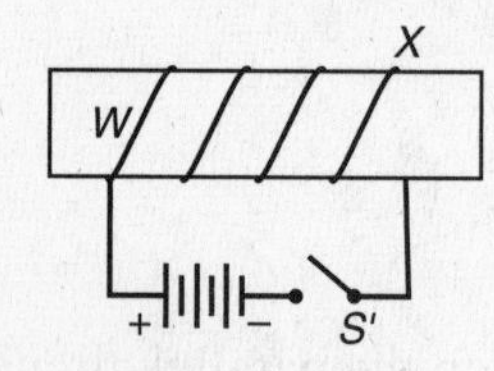

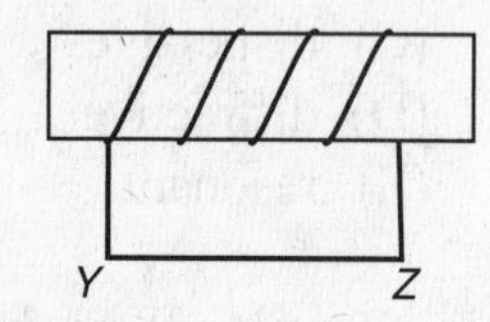

39. If switch S' is closed and kept closed, the end of the core that is marked *W*

(A) becomes an N-pole momentarily
(B) becomes and stays an N-pole
(C) becomes an S-pole momentarily
(D) becomes and stays an S-pole
(E) exhibits no effect because of the current

40. The instant after switch S' is closed

(A) there will be no current in wire *YZ*
(B) conventional current in wire *YZ* will be from *Y* to *Z*
(C) conventional current in wire *YZ* will be from *Z* to *Y*
(D) conventional current in wire *YZ* will be from *Z* to *Y* and then from *Y* to *Z*
(E) conventional current in wire *YZ* will be from *Y* to *Z* and then from *Z* to *Y*

---

41. The diagram below represents two equal negative point charges, *Y* and *Z*, that are a distance *d* apart. Where would the electric field intensity due to these two charges be zero?

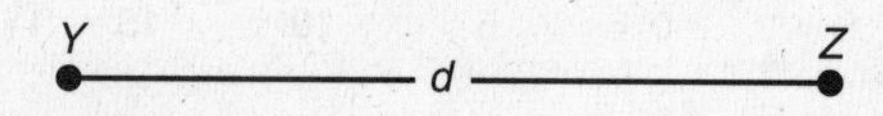

(A) On *Y*
(B) On *Z*
(C) On both *Y* and *Z*
(D) Midway between *Y* and *Z*
(E) None of the above

42. A positive charge is moving with constant speed at right angles to a uniform magnetic field. If the speed of the charge were doubled, the force exerted on the particle by the magnetic field would be

(A) unaffected
(B) quadrupled
(C) doubled
(D) halved
(E) reduced to one fourth of the original value

43. Of the following, the particle whose mass is closest to that of the neutron is the

(A) meson
(B) deuteron
(C) neutrino
(D) proton
(E) positron

GO ON TO THE NEXT PAGE ►

44. The time of one vibration of a simple pendulum may be decreased by

(A) increasing the length of the pendulum
(B) decreasing the length of the pendulum
(C) using a heavier bob
(D) using a lighter bob
(E) taking the pendulum up to the top of the Empire State Building

45. The gravitational force between two masses is equal to 36 newtons. If the distance between them is tripled, then the force of gravity will be

(A) 36 N
(B) 18 N
(C) 9 N
(D) 4N
(E) 27 N

46. A planet is orbiting a star in an elliptical orbit. On the basis of Kepler's second law, which of the following statements will be correct as the planet gets closer to the star?

(A) The kinetic energy increases, while the potential energy decreases.
(B) The kinetic energy decreases, while the potential energy increases.
(C) The sum of the kinetic and potential energies remain constant throughout the orbit.
(D) Both (A) and (C) are correct.
(E) Both (B) and (C) are correct.

47. It is certain that a rod is electrically charged if it

(A) repels a pith ball
(B) attracts a pith ball
(C) attracts the N-pole of a compass needle
(D) repels the N-pole of a compass needle
(E) points north

48. The rate of heat production of a wire immersed in ice water and carrying an electric current is proportional to

(A) the current
(B) the reciprocal of the current
(C) the reciprocal of the square of the current
(D) the square of the current
(E) the square root of the current

49. According to electromagnetic theory, Lenz's law can be explained best by which of the following laws of physics?

(A) Law of conservation of linear momentum
(B) Law of conservation of angular momentum
(C) Law of conservation of energy
(D) Law of universal gravitation
(E) None of the above

50. A planet has half the mass of Earth and half the radius. Compared to the acceleration due to gravity at the surface of Earth, the acceleration due to gravity at the surface of this other planet is

(A) the same
(B) halved
(C) doubled
(D) quartered
(E) quadrupled

51. Boyle's law describes the behavior of a gas when

(A) its pressure is kept constant
(B) its volume is kept constant
(C) its density is kept constant
(D) its mass is kept constant
(E) nothing is kept constant

DIAGNOSTIC TEST

GO ON TO THE NEXT PAGE ➤

52. Electrical appliances are usually grounded in order to

(A) maintain a balanced charge distribution
(B) prevent a buildup of heat
(C) run properly using household electricity
(D) prevent a buildup of static charges
(E) prevent an overload in the circuit

53. The force acting on a satellite in circular orbit around Earth is chiefly

(A) the satellite's inertia
(B) the satellite's mass
(C) Earth's mass
(D) Earth's gravitational pull
(E) the Sun's gravitational pull

54. An object that is black

(A) absorbs black light
(B) reflects black light
(C) absorbs all light
(D) reflects all light
(E) refracts all light

55. If the intensity of monochromatic light is increased while incident on a pair of narrow slits in a diffraction experiment, the spacing between maxima in the pattern will

(A) increase
(B) decrease
(C) remain the same
(D) increase or decrease depending on frequency
(E) not enough information provided

56. A 60-watt, 110-volt tungsten filament lamp is operated on 120 volts. Which of the following statements regarding the lamp is (are) true?

I. It will consume more than 60 watts while operating.
II. It will have a lower resistance than at 110 volts.
III. It will be brighter than at 110 volts.
IV. It will burn out after operating ½ hour or less.

(A) I, II, and III only
(B) I and III only
(C) II and IV only
(D) IV only
(E) None of the above

57. A man standing in an elevator is taken up by the elevator at constant speed. Which of the following is (are) true of the push that the man exerts on the floor of the elevator?

I. It is equal to his weight.
II. It is equal to less than his weight.
III. It is equal to more than his weight.
IV. It is dependent on the value of the constant speed.

(A) I, II, and III only
(B) I and III only
(C) II and IV only
(D) IV only
(E) None of the above

GO ON TO THE NEXT PAGE ➤

58. An object with a constant mass rests on a smooth and perfectly horizontal table. If a horizontal force **F** is applied, acceleration **a** results. If **F** is doubled without changing the direction, what will be the effect(s) on the acceleration?

I. The acceleration will remain the same.
II. The acceleration will be doubled
III. The acceleration will decrease.
IV. The acceleration will increase but not double.

(A) I, II, and III only
(B) II only
(C) II and IV only
(D) IV only
(E) None of the above

59. A projectile is launched from level ground with a velocity of 100 m/s at a 30° angle to the horizontal. If air resistance is neglected, approximately how high will the projectile rise?

(A) 50 m
(B) 250 m
(C) 125 m
(D) 98 m
(E) 500 m

60. A man pulls an object up an inclined plane with a force *F* and notes that the object's acceleration is 5 meters per second squared. He doubles the force without changing its direction. Which of the following will then be true of the acceleration?

I. It decreases.
II. It increases.
III. It remains the same.
IV. It is doubled.

(A) I, II, and III only
(B) I and III only
(C) II and IV only
(D) IV only
(E) None of the above

61. An object with mass *m* is moving at constant velocity **v**, in a horizontal circle with radius *r*.

I. The angular momentum is directly proportional to both **v** and *r*.
II. The kinetic energy varies inversely with $\mathbf{v}^2$.
III. The centripetal acceleration varies directly with *m*.

Which of the above statements is (are) correct?

(A) I only
(B) I and II
(C) I, II, and III
(D) II and III
(E) All of the above

62. Assume that you have two balls of identical volume, one weighing 2 newtons and the other 10 newtons. Both are falling freely after being released from the same point simultaneously. Which of the following will be true?

I. The 10-N ball falling freely from rest will be accelerated at a greater rate than the 2-N ball.
II. At the end of 4 s of free fall, the 10-N ball will have 5 times the momentum of the 2-N ball.
III. At the end of 4 s of free fall, the 10-N ball will have the same kinetic energy as the 2-N ball.
IV. The 10-N ball possesses greater inertia than the 2-N ball.

(A) I, II, and III only
(B) I and III only
(C) II and IV only
(D) IV only
(E) None of the above

DIAGNOSTIC TEST

GO ON TO THE NEXT PAGE ►

63. Lighted candle *X*, shown below is placed 20 centimeters from *Y*. An observer places her eye 45 centimeters on the other side of *Y* and, looking toward *X*, sees an image of *X*.

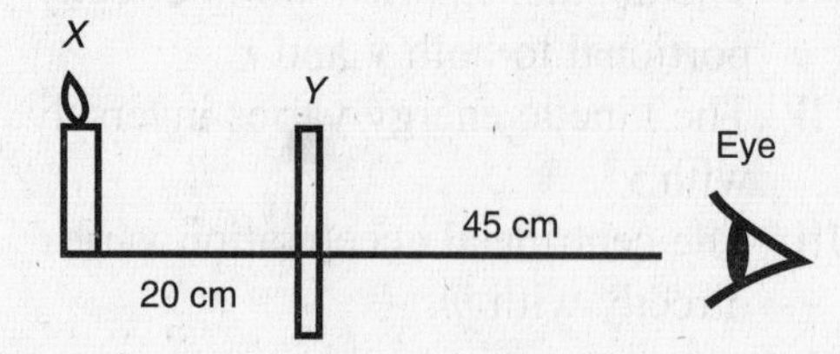

Which of the following may object *Y* be?

I. A new convex mirror of 10-cm focal length of the type often used in the laboratory.
II. A new convex lens of 10-cm focal length of the type often used in the laboratory.
III. A new concave mirror of 10-cm focal length of the type often used in the laboratory.
IV. A new concave lens of 10-cm focal length of the type often used in the laboratory.

(A) I, II, and III only
(B) I and III only
(C) II and IV only
(D) IV only
(E) None of the above

64. A lens is used to produce a sharp image on a screen. When the right half of the lens is covered with an opaque material, how will the image be affected?

I. The right half of the image will disappear.
II. The left half of the image will disappear.
III. The image size will become approximately half of the original size.
IV. The image brightness will become approximately half of the original brightness.

(A) I, II, and III only
(B) I and III only
(C) II and IV only
(D) IV only
(E) None of the above

65. When a beam of light goes from a rarer to a denser medium such as glass and has an angle of incidence equal to zero, which of the following properties of the beam of light does (do) NOT change?

I. Amplitude
II. Speed
III. Wavelength
IV. Direction

(A) I, II, and III only
(B) I and III only
(C) II and IV only
(D) IV only
(E) None of the above

66. Capacitor *P* is connected to a battery through switch *S* and wires *Y* and *Z*, as shown below. The capacitor's dielectric is marked *X*.

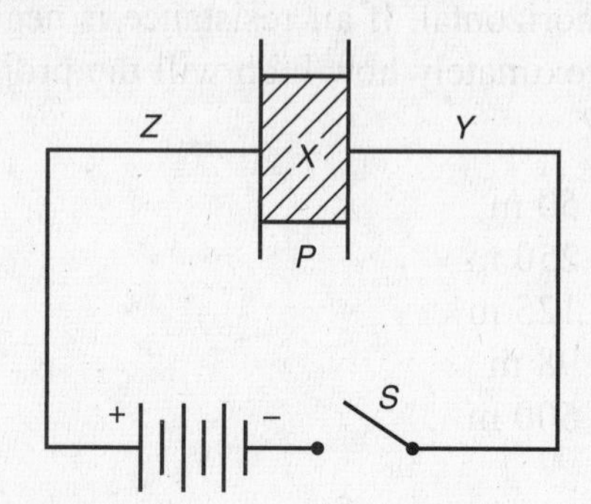

For a short time after the switch is closed, electrons will move through which of the following?

I. *Y*
II. *X*
III. *Z*
IV. *S*

(A) I, II, and III only
(B) I and III only
(C) II and IV only
(D) IV only
(E) I, III, and IV

GO ON TO THE NEXT PAGE ➤

67. A point source of light is placed at the principal focus of a convex lens. Which of the following will be true of the refracted light?

I. It will diverge.
II. It will be parallel to the principal axis.
III. It will seem to come from a point ½ of the radius of curvature from the lens.
IV. It will converge.

(A) I, II, and III only
(B) I and III only
(C) II only
(D) IV only
(E) None of the above

68. Which of the following statements is (are) correct about the photoelectric effect?

I. The number of electrons emitted is independent of the intensity of the incident light.
II. The stopping potential decreases with increased frequency.
III. In a graph of maximum kinetic energy versus incident frequency, all metals have different threshold frequencies, but all have the same slope.
IV. The maximum kinetic energy of the emitted electrons is independent of the intensity of the incident light.

(A) I only
(B) III only
(C) II and IV
(D) II, III, and IV
(E) III and IV

69. Which of the following is (are) true of an object starting from rest and accelerating uniformly?

I. Its kinetic energy is proportional to its displacement.
II. Its displacement is proportional to the square root of its velocity.
III. Its kinetic energy is proportional to the square of its speed.
IV. Its velocity is proportional to the square of elapsed time.

(A) I, II, and III only
(B) I and III only
(C) II and IV only
(D) IV only
(E) None of the above

70. Which of the following particles, all moving with the same velocity will have the longest de Broglie wavelength?

(A) An electron
(B) A proton
(C) A neutron
(D) An alpha particle
(E) A photon

Questions 71–73

Monochromatic light falls on a metal surface that has a work function of 6.7×10^{-19} joule. Each photon has an energy of 8.0×10^{-19} joule. (Planck's constant = 6.63×10^{-34} joule-second. One electron volt = 1.60×10^{-19} joule.)

71. What is the maximum kinetic energy, in joules, of the photoelectrons emitted by the surface?

(A) 1.3×10^{-19}
(B) 1.6×10^{-19}
(C) 2.6×10^{-19}
(D) 6.7×10^{-19}
(E) 8.0×10^{-19}

GO ON TO THE NEXT PAGE ➤

72. What is the energy of each photon, in electron volts?

(A) 1.6
(B) 1.6×10^{-19}
(C) 5.0
(D) 6.7
(E) 8.0

73. What is the frequency of each photon, in hertz?

(A) 3.7×10^{14}
(B) 4.2×10^{14}
(C) 1.2×10^{15}
(D) 3.7×10^{15}
(E) 7.0×10^{15}

---

74. What is the relationship between the atomic number, *Z*, the mass number, *A*, and the number of neutrons, *N*, in a nucleus?

(A) $Z = AN$

(B) $Z = \frac{A}{N}$

(C) $Z = \frac{N}{A}$

(D) $Z = A - N$
(E) $Z = N - A$

75. When lead, ${}^{214}_{82}\text{Pb}$, emits a beta particle, the resultant nucleus will be

(A) ${}^{214}_{83}\text{Bi}$
(B) ${}^{214}_{84}\text{Po}$
(C) ${}^{213}_{82}\text{Pb}$
(D) ${}^{214}_{81}\text{Tl}$
(E) ${}^{213}_{81}\text{Tl}$

STOP

Answer Key

1. **D**	16. **B**	31. **D**	46. **D**	61. **A**
2. **C**	17. **D**	32. **C**	47. **A**	62. **C**
3. **E**	18. **D**	33. **C**	48. **D**	63. **C**
4. **D**	19. **B**	34. **E**	49. **C**	64. **D**
5. **E**	20. **D**	35. **C**	50. **C**	65. **D**
6. **B**	21. **A**	36. **E**	51. **D**	66. **E**
7. **C**	22. **D**	37. **B**	52. **D**	67. **C**
8. **A**	23. **C**	38. **D**	53. **D**	68. **E**
9. **B**	24. **C**	39. **B**	54. **C**	69. **B**
10. **B**	25. **B**	40. **B**	55. **C**	70. **A**
11. **A**	26. **C**	41. **D**	56. **B**	71. **A**
12. **A**	27. **A**	42. **C**	57. **E**	72. **C**
13. **B**	28. **A**	43. **D**	58. **C**	73. **C**
14. **B**	29. **D**	44. **B**	59. **C**	74. **D**
15. **A**	30. **D**	45. **D**	60. **E**	75. **A**

Answer Explanations

Refer to the appropriate chapter for review of explanations.

1. D The velocity graph indicates constant speed when it is horizontal. This occurs during interval *DE*.

2. C The velocity graph indicates increasing speed when it is rising. This occurs during interval *CD*.

3. E The slope of a velocity graph represents the acceleration. When the graph is curved, the acceleration is changing. This occurs during interval *EF*.

4. D Acceleration is the rate of change in velocity. The correct unit is m/s^2.

5. E Torque is the product of force x lever arm. Torque is a vector quantity that seems to have the unit N · m (which appears to be J). But be careful! Energy is a scalar and a different physical quantity. Sometimes, we use the unit m · N instead for torque.

6. B Power is the rate of change of work measured in unit J/s or W.

7. C Momentum is equal to the product of mass x velocity, measured in the unit kg · m/s.

8. A Energy is a scalar quantity measured in the unit N · m, or Js.

9. B The kinetic energy of a body is given by the expression $\frac{1}{2}m\mathbf{v}^2$. Since the mass of the body is constant, its kinetic energy is proportional to $\mathbf{v}^2$, the square of its speed.

10. B The acceleration described in the question equals $\mathbf{v}^2/r$, where **v** is the speed and *r* is the radius of the circular path. Since *r* is constant, the acceleration is proportional to $\mathbf{v}^2$, the square of the speed of the object.

11. A The momentum of a body is equal to the product of its mass and its velocity. Since its mass remains constant, its momentum is proportional to its velocity.

12. A The acceleration of an object in equilibrium is zero. Its velocity is constant (and may be zero). Its displacement **d** = **v***t*. In this question, $t = 1$ s, a constant. Hence, the displacement is proportional to the velocity.

13. B Such motion is described by the relationship $\mathbf{v}^2 = 2\mathbf{ad}$. Since 2**a** is constant, $\mathbf{v}^2$ is proportional to **d**, and the displacement, **d**, is proportional to the square of the velocity, **v**.

14. B Copper has the least resistance of all the metals listed and therefore the greatest current.

15. A Nichrome has the highest resistance of all the metals listed. Since power is proportional to resistance, nichrome generates the greatest power.

16. B Two forces that act on an object in opposite directions tend to nullify each other. The resultant of such forces is their difference (in this case, 1 N), and the direction of the resultant is the direction of the larger force.

17. D If two forces, acting on the same point, are at right angles to each other, their resultant may be found by applying the Pythagorean theorem; $R^2 = 6^2 + 8^2$; $R = 10$ N. You should have recognized the 3-4-5 right triangle; the two sides, 6 and 8, are twice 3 and 4 respectively, and therefore the hypotenuse, giving the value of the resultant, is twice 5, or 10.

18. D Potential energy = weight × height
= 10 N × 5 m = 50 J

19. B The 10-N weight is not in free fall as it is also being pulled up by the string. The acceleration is therefore less than **g**.

20. D The 10-N weight is accelerating downward because of the action of a net force. Therefore, the tension in rope *A* must be less than 10 N. The 6-N weight is accelerating upward because of a net force, so the tension pulling it up must be more than 6 N.

21. A For a freely falling object, the distance covered is equal to the product of one-half its acceleration and the square of the time of fall:

$$\mathbf{d} = \tfrac{1}{2}\mathbf{g}t^2$$

$$= \tfrac{1}{2}(9.8 \text{ m/s}^2)(1 \text{ s})^2$$

$$= 4.9 \text{ m}$$

22. D Acceleration measures the rate of change of velocity. The fact that the acceleration is zero does not necessarily mean that the velocity is zero. Therefore (A) is correct. Additionally, since **F** = **ma** if the acceleration is zero, the net force acting on the object is zero. Therefore (C) is correct.

23. C That the rays will be parallel follows from the fact that all the angles of reflection will be equal.

24. C The boiling point is the temperature at which the gas changes into a liquid (80°C).

25. B The melting point is the temperature at which the liquid changes into a solid (60°C).

26. C The heat of vaporization occurs during the change from gas to liquid. At 6 kJ/min the graph shows that this takes 2 mins. The correct answer is 6 kJ/kg.

27. A The heat of fusion occurs during the change from liquid to solid. At 6 kJ/min the graph shows that this takes 1 min. The correct answer is 3 kJ/kg.

28. A The specific heat measures the change in heat energy per kilogram per degree Celsius. The graph shows that the specific heat as a solid is greater than it is as a gas.

29. D In the double-slit arrangement the light going through each slit is *diffracted:* It bends around behind the barrier; then it spreads out on the screen; the light from the two slits overlaps, and interference occurs.

30. D A change of 50°C is equivalent to a change of 50 K.

31. D Graph D shows an object uniformly accelerating from rest, maintaining constant speed, and then decelerating to a stop, all in the forward direction.

32. C You can obtain the total displacement (distance traveled) by adding the areas of the three segments. The first area, from 0 to 5 s, is equal to 50 m. The second area, from 5 to 10 s, is equal to 100 m. The third area, from 10 to 15 s, is also equal to 50 m. The total is 200 m.

33. C If the object is heading east, then a net force directed west would imply a negative acceleration. This occurs during interval 10–15 when the slope is negative.

34. E The amplitude of a sound wave is observed as its loudness.

35. C A medium is said to be dispersive if the velocity of light depends on its frequency while in that medium.

36. E The easy solution is to remember that, in a parallel circuit, the equivalent resistance is always lower than the resistance of the lowest resistor. The direct solution is that, for two resistors:

$$R_{eq} = R_1 R_2$$

$$R_1 + R_2 = 4\ \Omega$$

37. B Ohm's law states that $V = IR$. Thus

$$I = 12\text{ V}$$

$$4\ \Omega = 3\text{ A}$$

38. D In a parallel circuit, each branch has the same potential drop across it. Thus:

$$I = V$$

$$R = 12\text{ V}$$

$$20\ \Omega = 0.6\text{ A}$$

39. B The coil on the left becomes an electromagnet. Conventional current goes from plus to minus outside the battery. Grasp the coil with the right hand so that the fingers point in the direction of conventional current; the outstretched thumb points in the direction of the N-pole: in this case, *W.*

40. B A changing current in coil *X* induces a current in the other coil. Current in coil *X* builds up slowly to its maximum value because of the large self-induced emf in coil *X*. Until this maximum value is reached, current is induced in the second coil in a direction opposing the current buildup in coil *X*. For this to occur, the left end of the second coil should be an S-pole, which will tend to weaken the left magnet. Using the right-hand rule, conventional current will flow from *y* to *z*.

41. D The electric field intensity due to a point charge is proportional to the charge and inversely proportional to the square of the distance from the charge. In this question, charges *Y* and *Z* are equal. Therefore at the point midway between them the field intensity due to each charge is the same in magnitude and directed oppositely. (Recall that the direction of the field intensity is the direction of the force on an imagined positive charge.) Therefore, the two field intensities cancel each other completely at the point midway between the charges.

42. C The force acting on a charge moving perpendicularly through a magnetic field is proportional to the speed with which the charge moves (recall: $\mathbf{F} = q\mathbf{v}B$). Therefore, the force on the charge doubles if the speed doubles.

43. D The proton is slightly smaller in mass than the neutron. They differ by about 1 part in 2,000.

44. B The time of one complete vibration of a simple pendulum or its period is equal to $2\pi\sqrt{\frac{L}{\mathbf{g}}}$. If the length is decreased, the period is decreased. If **g** is decreased, by going up to the top of the Empire State Building, the period is increased. In the simple pendulum, the period is independent of the mass of the bob.

45. D If the distance is tripled, the inverse square law of gravitation reduces the force by a factor of 9: 36 N | 9 = 4 N.

46. D As a planet gets closer to the star, the force of gravity increases, and so it moves faster and its kinetic energy increases. Therefore (A) is correct. Since mechanical energy is constant, the potential energy must decrease and the sum of the kinetic and potential energies must remain constant. Therefore (C) is correct.

47. A The test to determine whether an object is electrically charged is to see if it will repel an electrically charged object. An uncharged rod will attract, and be attracted by, a charged pith ball.

48. D The rate of heat production of a wire = I^2R. The resistance of the wire is practically constant. Therefore, the rate of heat production is proportional to the square of the current.

49. C Lenz's law states that induced current flows in a direction such that its magnetic field opposes the magnetic field that induced it. This is essentially the law of conservation of energy.

50. C Half the mass will make the force reduce by half. However, reducing the distance by half will increase the force by four times. Therefore, the net effect is to double the acceleration due to gravity at the surface.

51. D Boyle's law applies to a definite mass of gas at constant temperature. Pressure, volume, and also density (= mass/vol) change.

52. D Grounding prevents the buildup of static charges on the surfaces of appliances.

53. D If it were not for Earth's gravitational pull, the satellite would tend to follow a straight line tangent to its orbit—until affected by the gravitational pull of some other object.

54. C A perfectly black object absorbs all light. There is no visible black light, but the term is sometimes applied to ultraviolet light.

55. C The distance between maxima in a diffraction pattern does not depend on the intensity of the light used.

56. B The rating of the lamp is 60 W at 110 V; that is, the lamp uses 60 W when used on 110 V. The lamp may operate on a lower voltage: It will not be as bright; at 60 V it will be quite dim. It may be used on somewhat higher voltages. Higher voltage results in higher current and greater power consumption. Of course, if a much higher than rated voltage is used, the bulb will burn out quickly. Therefore, statements I and III are correct.

57. E The man is moving with constant velocity; therefore, no unbalanced force is acting on him. Earth's pull on him (his weight) is balanced by an equal upward push on him by the floor of the elevator. In accordance with Newton's third law, his push on the floor of the elevator equals the floor's push on him: his weight. Therefore, none of statements I–IV is true.

58. C According to Newton's second law, the acceleration is proportional to the unbalanced force. If the unbalanced force is doubled, the acceleration is doubled. The term *smooth table* implies that friction is negligible. **F** is then the only horizontal force and is the net force.

59. C Assume that the magnitude of $\mathbf{g} = 10 \text{ m/s}^2$. The vertical component of the launch velocity is $\mathbf{v}_y = (100 \text{ m/s}) \sin 30° = 50 \text{ m/s}$. The time for the projectile to reach its maximum height is the time gravity takes to decelerate the vertical velocity to zero. Thus

$$t_{up} = 50 \text{ m/s}$$

$$10 \text{ m/s}^2 = 5 \text{ s}$$

Now, at the highest point, the object stops rising and begins to free-fall. The time down is equal to the time up (in this case, 5 s). Thus

$$y = \tfrac{1}{2}\mathbf{g}t^2 = 125 \text{ m}.$$

60. E The acceleration will be increased, but not doubled. This can be seen readily with a numerical example. Assume that F is parallel to the plane and equal to 6 N. Opposed to F is a component of the weight of the object, and there may also be friction; assume this adds up to 2 N. This leaves an unbalanced force of 4 N. Doubling F increases it to 12 N. Opposed to it will be the same 2 N. The unbalanced force now is 10 N, more than twice the 4 N that produced the acceleration in the first case. The new acceleration will be more than twice the original one. None of statements I–IV is correct.

61. A The angular momentum of the mass is given by $\mathbf{L} = m\mathbf{v}r$. The kinetic energy is directly proportional to $\mathbf{v}^2$, and the centripetal acceleration is independent of mass. The only correct statement is I.

62. C If air resistance is negligible, freely falling objects have the same acceleration. Then, after 4 s of free fall, both will have the same speed.

Their momentums (= $m\mathbf{v}$) will then be in the same ratio as their masses; this is true also of their kinetic energies ($\frac{1}{2}m\mathbf{v}^2$). The 10-N ball will have 5 times the momentum and 5 times the kinetic energies of the 2-N ball. Mass is a measure of an object's inertia; the greater the mass, the greater the inertia. Statements II and IV are correct.

63. C This question may be too tricky. Conventional mirrors do not transmit light, and nothing can be seen through them. If *Y* is a concave lens, then a virtual image of *X* is seen when the eye is placed on the other side of the lens; this is the conventional case. This is not obvious, however, for the case when *Y* is a convex lens. Here, *X* is at twice the focal length from the lens. A real image is brought to a focus on the other side of the lens and 20 cm from the lens. But there is no screen there, and the light keeps traveling on and will diverge as though an object were there. To the eye it will be the same as though an object were 25 cm away from it. The normal eye will see an image of *X*.

64. D In the formation of a real image with a lens, all the rays starting from one point on the object go, after refraction by the lens, to the same point of the image. All parts of the lens are usually used for this purpose. If, however, half of the lens is covered, the other half will still function to produce each point of the image, but only half as much light will get to each point. The image brightness will, consequently, be about half of the original brightness.

65. D If the angle of incidence is zero, the ray is normal to the surface. Such a ray is not deviated; in going from air to water it would not be bent at all. The frequency does not change either, but the wave is slowed down. Hence the wavelength must change: $\mathbf{v} = f\lambda$. There is some reflection of the wave at the new surface. Less energy means a lower amplitude of the wave. Only direction does *not* change.

66. E All parts of the capacitor are conductors except dielectric *X*. Electrons will move through the metallic conductors until the capacitor is charged.

67. C Light from the focal point wil refract parallel to the principle axis. No image will form as a consequence. Therefore, only statement II is correct.

68. E The number of photoelectrons emitted does depend on the intensity of the incident light, as verified by experiment. Only statements III and IV are true for the photoelectric effect.

69. B The object's kinetic energy is proportional to the square of its speed ($E_k = \frac{1}{2}m\mathbf{v}^2$). The square of its speed is proportional to its displacement ($\mathbf{v}^2 = 2a\mathbf{d}$). Hence, the kinetic energy is proportional to the displacement.

70. A The de Broglie wavelength applies to matter particles (this excludes the photon): $\lambda = h / m\mathbf{v}$, where h = Planck's constant. Since the electron has the smallest mass of the choices, it will exhibit the longest de Broglie wavelength.

71. A In the *photoelectric effect* the energy of the incident photon is used to pull the electron away from the metal; if the photon has more energy than is needed just to overcome the attractive force of the metal, the remaining energy gives the freed electron its kinetic energy:

$$\text{energy of photon} = \text{work function} + \text{kinetic energy}$$
$$8.0 \times 10^{-19}\text{ J} = 6.7 \times 10^{-19}\text{ J} + \text{kinetic energy}$$
$$\text{kinetic energy} = 1.3 \times 10^{-19}\text{ J}$$

72. C You are told that each photon has an energy of 8.0×10^{-19} J and that 1 eV volt = 1.60×10^{-19} J. Note by inspection that the energy of the photon is 5 times as great, or 5 eV.

73. C The energy of a photon is equal to the product of Planck's constant and its frequency:

$$E = hf$$
$$8.0 \times 10^{-19}\text{ J} = (6.63 \times 10^{-34}\text{ J-s})\,f$$
$$f = 1.2 \times 10^{15}\text{ Hz}$$

74. D The *mass number* of a nucleus, or of the atom of which it is a part, is equal to the number of protons (its atomic number) and the number of neutrons:

$$\text{mass number} = \text{atomic number} + \text{number of neutrons}$$
$$A = Z + N$$
$$Z = A - N$$

If you have any trouble with this, think of the mass number as approximately the total mass of the nucleus, where the unit of mass is taken as the mass of the proton. The mass of a neutron is approximately the same as the mass of a proton.

75. A The beta particle is an electron, which has a charge of –1. When a nucleus with a charge of +82 loses a –1 charge, it must become more positive by 1 more unit; that is, the charge must become +83. This charge is its atomic number. Of the given choices, only one has a subscript of 83, which is the atomic number. You don't have to do anything further; but at this point, just for review, note that the loss of an electron does not change the mass number of the nucleus. Choice (A) has not only the correct subscript, 83, but also the expected unchanged superscript, 214.

CHAPTER 2

Measurements and Relationships

Physics is an experimental science. This means that each theory must be based on experimental observations. Learning how to make observations properly is part of the laboratory experience in a physics class. Searching for patterns to establish relationships in physics often involves the use of graphs and algebra. A brief review of graphing and measurement techniques is presented below. A more extensive review of mathematics is provided in Appendix III.

Making Measurements

Relationships are patterns based on measurements. To begin, we ascribe to quantities of matter fundamental characteristics that can be used to distinguish them. Among these are mass, length, and electric current. Since the universe is in constant change, we add time to this list. These characteristics are called *fundamental.*

Quantitative comparisons are based on assigning a magnitude and a scale to each of these quantities. The *magnitude* indicates the relative size of the quantity, and the *scale* provides the rule for measurement. Another term we could use instead of *scale* is *unit* or *dimension.* Specifically, the units assigned to mass, length, and time (and several others) are called *fundamental units.* In physics, we use the International System (SI) of units; some of the more common units are presented in Tables 1 and 2.

TABLE 1 Fundamental SI Units Used in Physics

Quantity	Unit Name	Symbol
length	meter	m
mass	kilogram	kg
time	second	s
electric current	ampere	A
temperature	kelvin	K
amount of substance	mole	mol
plane angle	radian	rad

The units in Table 2 are called *derived* since they can be expressed in terms of the fundamental units shown in Table 1.

TABLE 2 Some Derived SI Units Used in Physics

Quantity Units	Unit Name	Symbol	Expression in Other SI
force	newton	N	$kg \cdot m/s^2$
frequency (cyclic)	hertz	Hz	s^{-1}
frequency (angular)			rad/s = 1/s
pressure	pascal	Pa	$N/m^2 = kg/ms^2$
energy (work, heat)	joule	J	$N \cdot m = kg \cdot m^2/s^2$
power	watt	W	$J/s = kg \cdot m^2/s^3$
electric charge	coulomb	C	As
electric potential (emf)	volt	V	$J/C = J/As = kg \cdot m^2/As^3$
capacitance	farad	F	$C/V = A^2s^4/kg \cdot m^2$
resistance	ohm	Ω	$V/A = kg \cdot m^2/A^2s^3$
magnetic flux	weber	Wb	$Vs = kg \cdot m^2/As^2$
magnetic flux density	tesla	T	$Wb/m^2 = kg/As^2$
inductance	henry	H	$Wb/A = kg \cdot m^2/A^2s^2$

Significant Figures and Standard Notation

We can count exactly, but we cannot measure exactly. If 20 students carefully count the number of baseballs in a box, we have the right to expect the same answer from each of the students. If they say that there are 45 baseballs, we do not think that maybe there are 45½ baseballs. However, if we ask the same 20 students to carefully measure the height of the box, we can expect different answers such as 62.3, 62.4, and 62.5 centimeters. Each of these measurements has three significant figures, but the last digit is doubtful. If, as far as we can tell, the last digit should be zero, we write the zero down and thus indicate the accuracy with which we tried to measure. In 62.0 centimeters the zero is significant, and we have three significant figures. In 0.0230 centimeters we still have only three significant figures. The first two zeros are used only to indicate the location of the decimal point. In *standard form,* the same number is written as 2.30×10^{-2} centimeter.

For more information on this concept, consult Appendix III.

Establishing Relationships

One way to establish a relationship between two physical quantities is by means of a graph. If the graph indicates that for one value of the independent quantity we assign one and only one value for the dependent quantity, we have a *functional relationship.* This usually means that we can represent the relationship by an algebraic equation.

Certain relationships produce graphs that are immediately recognizable. Usually, we plot a graph using *Cartesian coordinates* and represent the independent quantity along the horizontal or x-axis. If there is no change in the dependent quantity for any change in the independent quantity, we obtain a horizontal line, indicating a *constant relationship*.

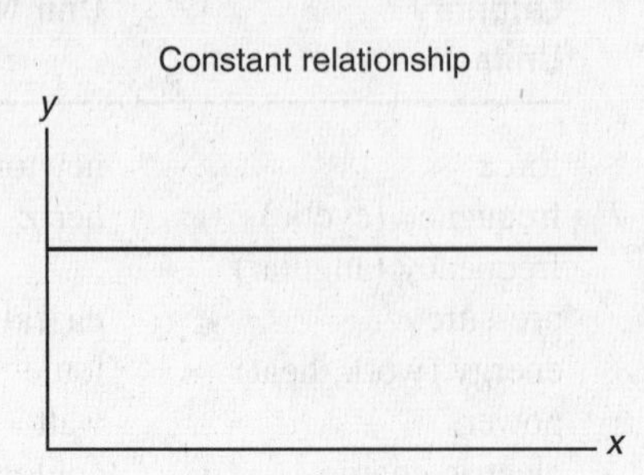

If there is a uniform change in the dependent quantity proportional to the independent quantity so that both increase (or decrease) simultaneously, we say that the relationship is a *direct relationship.* If the dependent quantity decreases when the independent quantity increases (or vice versa), we have an *inverse relationship*. Now these relationships can be either linear or nonlinear. Examples of linear direct and inverse relationships are presented below.

The graph of a linear direct relationship is a diagonal straight line that starts at the origin. This equation can be expressed as $y = kx$, where the constant k is called the *slope* of the line. Direct relationships that contain y-intercepts other than the origin can be rescaled to fit this form. One example is the absolute temperature (Kelvin) scale. Instead of plotting Celsius temperature versus pressure for an ideal gas (and getting –273°C) as the lowest possible temperature), we rescale the temperature axis so that the lowest possible temperature is defined as 0 K (called *absolute zero*). In this way we obtain the familiar gas-law relationships defined for absolute temperature only.

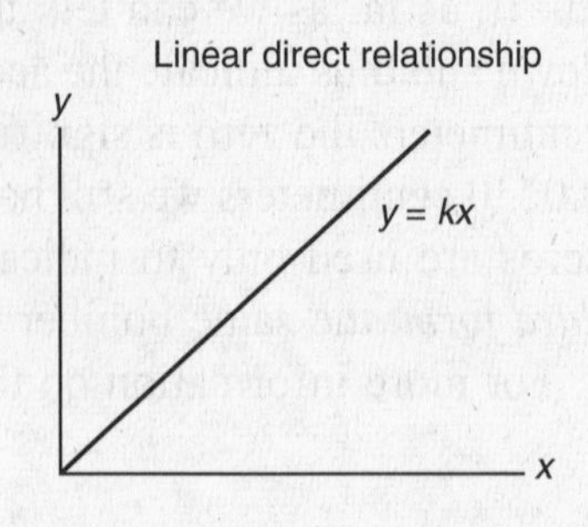

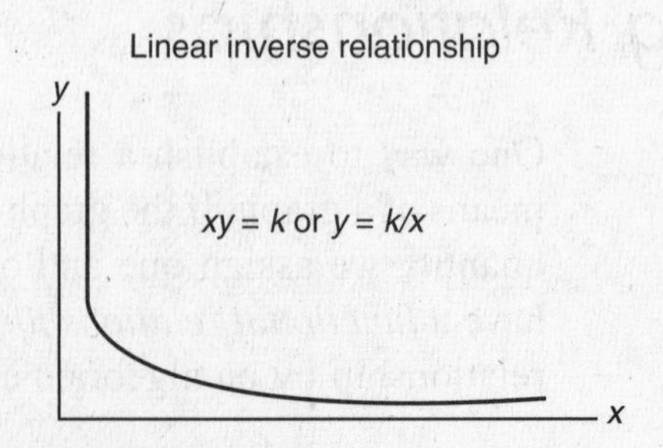

There are two typical nonlinear direct relationships that appear often in physics. The first is the parabolic or *squared relationship.* For example, in freely falling motion, the displacement of a mass is directly proportional to the square of the elapsed time. A graph of this relationship is shown below.

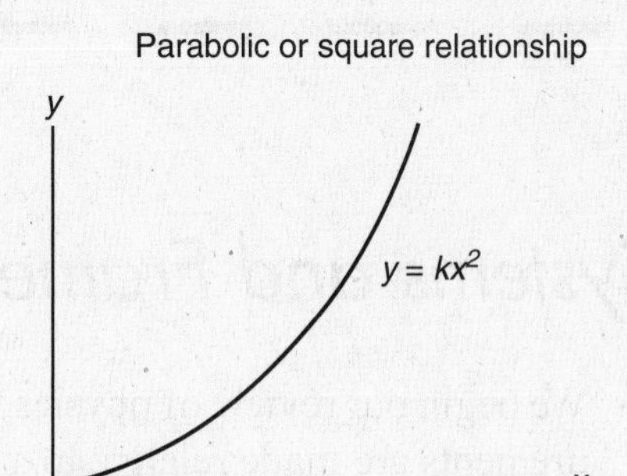

The other nonlinear direct relationship is the *square root relationship*. For example, the period of a swinging pendulum is directly proportional to the square root of its length. A graph of this relationship is shown below.

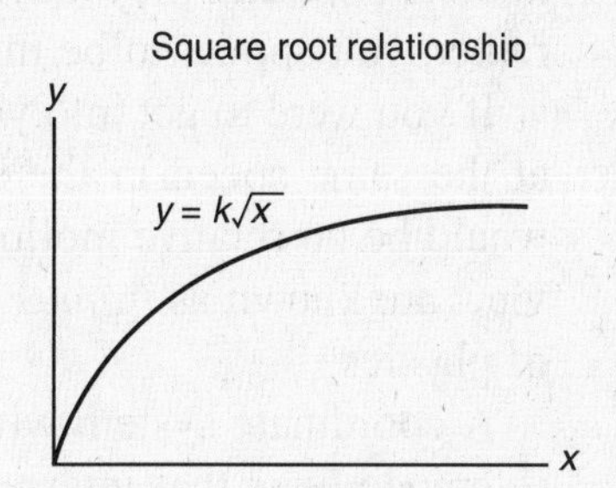

CHAPTER 3

Vectors

Coordinate Systems and Frames of Reference

We begin our review of physics with the idea that all observations and measurements are made relative to a suitably chosen frame of reference. In other words, when observations are made that will be the basis of future predictions, we must be careful to note from what point of view those observations are being made. For example, if you are standing on the street and see a car driving by, you observe the car and all of its occupants moving relative to you. However, to the driver and other occupants of the car, the situation is different: they may not appear to be in motion relative to themselves; rather, you appear to be moving backward relative to them.

If you were to get into your car, drive out to meet the other car, and travel at the same speed in the same direction, right next to the first car, there would be no relative motion between the two cars. These different points of view are known as *frames of reference*, and they are very important aspects of physics.

A coordinate system within a frame of reference is defined to be a set of reference lines that intersect at an arbitrarily chosen fixed point called the *origin*. In the Cartesian coordinate system, the reference lines are three mutually perpendicular lines designated as x, y, and z (see Figure 3.1). The coordinate system must provide a set of rules for locating objects within that frame of reference. In the three-dimensional Cartesian system, if we define a plane containing the x- and y-coordinates (let's say a horizontal plane), then the z-axis specifies direction up or down.

It is often useful to compare observations made in two different frames of reference. In the example above, the motion of the occupants of the car was reduced to zero if we transformed our coordinate system to the car moving at constant velocity. In this case we say that the car is an *inertial frame of reference*. In such a frame, it is impossible to observe whether or not the reference frame is in motion if the observers are moving with it.

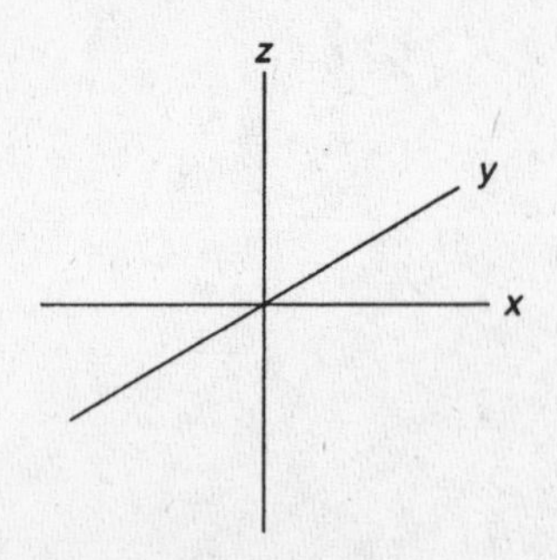

FIGURE 3.1

What Is a Vector?

Another way of locating the position of an object in the Cartesian system is use a directed line segment, or *vector*. If you draw an arrow, starting from the origin, to a point in space (see Figure 3.2), you have defined a position vector **R** whose magnitude is given by | **R** | and is equal to the linear distance between the origin and point (x, y). The direction of the vector is given by the acute angle, indentified by the Greek letter θ, that the arrow makes with the positive x-axis. Any quantity that has both magnitude and direction is called a *vector quantity*. Any quantity that has only magnitude is a *scalar quantity*. Examples of vector quantities are force, velocity, weight, and displacement. Examples of scalar quantities are mass, distance, speed, and energy.

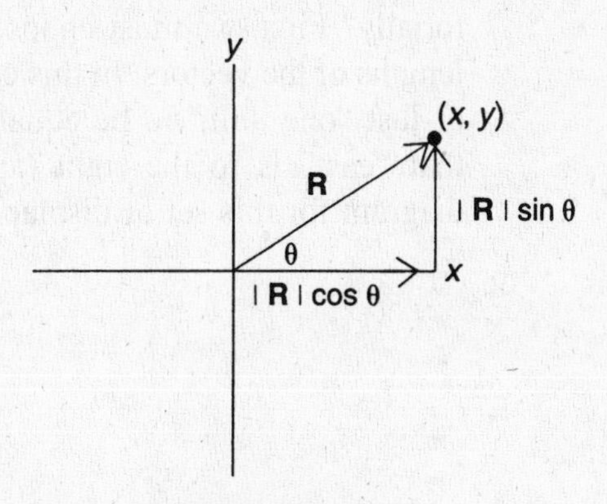

FIGURE 3.2

If we designate the magnitude of vector **R** as r, and the direction angle is given by θ, we have an alternative coordinate system called the *polar coordinate* system. In two dimensions, to locate the point in the Cartesian system (x, y) involves going x units horizontally and y units vertically. Additionally, we can show (see Figure 3.2) that

$$x = r \cos \theta \quad \text{and} \quad y = r \sin \theta$$

Having established this polar form for vectors, we can easily show that r and θ can be expressed in terms of x and y as follows:

$$r^2 = x^2 + y^2$$

$$\tan \theta = \frac{y}{x}$$

The magnitude of **R** is then given by

$$|\mathbf{R}| = r = \sqrt{x^2 + y^2}$$

and the direction angle θ is given by

$$\theta = \tan^{-1} \frac{y}{x}$$

Addition of Vectors

Geometric Considerations

The ability to combine vectors is a very important tool of physics. From a geometric standpoint, the "addition" of two vectors is not the same as the addition of two numbers. When we state that, given two vectors **A** and **B**, we wish to form the third vector **C** such that **A** + **B** = **C**, we must be careful to preserve the directions of the vectors relative to our chosen frame of reference.

One way to do this "addition" is to construct what is called a *vector diagram*. A vector is identified geometrically by a directed arrow that has a "head" and a "tail." We will consider a series of examples. First, suppose a girl is walking from her house a distance of five blocks east and then an additional two blocks east. How can we represent these displacements vectorially? First, we must choose a suitable scale to represent the magnitude lengths of the vectors. In this case, let us just call the scale "one vector unit" or just "one unit," to be equal to one block of distance. Let us also agree that "east" is to the right (and hence "north" is directed up). A vector diagram for this set of displacements is given in Figure 3.3.

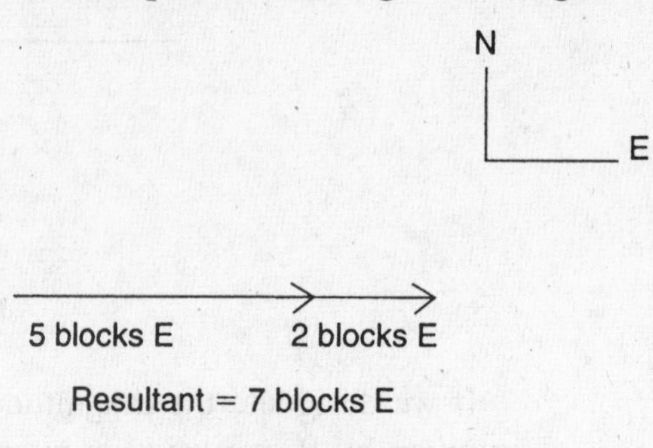

FIGURE 3.3

Now, for our second example, suppose the girl walks five blocks east and then two blocks northeast (that is, 45 degrees north of east). Since our intuition tells us that the final displacement will be in the general direction of north and east, we draw our vector diagram, using the same scale as before, so that the two vectors are connected head to tail. The vector diagram will look like Figure 3.4. The resultant is drawn from the tail of the first vector to the head of the second, forming a triangle (some texts use the "parallelogram" method of construction, which is equivalent). The direction of the resultant is measured from the tail of the vector inside the triangle.

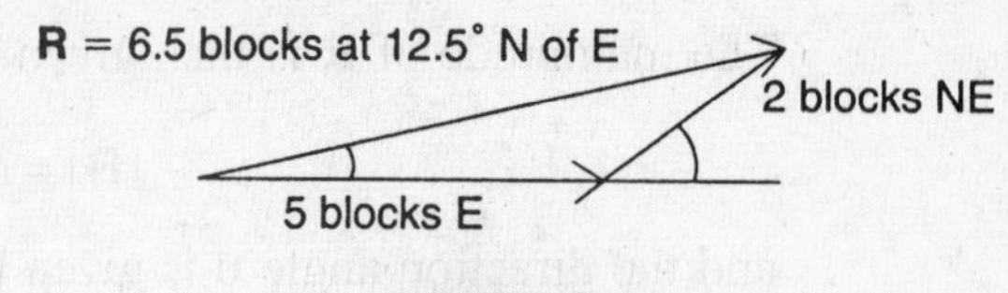

FIGURE 3.4

For our third example, suppose the girl walks five blocks east and then two blocks north. Figure 3.5 shows the vector diagram for this set of displacements.

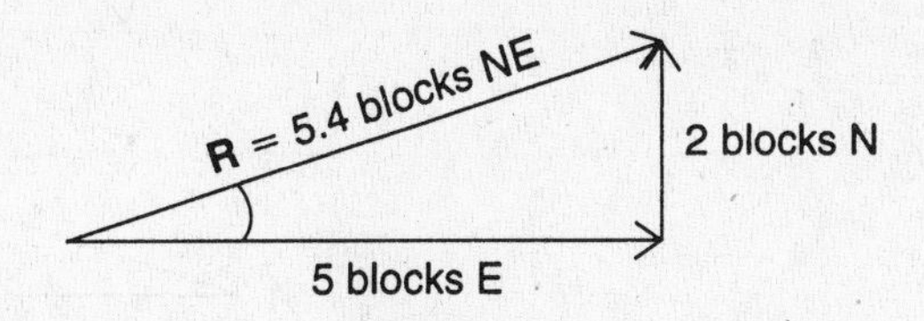

FIGURE 3.5

Finally, for our fourth example, suppose the girl walks five blocks east and then two blocks west. Her final displacement will be three blocks east of the starting point.

From these examples, we conclude that, as the angle between two vectors increases, the magnitude of their resultant decreases. Also, the magnitude of the resultant between two vectors is maximum when the vectors are in the same direction (at a relative angle of 0 degree), and minimum when they are in opposite directions (at a relative angle of 180 degrees). It is important to remember that the vectors must be constructed head to tail and a suitable scale chosen for the system.

Note that the two vectors do not form a closed figure. The "missing link" is the resultant. This fact suggests an important concept. If a third vector was given that was equal in magnitude but opposite in direction to the predicted resultant, the vector sum of that resultant and the third vector would be zero. In other words, the vector sum of those three vectors would form a closed triangle of zero resultant, meaning that the girl in our examples returned to her starting point leaving a zero displacement. In physics, that third vector, equal and opposite to the resultant, is called the *equilibrant*. Figure 3.6 illustrates this concept.

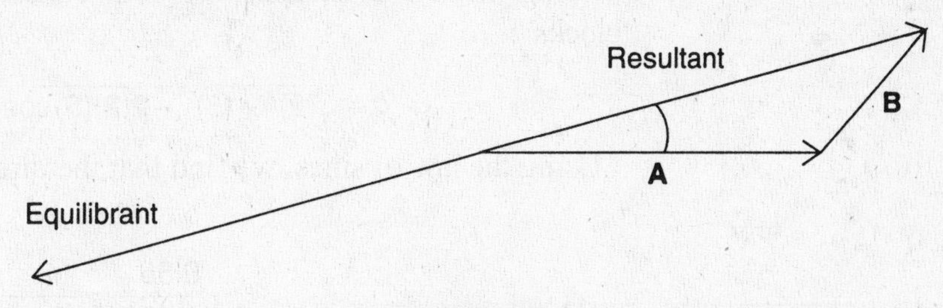

FIGURE 3.6

Algebraic Considerations

In the above examples, the resultant between two vectors was constructed using a vector diagram. The magnitude of the resultant was the measured length of the vector drawn from the tail of the first vector to the head of the second. If we sketch such a situation, we form a triangle whose sides are related by the *law of cosines* and whose angles are related by the *law of sines*.

Consider the vector triangle in Figure 3.7 with arbitrary sides a, b, and c and corresponding angles A, B, and C.

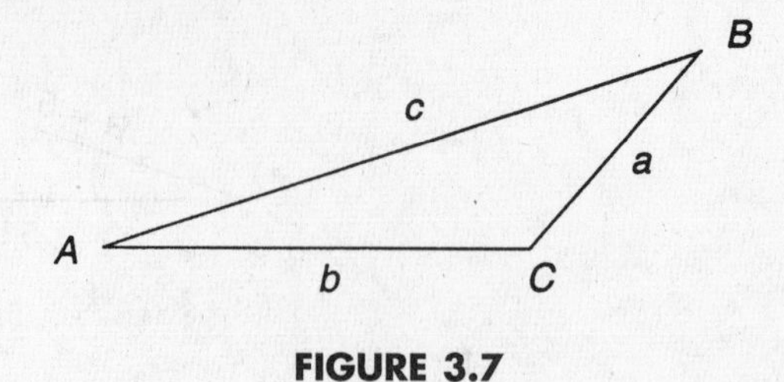

FIGURE 3.7

The law of cosines states that

$$c^2 = a^2 + b^2 - 2ab \cos C$$

The law of sines states that

$$\frac{a}{\sin A} = \frac{b}{\sin B} = \frac{c}{\sin C}$$

If we use the information given in our second example above, we see that in Figure 3.8 the vectors have the following magnitudes and directions:

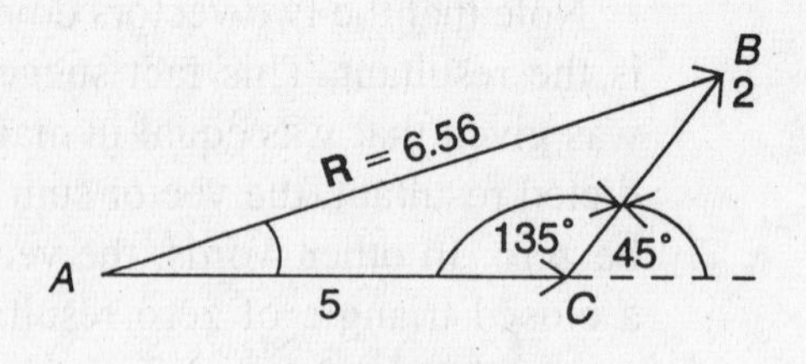

FIGURE 3.8

Using the law of cosines, we obtain the magnitude of the resultant in "blocks":

$$c = \sqrt{(2)^2 + (5)^2 - 2(2)(5)\cos 135} = 6.56$$

Using the law of sines, we find that the direction of the resultant is angle *A*:

$$\frac{6.56}{\sin 135} = \frac{2}{\sin A}$$

Angle *A* turns out to be equal to 12.45 degrees north of east.

Addition of Multiple Vectors

The discussion of the equilibrant, that is, the third vector, leads to the idea of the addition of multiple vectors. As long as the vectors are constructed head to tail, multiple vectors can be added in any order. If the resultant is zero, the diagram constructed will be a closed geometric figure, as shown in the case of three vectors forming a closed triangle in the example above. Thus, if the girl walks five blocks east, two blocks north, and five blocks west, the vector diagram will look like Figure 3.9, with the resultant displacement being equal to two blocks north.

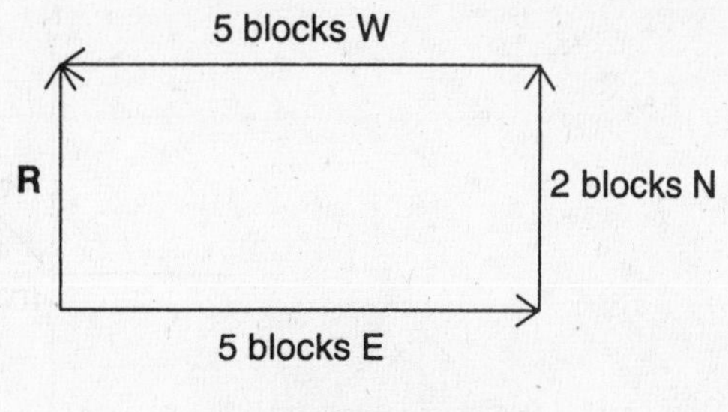

FIGURE 3.9

Subtraction of Vectors

The difference between two vectors can be understood by considering the process of displacement. Suppose that, in Figure 3.10, we identify two sets of points, (x, y) and (u, v), in a coordinate system as shown. Each point can be located within the coordinate system by a position vector drawn from the origin to that point. The displacement of an object from point A to point B would be represented by the vector drawn from A to B. This vector is defined to be the difference between the two position vectors and is sometimes called $\Delta\mathbf{R}$, so that $\Delta\mathbf{R} = \mathbf{R}_2 - \mathbf{R}_1$.

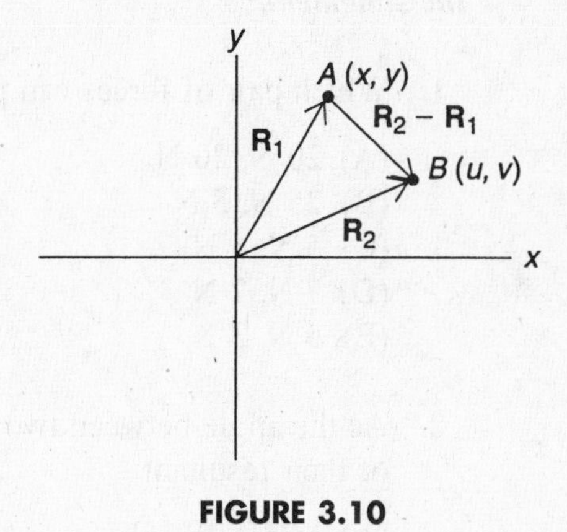

FIGURE 3.10

From our understanding of vector diagrams, we can see that $\mathbf{R}_1$ and $\Delta\mathbf{R}$ are connected head to tail, and thus that $\mathbf{R}_1 + \Delta\mathbf{R} = \mathbf{R}_2$!

Addition Methods Using the Components of Vectors

Any single vector in space can be *resolved* into two perpendicular components in a suitably chosen coordinate system. The methods of vector resolution have been discussed, but we can review them here. Given a vector (sketched in Figure 3.11) representing a displacement of 100 meters northeast (i.e., making a 45-degree angle with the positive x-axis), we can observe that it is composed of an x- and a y- (a horizontal and a vertical) component that can be geometrically or algebraically determined.

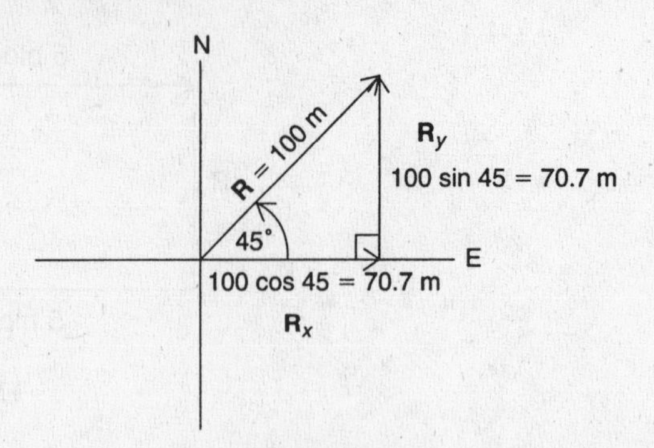

FIGURE 3.11

Geometrically, if we were to draw the 100-meter vector to scale at the correct angle, then projecting a perpendicular line down from the head of the given vector to the x-axis would construct the two perpendicular components. Algebraically, we see that, if we have a given vector **R**, then, in the x-direction, we have

$$\mathbf{R}_x = \mathbf{R}\cos\theta \quad \text{and} \quad \mathbf{R}_y = \mathbf{R}\sin\theta$$

where the magnitude of **R** is R.

Questions Chapter 3

In each case, select the choice that best answers the question or completes the statement.

1. Which pair of forces can produce a resultant of 15 newtons?

(A) 20 N, 20 N
(B) 25 N, 5 N
(C) 5 N, 5 N
(D) 7 N, 7 N
(E) 5 N, 3 N

2. As the angle between two concurrent forces increases, the magnitude of their resultant

(A) increases only
(B) decreases and then increases
(C) increases and then decreases
(D) remains the same
(E) decreases only

3. A girl walks 2 meters north, 4 meters west, and 2 meters south. Her final displacement is

(A) 4 m east
(B) 4 m west
(C) 2 m north
(D) 2 m south
(E) 5 m northwest

4. A cart is pulled by a rope making an angle of 45° to the horizontal. If 100 newtons of force are applied to the rope, the magnitude of the horizontal component force is approximately

(A) 45 N
(B) 141 N
(C) 100 N
(D) 80 N
(E) 71 N

5. The resultant of a 3-newton and a 4-newton force acting simultaneously on an object at right angles to each other is, in newtons,

(A) 0
(B) 1
(C) 3.5
(D) 5
(E) 7

6. Two forces act together on an object. The magnitude of their resultant is least when the angle between the forces is

(A) 0°
(B) 45°
(C) 60°
(D) 90°
(E) 180°

7. The resultant of a 5-newton and a 12-newton force acting simultaneously on an object in the same direction is, in newtons,

(A) 0
(B) 5
(C) 7
(D) 13
(E) 17

8. A vector is given by its components, $\mathbf{A}_x = 2.5$ and $\mathbf{A}_y = 7.5$. What angle does vector **A** make with the positive x-axis?

(A) 72°
(B) 18°
(C) 25°
(D) 50°
(E) 75°

9. Which of the following sets of displacements have equal resultants when performed in the order given?

I: 6 m east, 9 m north, 12 m west
II: 6 m north, 9 m west, 12 m east
III: 6 m east, 12 m west, 9 m north
IV: 9 m north, 6 m east, 12 m west

(A) I and IV
(B) I and II
(C) I, III, and IV
(D) I, II, IV
(E) II and IV

10. Which vector represents the direction of the two concurrent vectors shown below?

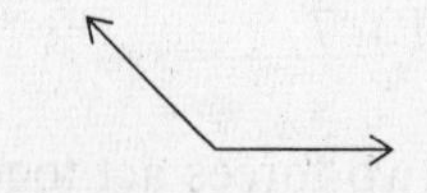

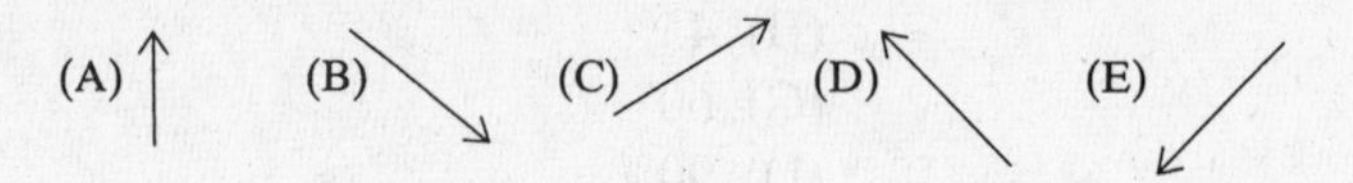

11. Three forces act concurrently on a point *P* as shown below. Which vector represents the direction of the resultant force on point *P*?

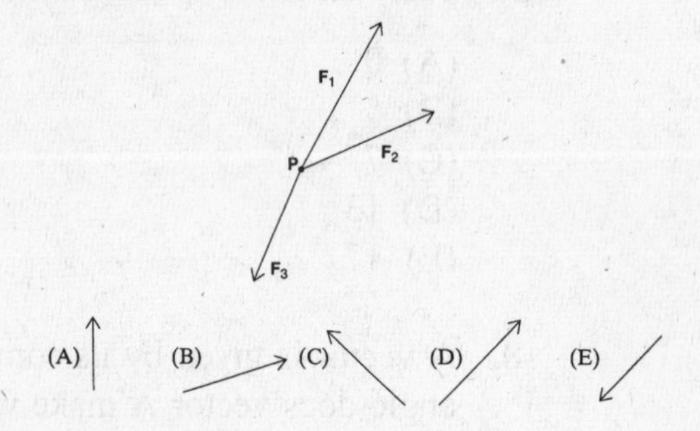

Explanations to Questions Chapter 3

Answers

1. (A)	**4.** (E)	**7.** (E)	**10.** (C)
2. (E)	**5.** (D)	**8.** (A)	**11.** (B)
3. (B)	**6.** (E)	**9.** (C)	

Explanations

1. **A** The answer can be determined by calculating the maximum and minimum resultants. These values are found by adding and subtracting the two vectors. If you try this for all five pairs, you will see that only 20 N has a maximum resultant that is more than 15 N and a minimum resultant that is less than 15 N.
2. **E** This is a recall question. As the angle between the two concurrent forces or vectors increases, the magnitude of their resultant decreases.
3. **B** The final displacement of the girl is the vector from her starting point to her ending point. If you follow the displacements in order, you will see that she ends up 4 m west of her starting point.
4. **E** To solve this problem, note that the 100-N force is the hypotenuse of a right triangle. The horizontal component force is therefore given by **F** = (100 N) cos 45 = 70.7 N, which is approximately equal to 71 N of force.
5. **D** You can draw the two forces to *scale,* letting 1 cm = 1 N. If you complete the parallelogram and draw the diagonal, as shown below, you notice that the resultant is 5 N. You should also recognize the *3-4-5 right triangle*; the resultant forms the hypotenuse of a right triangle whose arms are, respectively, 3 and 4. Therefore the hypotenuse is 5.

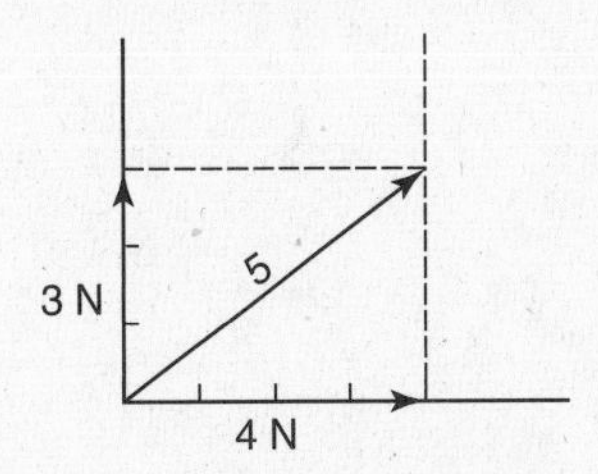

6. **E** When two vectors act in opposite directions (an angle of 180°), their resultant is the difference between the two vectors. When two vectors act in the same direction (an angle of 0°), their resultant is the sum of the two vectors. As the angle between the two vectors increases from 0° to 180°, their resultant decreases from the sum to the difference of the two vectors. To determine the actual value of any of these angles, use a parallelogram.
7. **E** When two forces act in the same direction, the resultant of the two forces is their sum. If you said 13, you were too hasty. You should recognize the *5-12-13 right triangle,* but the two given forces are not at right angles to each other. Always read the question carefully.
8. **A** The angle for vector **A** to the positive *x*-axis is given by

$$\tan \theta = \frac{\mathbf{A}_y}{\mathbf{A}_x} = \frac{7.5}{2.5}$$

Thus tan θ = 3 and θ = 71.5° or, rounded off, 72°.

9. **C** Vectors can be added in any order. The only requirement is that the vectors have the same magnitude and direction as they are shuffled. A look at the four sets of displacements indicates that I, III, and IV consist of the same vectors listed in different orders. These three sets will produce equal resultants.

10. C Sketch the vectors head to tail as if forming a vector triangle in construction. The resultant is drawn from the tail of the first vector to the head of the second. Choose the horizontal vector as the first one; then choice (C) represents the general direction of the resultant.

11. B Sketch the vectors in any order, as shown below. Draw the resultant from the tail of the first vector to the head of the last.

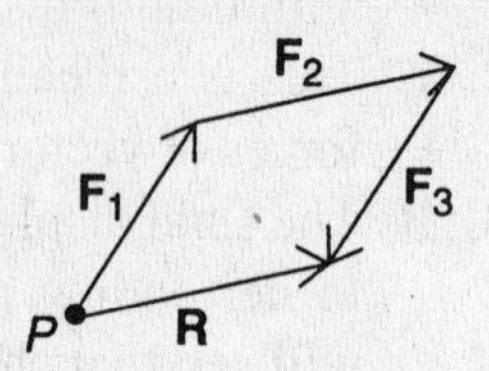

CHAPTER 4

One-Dimensional Motion

Introduction

Motion involves a change in position of an object over time. When we observe an object moving, it is always with respect to a frame of reference. Since motion occurs in a particular direction, it is dependent on the choice of a coordinate system for that frame of reference. This fact leads us to the conclusion that there is a vector nature to motion that must be taken into account when we want to analyze how something moves.

In this chapter, we shall confine our discussions to motion in one dimension. In physics, the study of motion is called *kinematics*. No hypotheses are made in kinematics concerning why something moves. Kinematics is a completely descriptive study of *how* something moves.

Average and Instantaneous Motion

Motion can take place either uniformly or nonuniformly. In other words, if an object is moving, relative to a frame of reference, the displacement changes, over a period of time, may be equal or unequal.

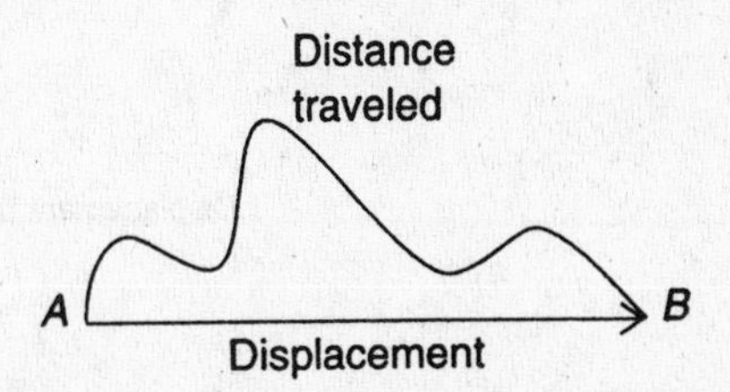

FIGURE 4.1 Displacement Versus Distance

If, in Figure 4.1, we consider the actual distance traveled by an object along some arbitrary path, we are dealing with a scalar quantity. The displacement vector **AB**, however, is directed along the line connecting points *A* and *B* (whether or not this is the actual route taken). Thus, when a baseball player hits a home run and runs around the bases, he or she may have traveled a distance of 360 feet (the bases are 90 feet apart), but the player's final displacement is zero (having started and ended up in the same place)!

If we are given the displacement vector of an object for a period of time Δt, we define the *average velocity*, $\bar{\mathbf{v}}$, to be equal to

$$\bar{\mathbf{v}} = \frac{\Delta \mathbf{x}}{\Delta t}$$

This is a vector quantity since directions are specified. Numerically, we think of the average speed as being the ratio of the total distance traveled to the total elapsed time. The units of velocity are meters per second (m/s).

If we are interested in the velocity at any instant in time, we can define the *instantaneous velocity* to be the velocity, $\mathbf{v}$, as determined at any precise time period. In a car, the speedometer registers instantaneous speed, which can become velocity if we take into account the direction of motion. If the velocity is constant, the average and instantaneous velocities are equal, and we can write simply

$$\Delta\mathbf{x} = \mathbf{v}\,\Delta t$$

It is possible that the observer making the measurements of motion is likewise in motion relative to Earth. In this case, we consider the relative velocity between the two systems. If two objects, a and b, are moving in the same direction, the relative velocity between the objects is given by

$$\mathbf{v}_{rel} = \mathbf{v}_a - \mathbf{v}_b$$

In a similar way, if two objects, a and b, are approaching each other in opposite directions, the relative velocity between them is given by

$$\mathbf{v}_{rel} = \mathbf{v}_a + \mathbf{v}_b$$

Uniformly Accelerated Motion

If an object is moving with a constant velocity such that its position is taken to be zero when it is first observed, a graph of the expression $\mathbf{x} = \mathbf{v}t$ will represent a direct relationship between position and time (see Figure 4.2).

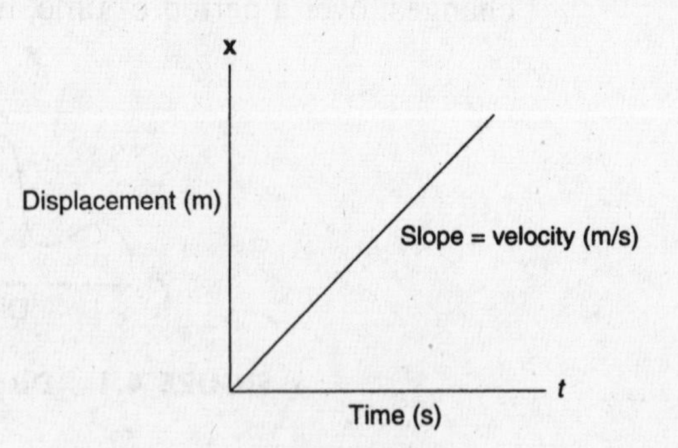

FIGURE 4.2

Since the line is straight, the constant slope (in which $\mathbf{v} = \Delta\mathbf{x}/\Delta t$) indicates that the velocity is constant throughout the time interval. If we were to plot velocity versus time for this motion, the graph might look like Figure 4.3.

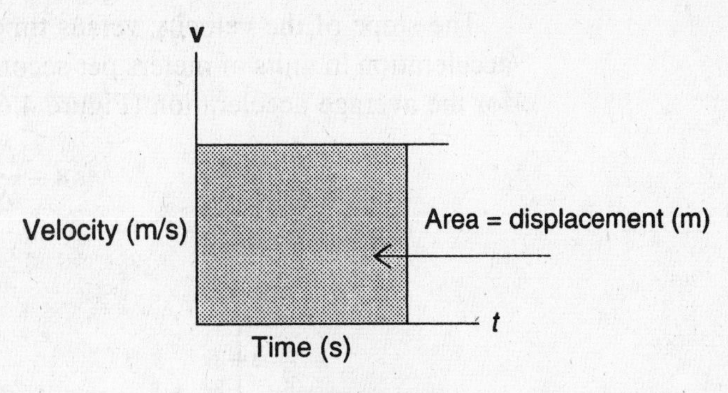

FIGURE 4.3

Notice that, for any time t, the area under the graph equals the displacement.

It is possible that the object is changing direction while maintaining a constant speed (uniform circular motion is an example of this situation), or that it is changing both speed and direction. In any case, if the velocity is changing, we say that the object is *accelerating*. If the velocity is changing uniformly, the object has uniform acceleration. In this case, a graph of velocity versus time will look like Figure 4.4.

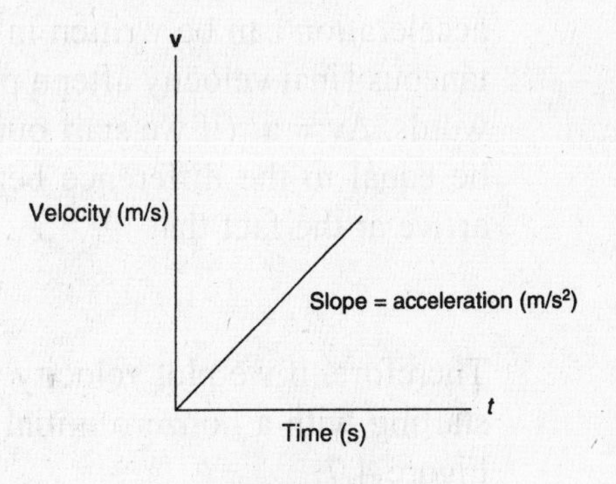

FIGURE 4.4

The displacement from $t = 0$ to any other time is equal to the area of the triangle formed. However, between any two intermediary times, the resulting figure is a trapezoid. If we make several measurements, the displacement versus time graph for uniformly accelerated motion is a parabola starting from the origin (if we make the initial conditions that, when $t = 0$, $x = 0$ and $\mathbf{v} = 0$; see Figure 4.5).

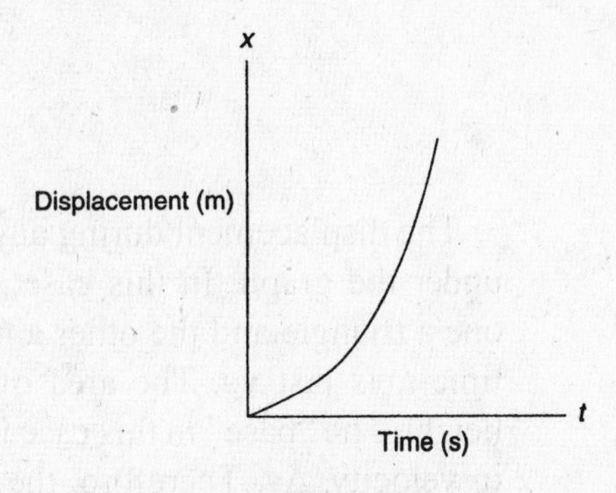

FIGURE 4.5

The slope of the velocity versus time graph is defined to be the average acceleration in units of meters per second squared (m/s^2). We can now write, for the average acceleration (Figure 4.6),

$$\bar{\mathbf{a}} = \frac{\Delta v}{\Delta t}$$

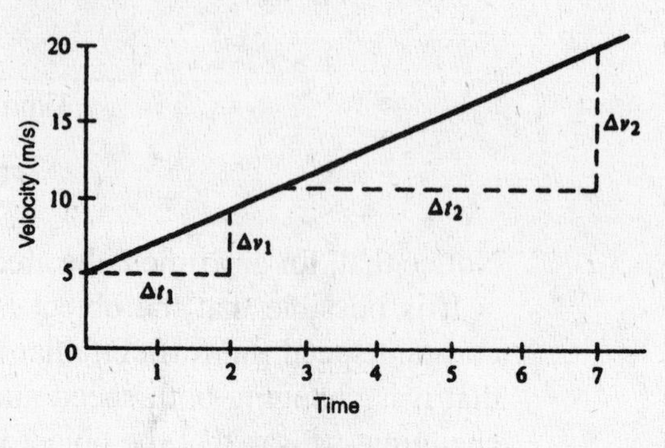

FIGURE 4.6

If the acceleration is taken to be constant in time, our expression for average acceleration can be written in a form that allows us to calculate the instantaneous final velocity after a period of acceleration has taken place. In other words, $\Delta\mathbf{v} = \mathbf{a}t$ (if we start our time interval from zero). If we define $\Delta\mathbf{v}$ to be equal to the difference between an initial and a final velocity, we can arrive at the fact that

$$\mathbf{v}_f = \mathbf{v}_i + \mathbf{a}t$$

Therefore, if we plot velocity versus time for uniformly accelerated motion starting with a nonzero initial velocity, the result is a graph that looks like Figure 4.7.

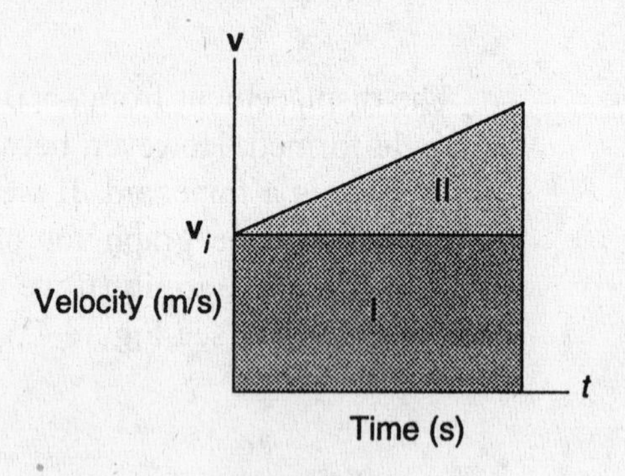

FIGURE 4.7

The displacement during any period of time will be equal to the total area under the graph. In this case, the total area will be the sum of two areas, one a triangle and the other a rectangle. The area of the rectangle, for some time t, is just $\mathbf{v}_i t$. The area of the triangle is one-half the base times the height. The "base" in this case is the time period, t; the "height" is the change in velocity, $\Delta\mathbf{v}$. Therefore, the area of the triangle is $(1/2)\Delta\mathbf{v}t$. If we recall the definition of $\Delta\mathbf{v}$ and the fact that $\mathbf{v}_f = \mathbf{v}_i + \mathbf{a}t$, we obtain the following formula for the distance traveled during uniformly accelerated motion starting with an initial velocity (assuming we start from the origin):

$$x = \mathbf{v}_i t + \frac{1}{2}\mathbf{a}t^2$$

This analysis suggests an alternative method of determining the average velocity of an object during uniformly accelerated motion. If we start, for example, with a velocity of 10 meters per second and accelerate uniformly for 5 seconds at a rate of 2 meters per second squared, the average velocity during that interval is the average of 10, 12, 14, 16, 18, and 20 meters per second, which is just 15 meters per second. Therefore, we can simply write

$$\overline{\mathbf{v}} = \frac{\mathbf{v}_i + \mathbf{v}_f}{2}$$

Occasionally, a problem in kinematics does not explicitly mention the time involved. For this reason it would be nice to have a formula for velocity that does not involve the time factor. We can derive one from all the other formulas. Since the above equation relates the average velocity to the initial and final velocities (for uniformly accelerated motion), we can write our displacement formula as

$$\mathbf{x} = \frac{\mathbf{v}_i + \mathbf{v}_f}{2}t$$

Now, since $\mathbf{v}_f = \mathbf{v}_i + \mathbf{a}t$, we can express the time as $t = (\mathbf{v}_f - \mathbf{v}_i)/\mathbf{a}$. Therefore

$$x = \left(\frac{\mathbf{v}_i + \mathbf{v}_f}{2}\right)\left(\frac{\mathbf{v}_f - \mathbf{v}_i}{\mathbf{a}}\right)$$

$$= \frac{\mathbf{v}_f^2 - \mathbf{v}_i^2}{2\mathbf{a}}$$

Accelerated Motion Due to Gravity

Since velocity and acceleration are vector quantities, we need to consider the algebraic conventions accepted for dealing with various directions. For example, we usually agree to consider motion up or to the right as positive, and motion down or to the left as negative. In this way, negative velocity implies backward motion (since negative speed doesn't make sense). Negative acceleration, on the other hand, implies that the object is slowing down.

In the preceding section, we considered the case of uniformly accelerated motion. A naturally occurring situation in which acceleration is constant involves motion due to gravity. Since gravity acts in the downward direction, its acceleration, common to all masses, is likewise taken to be in the downward direction. Notice that this convention may or may not imply deceleration. An object dropped from rest is accelerating as time passes, but in the downward or negative direction.

In physics, the acceleration due to gravity is generally taken to be −9.8 meters per second squared and is represented by the symbol **g**. As a vector quantity, **g** is negative and the direction of motion is reflected in either the position or the velocity equation. For example, if an object was dropped, the displacement in the −**y** direction would be given by

$$\mathbf{y} = -\frac{1}{2}\mathbf{g}t^2$$

If, however, the object was thrown down, with an initial velocity, we could write

$$\mathbf{y} = -\mathbf{v}_i t - \frac{1}{2}\mathbf{g}t^2$$

Finally, if the object was thrown upward, with some initial velocity, we would have

$$\mathbf{y} = \mathbf{v}_i t - \frac{1}{2}\mathbf{g}t^2$$

Graphical Analysis of Motion

We have seen that much information can be obtained if we consider the graphical analysis of motion. If complex changes in motion are taking place, visualization may provide a better understanding of the physics involved than algebra. The techniques of graphical analysis are as simple as slopes and areas. For example, we already know that, for uniformly accelerated motion, the graph of distance versus time is a parabola. Since the slope is changing, the instantaneous velocity can be approximated at a point **P** by finding the slope of a tangent line drawn to a given point on the curve (see Figure 4.8).

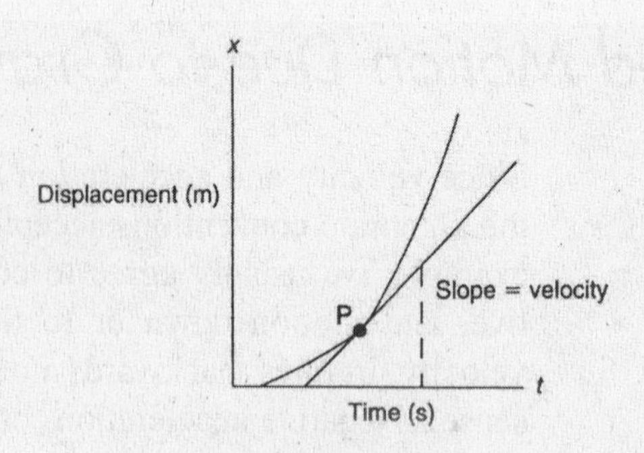

FIGURE 4.8

What would happen if an object accelerated from rest maintained a constant velocity for a while, and then slowed down to a stop? Using what we know about graphs of velocity and acceleration in displacement versus time, we might represent the motion as shown in Figure 4.9.

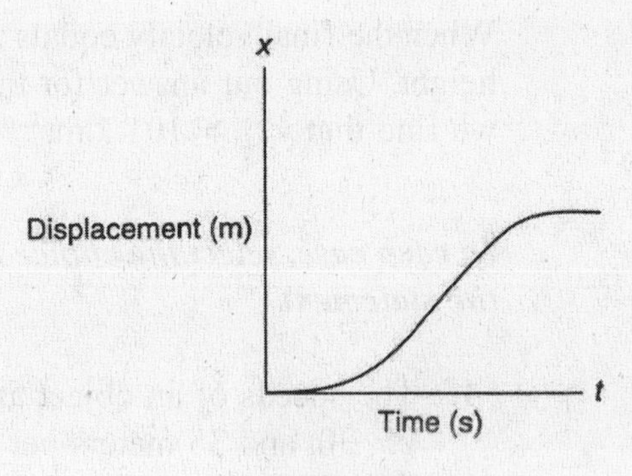

FIGURE 4.9

We can apply many instances of motion to graphs. In the case of changing velocity, consider the graph of velocity versus time for an object thrown upward into the air, reaching its highest point, changing direction, and then accelerating downward. This motion has a constant downward acceleration that, at first, acts to slow the object down, but later acts to speed it up. A graph of this motion is seen in Figure 4.10a and b.

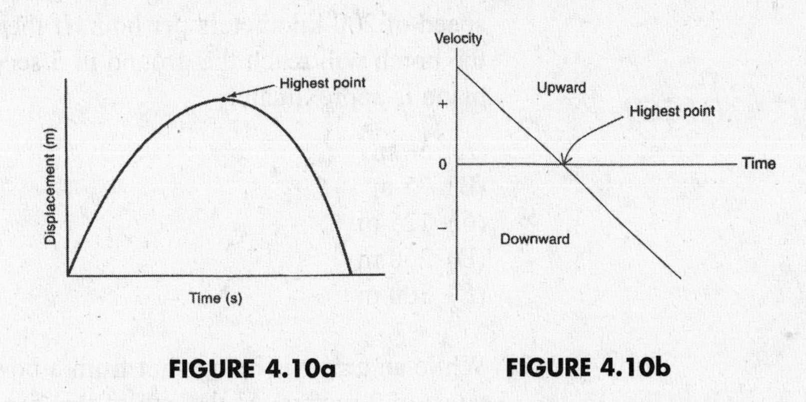

FIGURE 4.10a FIGURE 4.10b

Example: A ball is thrown straight up into the air and, after being in the air for 9 seconds, is caught by a person 5 meters above the ground. To what maximum height did the ball go?

The equation

$$\mathbf{y} = \mathbf{v}_i t - \frac{1}{2}\mathbf{g}t^2$$

represents the vertical position of the ball above the ground for any time t. Since the ball is thrown upward, the initial velocity is positive, while the acceleration of gravity is always directed downward (and hence is algebraically negative). Thus, we can use the above equation to find the initial velocity when $\mathbf{y}$ = 5 m and t = 9 s. Substituting these values, as well as the magnitude of $\mathbf{g}$ (9.8 m/s^2) in the equation, we get $\mathbf{v}_i$ = 44.65 m/s.

Now, to find the maximum height, we note that, when the ball rises to its maximum, its speed becomes zero since its velocity is changing direction. Since we do not know how long the ball takes to rise to its maximum height (we could determine this value if desired), we can use the formula

$$\mathbf{v}_f^2 - \mathbf{v}_i^2 = -2\mathbf{gy}$$

When the final velocity equals zero, the value of **y** is equal to the maximum height. Using our answer for the initial velocity and the known value of **g**, we find that $\mathbf{y}_{max} = 101.7$ m.

Questions Chapter 4

In each case, select the choice that best answers the question or completes the statement.

1. The speeds of an object at the ends of 4 successive seconds are 20, 25, 30, and 35 meters per second, respectively. The acceleration of this object is

(A) 5 m/s
(B) 5 m/s^2
(C) 5 s/m^2
(D) 5 s/m
(E) 20 m/s

2. A bomb is dropped from an airplane moving horizontally with a speed of 200 kilometers per hour. If the air resistance is negligible, the bomb will reach the ground in 5 seconds when the altitude of the plane is approximately

(A) 50 m
(B) 75 m
(C) 125 m
(D) 250 m
(E) 300 m

3. While an arrow is being shot from a bow, it is accelerated over a distance of 2 meters. At the end of this acceleration it leaves the bow with a speed of 200 meters per second. The average acceleration imparted to the arrow is

(A) 200 m/s^2
(B) 400 m/s^2
(C) 500 m/s^2
(D) 1,000 m/s^2
(E) 10,000 m/s^2

Questions 4–6 refer to the velocity versus time graph shown below.

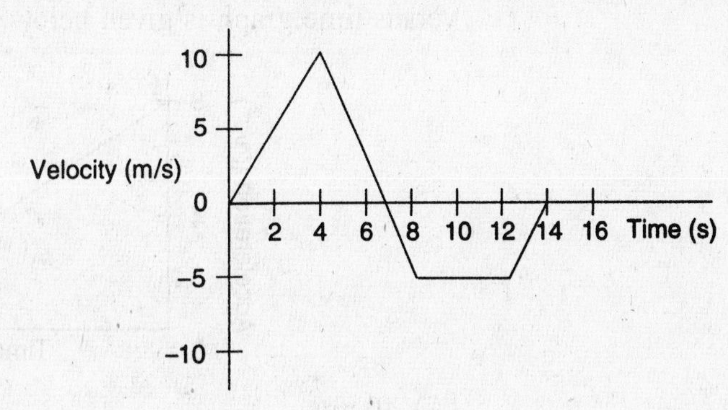

4. The total distance traveled by the object during the indicated 14 seconds is

(A) 7.5 m
(B) 25 m
(C) 62.5 m
(D) 77.5 m
(E) 82.1 m

5. The total displacement of the object during the 14 seconds indicated is

(A) 7.5 m
(B) 25 m
(C) 62.5 m
(D) 77.5 m
(E) 82.1 m

6. The average velocity, in meters per second, of the object is

(A) 0
(B) 0.5
(C) 2.5
(D) 4.5
(E) 5.6

7. What is the total change in velocity for the object whose acceleration versus time graph is given below?

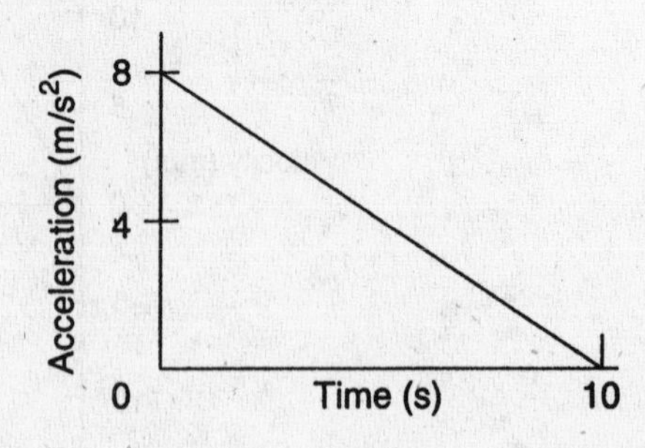

(A) 40 m/s
(B) –40 m/s
(C) 80 m/s
(D) –80 m/s
(E) 0 m/s

8. A particle moves along the x-axis subject to the following position function:

$$x(t) = 2t^2 + 3t - 1$$

What was its average velocity during the interval $t = 0$ to $t = 3$?

(A) 2.0 m/s
(B) 3.3 m/s
(C) 5 m/s
(D) 8.3 m/s
(E) 9.0 m/s

9. An object has an initial velocity of 15 meters per second. How long must it accelerate at a constant rate of 3 meters per second squared before its average velocity is equal to twice its initial velocity?

(A) 5 s
(B) 10 s
(C) 15 s
(D) 20 s
(E) 25 s

10. A handball is tossed vertically upward with a velocity of 19.6 meters per second. Approximately how high will it rise?

(A) 15 m
(B) 20 m
(C) 25 m
(D) 30 m
(E) 60 m

Explanations to Questions Chapter 4

Answers

1. (B)	**4.** (C)	**7.** (A)	**10.** (B)
2. (C)	**5.** (A)	**8.** (E)	
3. (E)	**6.** (B)	**9.** (B)	

Explanations

1. B Acceleration = $\frac{\text{change in velocity}}{\text{time in which change takes place}}$

$$= \frac{5 \text{ m/s}}{1 \text{ s}}$$

$$= \frac{5 \text{ m/s}}{\text{s}}$$

This means that each second the speed is changing 5 m/s.

2. C The horizontal velocity does not affect the vertical velocity. A suitable expression for free fall, since we know the time of falling and want to calculate the distance covered, is $\mathbf{d} = \frac{1}{2}\mathbf{g}t^2$. Here,

$$\mathbf{d} = \tfrac{1}{2}(9.8 \text{ m/s}^2)(5 \text{ s})^2 = 123 \text{ m}.$$

3. E The arrow is accelerated from rest. Its final speed is 200 m/s; it is accelerated over a known distance, 2 m; $\mathbf{v}^2 = 2\,\mathbf{ad}$; $(200 \text{ m/s})^2 = 2a(2 \text{ m})$. Then

$$\mathbf{a} = \frac{40{,}000}{4} = 10{,}000 \text{ m/s}^2.$$

The term *average acceleration* is used to take care of fluctuations; don't divide by 2.

4. C The total distance traveled by the object in 14 s is equal to the total area under the graph. Break the figure up into triangles and rectangles, and note that their areas add up to 62.5 m (recall that distance is a scalar).

5. A The total displacement of the object is the sum of the positive (forward) and negative (backward) areas, representing the fact that the object moves away from the origin and then back (as indicated by the velocity vs. time graph) going below the x-axis. Add and subtract the proper areas:

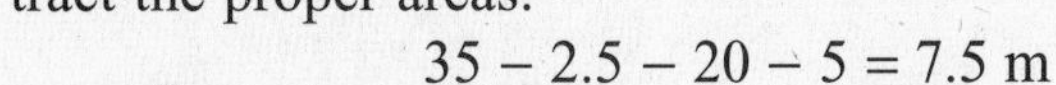

$$35 - 2.5 - 20 - 5 = 7.5 \text{ m}$$

for the final displacement.

6. **B** Average velocity is the change in displacement for the object over the change in time (not distance). Since from question 5 you know the displacement change to be 7.5 m in 14 s, the average velocity is therefore 0.5 m/s.

7. **A** The change in velocity for the object is equal to the area under the graph: 1/2 (+8)(10) = 40 m/s.

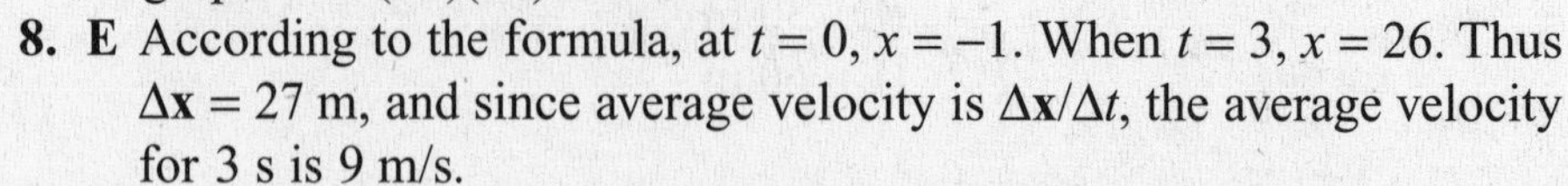

8. **E** According to the formula, at $t = 0$, $x = -1$. When $t = 3$, $x = 26$. Thus $\Delta \mathbf{x} = 27$ m, and since average velocity is $\Delta \mathbf{x}/\Delta t$, the average velocity for 3 s is 9 m/s.

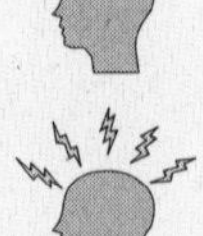

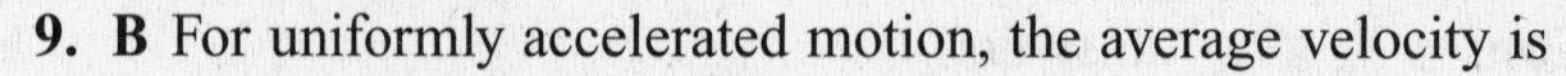

9. **B** For uniformly accelerated motion, the average velocity is

$$\bar{\mathbf{v}} = \frac{\mathbf{v}_f + \mathbf{v}_i}{2}$$

The question requires that the average velocity be equal to twice the initial velocity; thus, if the average velocity is 30 m/s, the final velocity attained must be 45 m/s. Now the question is: How long must the object accelerate from 15 m/s to achieve a speed of 45 m/s? A change in velocity of 30 m/s at a rate of 3 m/s^2 for 10 s is implied.

10. **B** This is a problem in motion with constant acceleration equal to 9.8 m/s^2 downward, in which the object has an initial velocity (19.6 m/s), but its final velocity at the moment in question is zero.

$$\mathbf{v}_1^2 = 2\ \mathbf{gd}$$

$$-(19.6 \text{ m/s})^2 = 2\ (-9.8 \text{ m/s}^2)\ (s)$$

$$\mathbf{d} = 19.6 \text{ m}$$

CHAPTER 5

Two-Dimensional Motion

Relative Motion

In Chapter 4, we reviewed the basic elements of one-dimensional rectilinear motion. In this chapter, we will consider only two-dimensional motion. In one sense, one-dimensional motion can be viewed as two-dimensional motion by a suitable transformation of coordinate systems.

For example, consider the definition of displacement from Chapter 4. If we define a coordinate system for a reference frame, the location of a point in that frame is determined by a position vector drawn from the origin of that coordinate system to the point. If the point is displaced, the one-dimensional vector drawn from point A to point B is called the *displacement vector* and is designated as $\mathbf{B} - \mathbf{A}$ (see Figure 5.1).

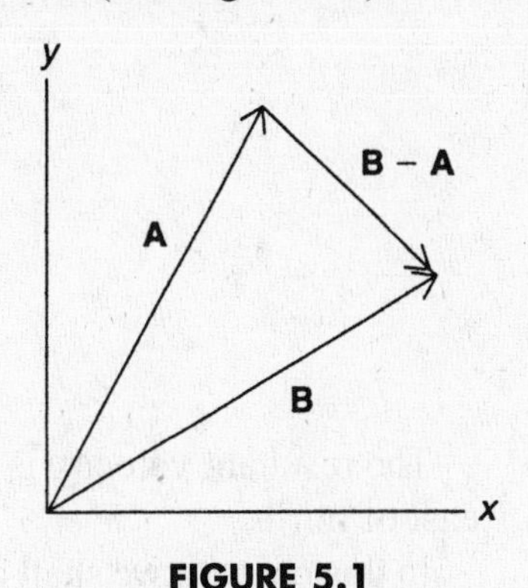

FIGURE 5.1

Even though the vector $\mathbf{B} - \mathbf{A}$ is one dimensional, we can resolve it into two components that represent mutually perpendicular and independent simultaneous motions. An example of this type of motion can be seen when a boat tries to cross a river or an airplane meets a crosswind. In the case of the boat, its velocity, relative to the river, is based on the properties of the engine and is measured by the speedometer on board. However, to a person on the shore, its *relative velocity* (or effective velocity) is different from what the speedometer in the boat may report. In Figure 5.2, we see such a situation with the river moving to the right at 4 meters per second and the boat moving upward across the river at 10 meters per second. If you like, we can call to the right "eastward" and up the river "northward."

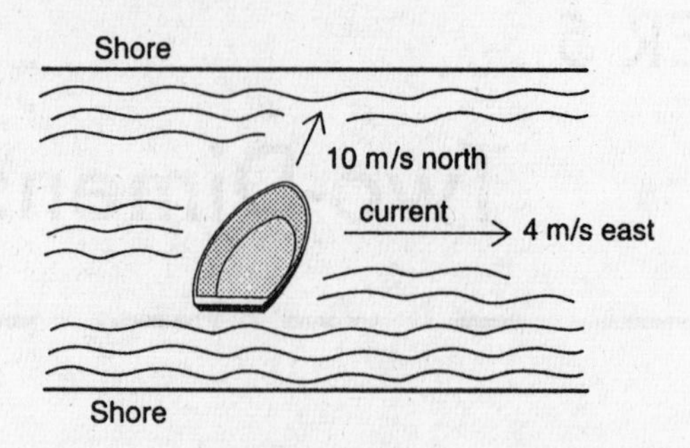

FIGURE 5.2

By vector methods, the resultant velocity relative to the shore is given by the Pythagorean theorem. The direction is found by means of a simple sketch (Figure 5.3) connecting the vectors head to tail to preserve the proper orientation. Numerically, we can use the tangent function or the law of sines.

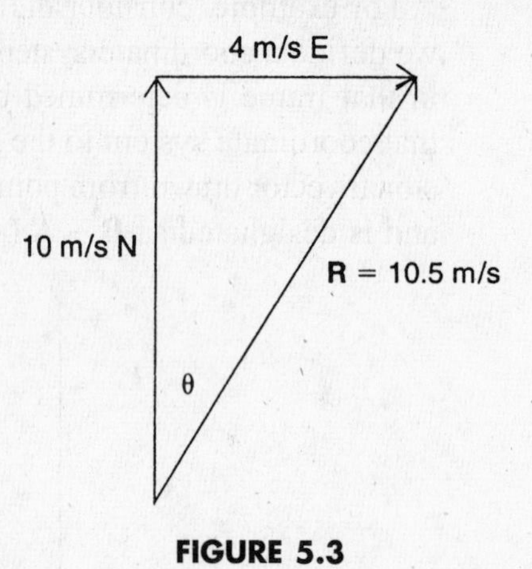

FIGURE 5.3

The resultant velocity is 10.5 meters per second at an angle of 22 degrees east of north.

In this chapter, we shall consider two-dimensional motion as viewed from the Earth frame of reference. Typical of this kind of two-dimensional motion is that of a projectile. Galileo Galilei was one of the pioneers in studying the mechanics in flying projectiles and the first to discover that the path of a projectile is a parabola.

Horizontally Launched Projectiles

If you roll a ball off a smooth table, you will observe that it does not fall straight down. With trial and error, you might observe that how far it falls will depend on how fast it is moving forward. Initially, however, the ball has no vertical velocity. The ability to "fall" is given by gravity, and the acceleration due to gravity is –9.8 meter per second squared. Since gravity acts in a direction perpendicular to the initial horizontal motion, the two motions are simultaneous and independent. Galileo demonstrated that the trajectory is a parabola.

We know that the distance fallen by a mass dropped from rest is given by the equation

$$|\mathbf{y}| = -\frac{1}{2}\mathbf{g}t^2$$

Since the ball that you rolled off the table is moving horizontally, with some initial constant velocity, it covers a distance (called the *range*) of $x = |\mathbf{v}_{ix}|t$. Since the time is the same for both motions, we can first solve for the time, using the x equation, and then substitute the result into the y equation. In other words,

$$t = \frac{x}{|\mathbf{v}_{ix}|}$$

and therefore

$$|\mathbf{y}| = -\frac{1}{2}\mathbf{g}\left(\frac{x}{\mathbf{v}_{ix}}\right)^2 = -\frac{\mathbf{g}x^2}{2\mathbf{v}_{ix}^2}$$

which is of course the equation of an inverted parabola. This equation of y in terms of x is called the *trajectory* of the projectile, while the two separate equations for x and y as functions of time are called *parametric equations*. Figure 5.4 illustrates this trajectory as well as a position vector **R**, which locates a point at any given time in space.

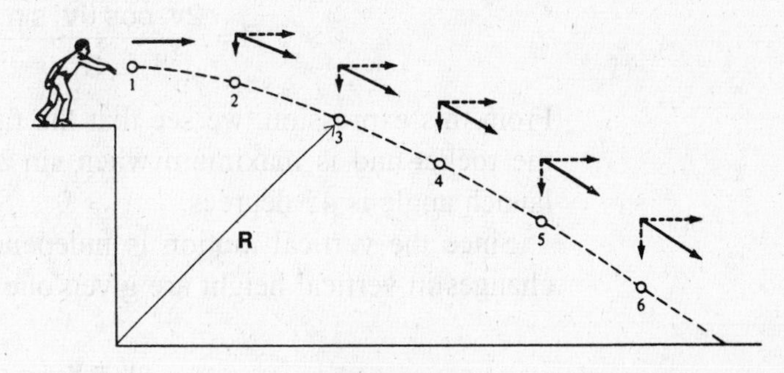

FIGURE 5.4

If the height from which a projectile is launched is known, the time to fall can be calculated from the equation for free fall. For example, if the height is 49 meters, the time to fall is 3.16 seconds. If the horizontal velocity is 10 meters per second, the maximum range will be 31.6 meters. We can also follow the trajectory by determining how far the object has fallen when it is 10 meters away from the base. Using the trajectory formula and the velocity given, we find that the answer is –4.9 meter.

Projectiles Launched at an Angle

Suppose that a rocket on the ground is launched with some initial velocity at some angle θ. The vector nature of velocity allows us to immediately write the equations for the horizontal and vertical components of initial velocity:

$$\mathbf{v}_{ix} = \mathbf{v}\cos\theta \quad \text{and} \quad \mathbf{v}_{iy} = \mathbf{v}\sin\theta$$

Since each motion is independent, we can consider the fact that, in the absence of friction, the horizontal velocity will be constant while the y velocity will decrease as the rocket rises. When the rocket reaches its maximum height, its vertical velocity will be zero; then gravity will accelerate the rocket back down. It will continue to move forward at a constant rate. How long will the rocket take to reach its maximum height? From the definition of acceleration and the equations in Chapter 4, we know that this will be the time needed for gravity to decelerate the vertical velocity to zero; that is,

$$t_{\text{up}} = \frac{|\,\mathbf{v}_{iy}\,|}{|\,\mathbf{g}\,|} = \frac{|\,\mathbf{v}_i\,|\sin\theta}{|\,\mathbf{g}\,|}$$

The total time of flight will be just twice this time. Therefore, the range is the product of the initial horizontal velocity (which is constant) and the total time. In other words,

$$\mathbf{R} = \text{Range} = 2\mathbf{v}_{ix}t_{\text{up}} = 2\mathbf{v}\cos\theta t_{\text{up}}$$

If we now substitute into this expression the time to go up, we obtain

$$\mathbf{R} = \frac{2\mathbf{v}_i\cos\theta\mathbf{v}_t\sin\theta}{\mathbf{g}} = \frac{\mathbf{v}_i^2\sin 2\theta}{\mathbf{g}}$$

From this expression, we see that the range is independent of the mass of the rocket and is maximum when sin 2θ equals 1. This occurs when the launch angle is 45 degrees.

Since the vertical motion is independent of the horizontal motion, the changes in vertical height are given one dimensionally as

$$\mathbf{y} = \mathbf{v}_{iy}t - \frac{1}{2}\mathbf{g}t^2$$

If we want to know the maximum height achieved, we simply use the value for the time to reach the highest point. To find the trajectory of the rocket, we first substitute for the value of the initial vertical velocity and then substitute for time. The time, t, is found from the fact that at all times $x = \mathbf{v}_i\cos\theta t$. Solving for and then substituting, we obtain the following trajectory:

$$\mathbf{y} = \mathbf{v}_i\sin\theta\left(\frac{x}{\mathbf{v}_i\cos\theta}\right) - \frac{1}{2}\mathbf{g}\left(\frac{\mathbf{x}}{\mathbf{v}_i^2\cos^2\theta}\right)$$

This simplifies to the final equation for the trajectory of the projectile:

$$\mathbf{y} = (\tan\theta)x - \left(\frac{\mathbf{g}}{2\mathbf{v}_i^2\cos^2\theta}\right)x^2$$

Of course, this equation also represents an inverted parabola. This trajectory is seen in Figure 5.5.

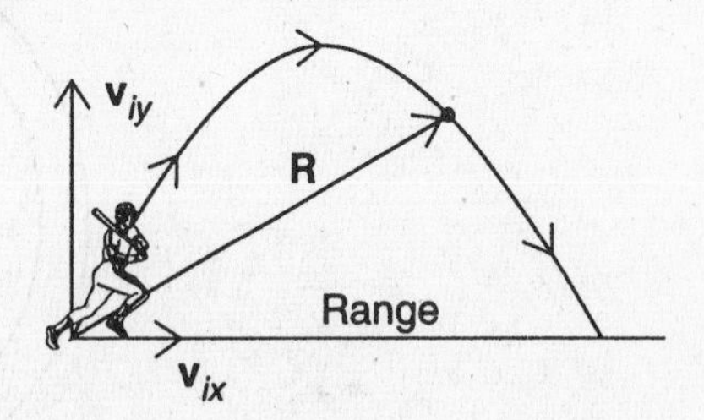

FIGURE 5.5

As an example, if a projectile is launched with an initial velocity of 100 meters per second at an angle of 30°, the maximum range will be equal to 883.7 meters. To find the maximum height, we could first find the time required to reach that height, but a little algebra will give us a formula for the maximum height independent of time. The maximum height can also be expressed as

$$\mathbf{y}_{max} = \frac{\mathbf{v}_i^2 \sin^2\theta}{2\mathbf{g}}$$

Using the known numbers, we find that the maximum height reached is 127.55 meters.

Uniform Circular Motion

Velocity is a vector. When velocity changes, magnitude or direction, or both, can change. When direction is the only quantity changing, as the result of a centrally directed deflecting force, the result is uniform circular motion.

Consider here an object already undergoing periodic, uniform circular motion. By this description we mean that the object maintains a constant speed as it revolves around a circle of radius R, in a period of time T. The number of revolutions per second is called the *frequency*, f. This is illustrated in Figure 5.6.

Clearly, if the total distance traveled around the circle is its circumference, then the constant average velocity is given by $2\pi R/T$. Also, if the circle maintains a constant radius, the quantity $2\pi/T$ is called the *angular frequency* or *angular velocity*, $\boldsymbol{\omega}$. The units of $\boldsymbol{\omega}$ involve units related to arc length. Before going any further with this topic, we should digress a moment to discuss these units.

Imagine a circle of radius R as shown in Figure 5.7.

A point particle is moving counterclockwise around this circle, making an angle ϕ with the positive x-axis. A corresponding arc length s is associated with angle ϕ. From geometry, we know that the length of an arc s is equal to

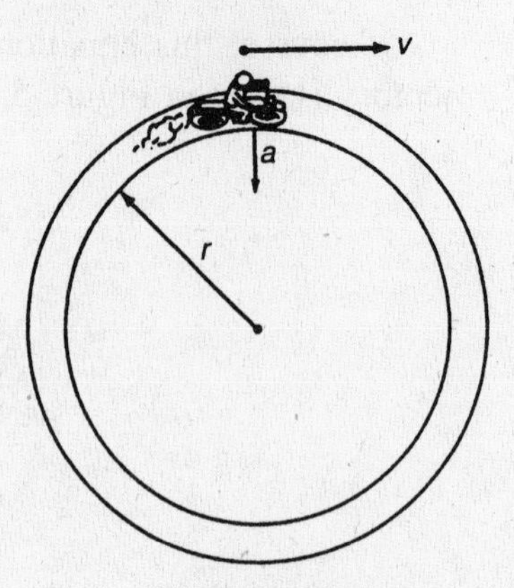

FIGURE 5.6

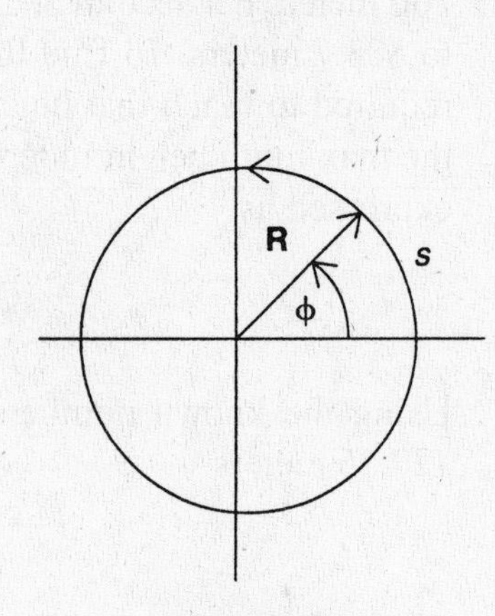

FIGURE 5.7

the circumference of the circle multiplied by the ratio of angle ϕ to the whole angle of the circle (360 degrees):

$$\mathbf{s} = 2\pi\mathbf{R}\left(\frac{\phi}{\text{whole-circle angle}}\right)$$

If we define the whole-circle angle to be equal to 2π and call the new unit "radians," we have

$$\mathbf{s} = \mathbf{R}\phi$$

which is a common formula from rotational dynamics in physics. The units of the angle are now such that 360 degrees corresponds to 2π radians. Thus, we have, for a uniformly changing angle in radians, the average angular velocity defined to be $\Delta\phi/\Delta t$. This is related to the linear velocity along the arc length by the formula

$$\frac{\Delta\mathbf{s}}{\Delta t} = \mathbf{R}\frac{\Delta\phi}{\Delta t} = \mathbf{R}\omega = \mathbf{v}$$

The units of angular velocity (or frequency) are now radians per second (rad/s). If we define the angular velocity as $\Delta\phi/\Delta t$, we can also write that $\Delta\phi = \omega\,\Delta t$, which looks remarkably similar to the linear formula $\Delta\mathbf{x} = \mathbf{v}\,\Delta t$.

Now, let us return to the idea of uniform circular motion. If the object is moving around a circle at a constant rate, its linear (tangential) and angular velocities are related by the angle swept out per second and the radius of the circle. We can see that, if we have an object on a rotating platform, or have a rotating solid, then the formula above informs us that all points have the same angular velocity but the linear velocity is directly proportional to the radius.

If the velocity is changing (as it is by virtue of its changing direction), what kind of acceleration does the object have? If we consider the change in velocity, $\Delta \mathbf{v}$, as a change in the quantity $R\omega$, then clearly

$$\frac{\Delta \mathbf{v}}{\Delta t} = R\frac{\Delta \omega}{\Delta t} = R\alpha$$

where α is called the *angular acceleration* and has units of radians per second squared (rad/s^2). Notice that, analogously to linear motion, we could write that, for constant angular acceleration, $\Delta\omega = \alpha\,\Delta t$. However, if we have uniform circular motion, the angular frequency is not changing, so the angular acceleration is zero! Thus, we are still left with the question concerning the nature of the acceleration involved.

This confusion is cleared up easily if we recognize that, as angle ϕ changes, so do the position vector and the arc length. Consider Figure 5.8.

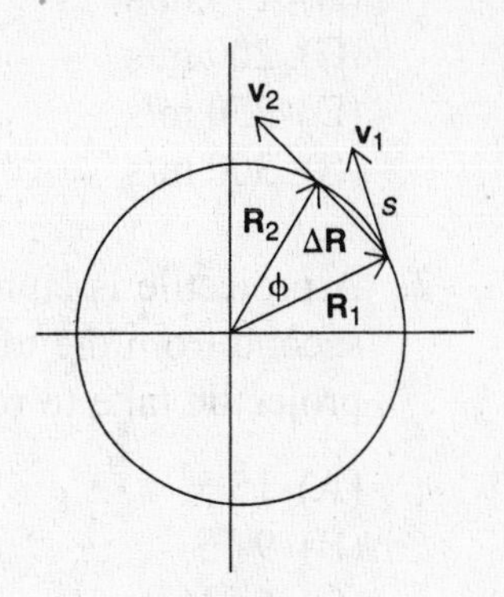

FIGURE 5.8

For small changes in angle, the arc length Δs is approximately equal to the displacement $\Delta \mathbf{R}$. Thus, we can write, for small approximations, that

$$\frac{\Delta \mathbf{R}}{\Delta t} = \frac{R\,\Delta\phi}{\Delta t}$$

This means that $\Delta \mathbf{R}/\mathbf{R} = \Delta\phi$. Since velocity is a vector, $\Delta \mathbf{v}$ is swept out in the same time as Δt and at the same angle, $\Delta\phi$. Thus

$$\frac{\Delta \mathbf{R}}{\mathbf{R}} = \frac{\Delta \mathbf{v}}{\mathbf{v}}$$

and so we can write

$$\frac{\Delta \mathbf{v}}{\Delta t} = \left(\frac{\mathbf{v}}{R}\right)\left(\frac{\Delta \mathbf{R}}{\Delta t}\right) = \frac{\mathbf{v}^2}{R} = \mathbf{a}_c$$

This quantity is known as the *centripetal acceleration*, and from the construction it is directed radially inward toward the center of the circle. The units are standard acceleration units. This quantity is different from the angular acceleration **a**, but is related to the angular velocity in the following way:

Since $\boldsymbol{\omega} = \mathbf{v}/R$, we have

$$\mathbf{a}_c = \frac{\mathbf{v}^2}{R} = \left(\frac{\mathbf{v}}{R}\right)\mathbf{v} = (\boldsymbol{\omega})(\boldsymbol{\omega}R) = \boldsymbol{\omega}^2 R$$

Finally, we can express the centripetal acceleration in terms of the period, T. Since we defined $\mathbf{v} = 2\pi R/T$, it immediately follows that

$$\mathbf{a}_c = \frac{4\pi^2 R}{T^2}$$

Questions Chapter 5

In each case, select the choice that best answers the question or completes the statement.

1. A projectile is launched at an angle of 45° with a velocity of 250 meters per second. If air resistance is neglected, the magnitude of the horizontal velocity of the projectile at the time it reaches maximum altitude is equal to

(A) 0 m/s
(B) 175 m/s
(C) 200 m/s
(D) 250 m/s
(E) 300 m/s

2. A projectile is launched horizontally with a velocity of 25 meters per second from the top of a 75 meter cliff. How many seconds will the projectile take to reach the bottom?

(A) 15.5
(B) 9.75
(C) 6.31
(D) 4.27
(E) 3.91

3. An object is launched from the ground with an initial velocity and angle such that the maximum height achieved is equal to the total range of the projectile. The tangent of the launch angle is equal to

(A) 1
(B) 2
(C) 3
(D) 4
(E) 5

4. At a launch angle of 45°, the range of a launched projectile is given by

(A) $\frac{\mathbf{v}_i^2}{\mathbf{g}}$

(B) $\frac{2\mathbf{v}_i^2}{\mathbf{g}}$

(C) $\frac{\mathbf{v}_i^2}{2\mathbf{g}}$

(D) $\sqrt{\frac{\mathbf{v}_i^2}{2\mathbf{g}}}$

(E) $\frac{2\mathbf{v}_i}{\mathbf{g}}$

5. A projectile is launched at a certain angle. After 4 seconds, it hits the top of a building 500 meters away. The height of the building is 50 meters. The projectile was launched at an angle of

(A) 14°
(B) 21°
(C) 37°
(D) 76°
(E) 85°

6. A projectile is launched at a velocity of 125 meters per second at an angle of 20°. A building is 200 meters away. At what height above the ground will the projectile strike the building?

(A) 50.6 m
(B) 90.2 m
(C) 104.3 m
(D) 114.6 m
(E) 125.6 m

7. The operator of a boat wishes to cross a 5-kilometer-wide river that is flowing to the east at 10 meters per second. He wishes to reach the exact point on the opposite shore 15 minutes after starting. At what speed and in what direction should the boat travel?

(A) 11.2 m/s at 26.6° E of N
(B) 8.66 m/s at 63.4° W of N
(C) 11.2 m/s at 63.4° W of N
(D) 8.66 m/s at 26.6° E of N
(E) 5 m/s due N

8. An object is moving around a circle of radius 1.5 meters at a constant velocity of 7 meters per second. The frequency of the motion, in revolutions per second, is

(A) 0.24
(B) 0.53
(C) 0.67
(D) 0.74
(E) 0.98

9. A stereo turntable has a frequency of 33 revolutions per minute. A record of radius 15.25 centimeters is placed on the turntable and begins to rotate. The velocity of a point on the edge of the record is

(A) 2.16 m/s
(B) 22.23 m/s
(C) 0.53 m/s
(D) 7.62 m/s
(E) 13.5 m/s

10. In 3 seconds, an object moving around a circle sweeps out an angle of 9 radians. If the radius of the circle is 0.5 meter, the centripetal acceleration of the object is

(A) 4.5 m/s^2
(B) 3 m/s^2
(C) 13.5 m/s^2
(D) 15 m/s^2
(E) 18 m/s^2

Explanations to Questions Chapter 5

Answers

1. (B)	**4.** (A)	**7.** (C)	**10.** (A)
2. (E)	**5.** (A)	**8.** (D)	
3. (D)	**6.** (A)	**9.** (C)	

Explanations

1. B The horizontal component of velocity remains constant in the absence of resistive forces and is equal to $\mathbf{v}_i \cos \theta$. Substituting the known numbers, we get $\mathbf{v}_x = (250)(0.7) = 175$ m/s.

2. E The time to fall is given by the free-fall formula:

$$t = \sqrt{\frac{2\mathbf{y}}{\mathbf{g}}}$$

If we substitute the known numbers, we get 3.91 s for the time.

3. D Since the maximum height is equal to the range, we set these two equations equal to each other:

$$\frac{\mathbf{v}_i^2 \sin^2 \theta}{2\mathbf{g}} = \frac{\mathbf{v}_i^2 \sin 2\theta}{\mathbf{g}}$$

Recalling that $\sin 2\theta = 2 \sin\theta \cos\theta$, we solve for $\tan\theta$ and find that it equals 4.

4. A From the formula for range, we see that, at 45°, $\sin 2\theta = 1$, and so

$$R = \frac{\mathbf{v}_i^2}{\mathbf{g}}$$

5. A We know that, after 4 s, the projectile has traveled horizontally 500 m. Therefore, the horizontal velocity was a constant 125 m/s and is equal to $\mathbf{V}_i \cos\theta$. We also know that, after 4 s, the y-position of the projectile is 50 m. Thus we can write:

$$50 = 4\mathbf{v}_i \sin\theta - 4.9(4)^2 = 4\mathbf{v}_i \sin\theta - 78.4$$

Therefore, $\mathbf{v}_i \cos\theta = 125$, $\mathbf{v}_i \sin\theta = 32.1$, and $\tan\theta = 0.2568$, so $\theta = 14.4°$.

6. A The x-component of velocity is 93.97 m/s, and the y-component is 34.2 m/s. Since the projectile travels 200 m horizontally at 93.97 m/s, the time of flight $t = 2.128$ s. Substituting this time into our equation for the y position gives $y = 50.6$ m.

7. C The river is flowing at 10 m/s to the right (east) and the resultant desired velocity is 5 m/s up (north). Therefore, the actual velocity, relative to the river, is heading W of N. By the Pythagorean theorem, the velocity of the boat must be 11.2 m/s. The angle is given by the tangent function. In the diagram below, not drawn to scale but correct for orientation, $\tan\theta = 10/5 = 2$. Therefore, $\theta = 63.4°$ W of N.

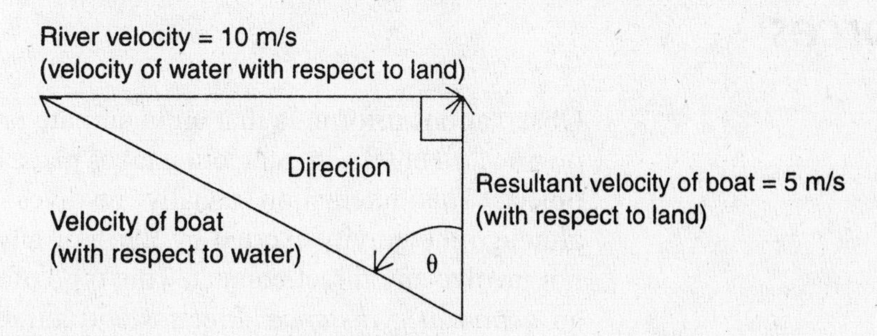

8. D The velocity is given by $\mathbf{v} = 2\pi R/T$, where T is the period, but the frequency f is just the reciprocal of the period. Thus, $\mathbf{v} = 2\pi Rf$. We substitute the known numbers to obtain $f = 0.74$ rev/s.

9. C First we convert 33 rpm into radians per second. We note that there are 2π rad/rev and 60 s in 1 min. The conversion gives an angular frequency of 3.454 rad/s. Now we convert the radius to meters so that we are in SI units (MKS). Thus, $R = 0.1525$ m. Now we know that, for uniform circular motion, $\mathbf{v} = R\omega$, so

$$\mathbf{v} = (0.1525\text{ m})(3.454\text{ rad/s}) = 0.53\text{ m/s}$$

10. A If the object sweeps out 9 rad in 3 s, its angular velocity is 3 rad/s. Since the radius is 0.5 m, the centripetal acceleration is given by

$$a_c = \omega^2 R = (9)(0.5) = 4.5\text{ m/s}^2$$

CHAPTER 6

Forces and Newton's Laws of Motion

Introduction

From our discussion in Chapter 3 of frames of reference, you should be able to convince yourself that, if two objects have the same velocity, the relative velocity between them is zero, and therefore to an observer one object looks as though it is at rest with respect to the other. In fact, it would be impossible to decide whether or not such an *"inertial"* observer was moving! Therefore, accepting this fact, we state that an object that appears to be at rest in the Earth frame of reference will simply be stated to be "at rest" relative to us (the observers).

Forces

Observations inform us that an inanimate object will not move freely of its own accord unless an interaction takes place between it and at least one other object. This interaction usually involves contact between the objects, although the gravity exerted by Earth or any other body on the object does not involve any direct contact. This type of interaction is sometimes called an *action at a distance*. Electrostatic attraction and repulsion, as well as magnetic attraction and repulsion, are other examples of this effect.

When interaction occurs, we say that a *force* has been created between the objects. If an object is moving (relative to us), then the force may change the direction of the motion or it may change its speed. In other words, there will be a change in the velocity, which is a vector quantity; the magnitude and/or direction of the velocity will change. If no change occurs, we must conclude the presence of another force that resists the changes induced by the applied one. Friction is an example of an opposing force of this type.

In contrast to deflecting forces, other forces may be restorative. An example is a spring or other elastic material that has been stretched. Once the material has been elongated, it will snap back in an attempt to return to its original status.

If all the forces acting on an object produce no net change, the object is in a state of *equilibrium*. If the object is moving relative to us, we say that it is in a state of *dynamic equilibrium*. If the object is at rest relative to us, we say that it is in a state of *static equilibrium*. Figure 6.1 illustrates these ideas.

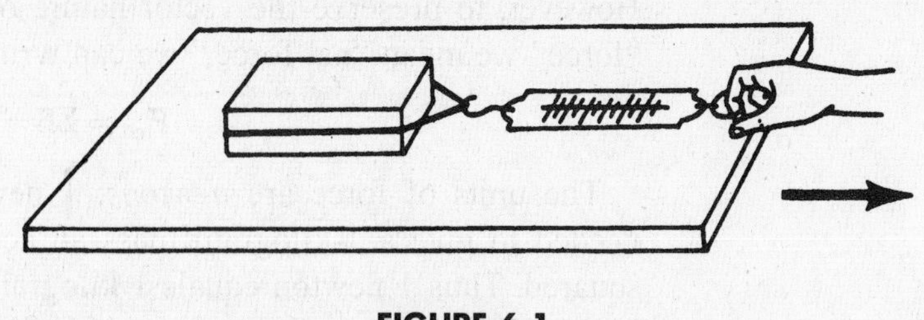

FIGURE 6.1

Newton's Laws of Motion

In 1687, at the urging of his friend Edmund Halley, Isaac Newton published his greatest work. It was titled the *Mathematical Principles of Natural Philosophy*, but is more widely known by a shortened version, *Principia*. In this book, Isaac Newton revolutionized the rational study of mechanics by the introduction of mathematical principles that all of nature was considered to obey. Using his newly developed ideas, Newton set out to explain the observations and analyses of Galileo Galilei and Johannes Kepler.

The ability of an object to resist a change in its state of motion is called *inertia*. This concept is the key to *Newton's first law of motion*:

> Every body continues in its state of rest, or of uniform motion in a straight line, unless it is compelled to change that state by forces acting on it.

In other words, an object at rest will tend to stay at rest, and an object in motion will tend to stay in motion, unless acted upon by an external force. By "rest," we of course mean the observed state of rest in a particular frame of reference. As stated above, the concept of "inertia" is taken to mean the ability of an object to resist a force attempting to change its state of motion. As we will subsequently see, this concept is covered under the new concept of *mass* (a scalar, as opposed to *weight*, a vector force).

If a mass has an unbalanced force incident upon it, the velocity of the mass is observed to change. The magnitude of this velocity change depends inversely on the amount of mass. In other words, a force acting along the direction of motion will cause a smaller mass to accelerate more than a larger mass. Newton's second law of motion expresses these observations as follows:

> The change of motion is proportional to the applied force, and is in the direction which that force acts.

The acceleration produced by the force is in the same direction as the force. This does not mean that the object's direction must remain the same. Mathematically, the second law is sometimes expressed as:

$$\mathbf{F} = m\mathbf{a}$$

However, to preserve the vector nature of the forces, and the fact that by "force" we mean "net force," we can write the second law as:

$$\mathbf{F}_{\text{net}} = \Sigma\mathbf{F} = m\mathbf{a}$$

The units of force are *newtons*; 1 newton (N) is defined as the force needed to give a 1-kilogram mass an acceleration of 1 meter per second squared. Thus 1 newton equals 1 kilogram-meter per second squared. This being the case, the weight of an object is given by

$$\mathbf{W} = m\mathbf{g}$$

where **g** is the acceleration due to gravity and the units of weight are newtons. In the SI system of units, the kilogram is taken as the standard unit of mass and corresponds to 9.8 newtons of weight, or about 2.2 English pounds.

Newton's third law of motion is crucial for understanding the conservation laws we will discuss later. It stresses the fact that forces are the result of mutual interactions and are thus produced in pairs. The third law is usually stated as follows (Figure 6.2):

For every action, there is an equal but opposite reaction.

FIGURE 6.2

When Newton wrote his second law of motion, his concept of "quantity of motion" was defined as the product of an object's mass and velocity. In modern times, this concept is known as *momentum*.

Static Applications of Newton's Laws

If we look more closely at Newton's second law of motion, we see an interesting implication. If the net force acting on an object is zero, the acceleration of the object will likewise be zero. Notice, however, that kinematically zero acceleration does not imply zero motion! It simply indicates that the velocity of the object is not changing. If the object is in a state of rest and remains at rest (because of zero net forces), the object is in static equilibrium. Some interesting problems in engineering deal with the static stability of structures. Let's look at a simple example.

Place this book on a table. You will observe that it is not moving relative to you. Its state of rest is provided by the zero net force between the downward force of gravity and the upward reactive force of the table (pushing on the floor, which in turn pushes up on the table, etc.). Figure 6.3 shows this setup. The upward reactive force of the table, sometimes called the *normal force*, always acts perpendicularly to the surface.

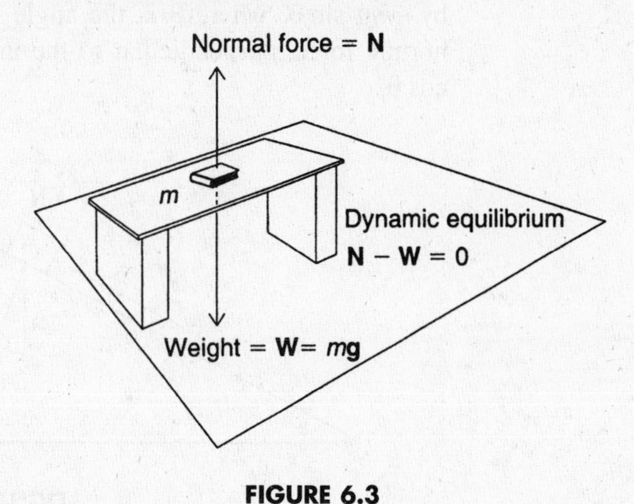

FIGURE 6.3

In this example, we have used both the second and third laws of motion. Remember: The second law is a vector equation, and so we must treat separately the vector forces that comprise the sum of all vector forces in each direction! In this case, we have:

$$\Sigma \mathbf{F}_y = \mathbf{N} - \mathbf{W} = 0$$

Another static situation occurs when a mass is hung from an elastic spring. It is observed that, when the mass is attached vertically to a spring that has a certain natural length, the amount of stretching, or elongation, is directly proportional to the applied weight. This relationship, known as *Hooke's law*, supplies a technique for measuring static forces.

Mathematically, Hooke's law is given as

$$\mathbf{F} = -k\mathbf{x}$$

where k is the spring constant in units of newtons per meter (N/m), and $\mathbf{x}$ is the elongation beyond the natural length. The negative is used to indicate that the applied force is restorative so that, if allowed, the spring will accelerate back in the opposite direction. As a static situation, a given spring can be calibrated for known weights or masses, and thus used as a "scale" for indicating weight or other applied forces.

Dynamic Applications of Newton's Laws

If the net force acting on an object is not zero, Newton's second law implies the existence of an acceleration in the direction of the net force. Therefore, if a 10-kilogram mass is acted on by a 10-newton force from rest, the accel-era-tion will be 1 meter per second squared.

In order to resolve the force of gravity into components parallel and perpendicular to the incline, we must first identify the relevant angle in the geometry. This procedure is outlined in Figure 6.4, and we can see that the magnitude of the downward component of weight along the incline is given by $-m\mathbf{g}\sin\theta$, where θ is the angle of the incline. The magnitude of the normal force, perpendicular to the incline, is therefore given by $|\mathbf{N}| = |m\mathbf{g}\cos\theta|$.

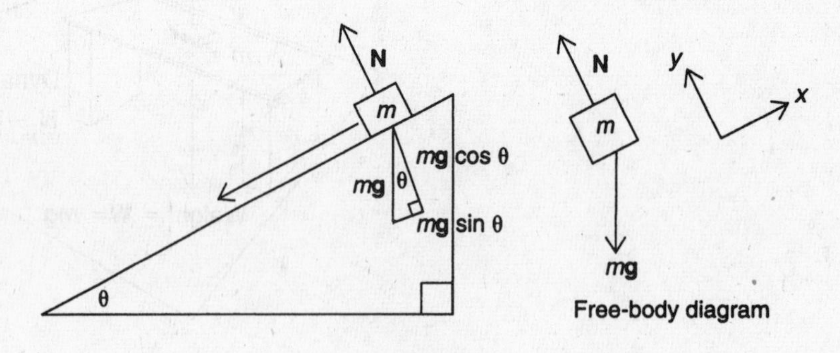

FIGURE 6.4

The direction of $m\mathbf{g}\cos\theta$ can be taken as inward into the incline's surface and provides the force necessary to keep the mass in contact with the surface. If friction were present, this force would be the main contributor to the frictional force, which would be directed opposite the direction of motion.

If, for example, the mass was 10 kilograms, the weight would be 98 newtons. If θ = 30°, we would write for the *x*-forces:

$$\Sigma F_x = m\mathbf{a} = -m\mathbf{g}\sin\theta = -98\sin 30 = 49\text{ N}$$

and since there is no acceleration in the *y*-direction, we would have:

$$\Sigma \mathbf{F}_y = 0 = \mathbf{N} - m\mathbf{g}\cos\theta = \mathbf{N} - 98\cos 30 = \mathbf{N} - 8.49$$

Thus the normal force is equal to 8.49 newtons.

Central Forces

Consider a point mass of 10 kilograms moving around a circle, supported by a string making a 45-degree angle to a vertical post (the so-called *conical pendulum*). Let's analyze this situation, which is shown in Figure 6.5.

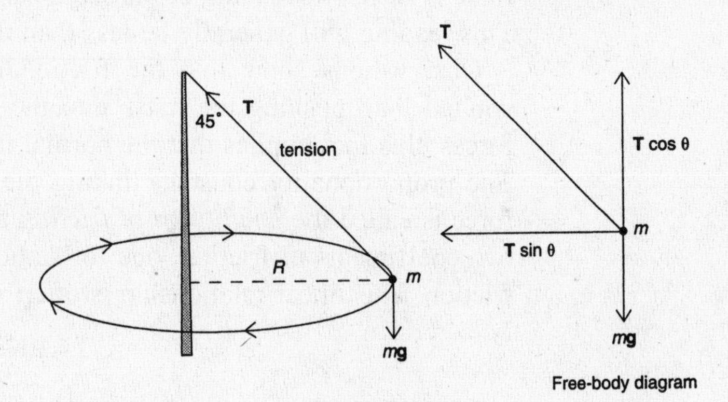

FIGURE 6.5

In this case, the magnitude of the weight, $m\mathbf{g}$, is balanced by the upward component of the tension in the string, given by $\mathbf{T}\cos\theta$. The inward component of tension, $\mathbf{T}\sin\theta$, is responsible for providing a *centripetal force*. However, from the frame of reference of the mass, the force experienced is "outward" and is referred to as the *centrifugal force*. This force is often called "fictitious" since it is perceived only in the mass's frame of reference.

If we specify that the radius is 1.5 meters and the velocity of the mass is unknown, we see that, vertically, $\mathbf{T}\cos 45 = m\mathbf{g}$. This fact implies that, if $m\mathbf{g} = 98$ newtons, then $\mathbf{T} = 138.6$ newtons.

Horizontally, we see that

$$\mathbf{T}\sin 45 = \frac{M\mathbf{v}^2}{R}$$

where the expression for centripetal force is given by

$$\mathbf{F}_c = \frac{m\mathbf{v}^2}{R}$$

using the logic of $\mathbf{F} = m\mathbf{a}$. Using our known information, we solve for velocity and find that $\mathbf{v} = 3.83$ meters per second. It is important to remember that the components of forces must be resolved along the principal axes of the chosen coordinate system.

Friction

Friction is a contact force between two surfaces that is responsible for opposing sliding motion. Even the smoothest surfaces are microscopically rough, with peaks and valleys like a mountain range. When an object is first moved, this friction plus inertia must be overcome. If a spring balance is attached to a mass and then pulled, the reading of the force scale when the mass first begins to move provides a measure of the static friction. Once the mass is moving, if we maintain a steady enough force, the velocity of movement will be constant. Thus, the acceleration will be zero, indicating that the net force is zero. The reading of the scale will then measure the kinetic friction. This reading will generally be less than the starting reading.

Observations show that the frictional force is directly related to the applied load pushing the mass into the surface. From our knowledge of forces, this fact implies that the normal force is responsible for this action. The proportionality constant linking the normal force with the frictional force is called the *coefficient of friction* and is symbolized by μ. There are two coefficients of friction, one for static friction and the other for kinetic friction. This linear relationship is often written as

$$f = \mu N$$

and the analysis of a situation involves identifying the normal force. For example, in Figure 6.6 a mass is being pulled along a horizontal surface by a string, making an angle θ.

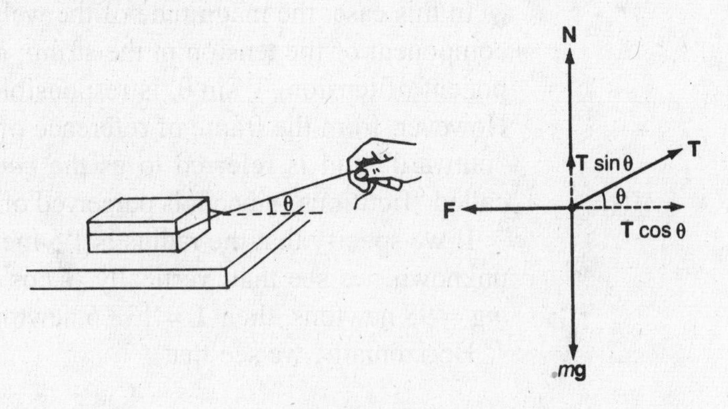

FIGURE 6.6

Questions Chapter 6

In each case, select the choice that best answers the question or completes the statement.

1. What is the approximate weight of a 7.5-kilogram mass sitting on the surface of Earth?

(A) 7.5 N
(B) 0.75 N
(C) 75 N
(D) 17.5 N
(E) 27.5 N

2. A 15-kilogram mass is sliding along a frictionless floor with an acceleration of 5 meters per second squared. What is the magnitude of the net force acting on the mass?

(A) 15 N
(B) 20 N
(C) 3 N
(D) 75 N
(E) 10 N

3. A 3-kilogram mass is traveling at 15 meters per second when friction begins to decelerate it to a stop in 5 seconds. What is the magnitude of the force of friction?

(A) 9 N
(B) 3 N
(C) 45 N
(D) 60 N
(E) 15 N

4. A person is standing on a bathroom scale in an elevator and notices that her weight is less than normal. This means that the elevator:

(A) must be moving upward
(B) must be moving downward
(C) must be at rest
(D) must have positive acceleration
(E) must have negative acceleration

5. A mass is sliding down an incline. As the angle of elevation for the incline increases, the magnitude of the component of its weight perpendicular to the incline:

(A) increases
(B) decreases
(C) increases, then decreases
(D) decreases, then increases
(E) remains the same

6. A 4-kilogram mass is accelerated by a horizontal force of 15 newtons. If the magnitude of the acceleration is 3 meters per second squared, how much friction is acting on the mass?

(A) 5 N
(B) 15 N
(C) 0 N
(D) 27 N
(E) 3 N

7. In the situation shown below, what is the tension in string 1?

(A) 69.3 N
(B) 98 N
(C) 138.6 N
(D) 147.6 N
(E) 155 N

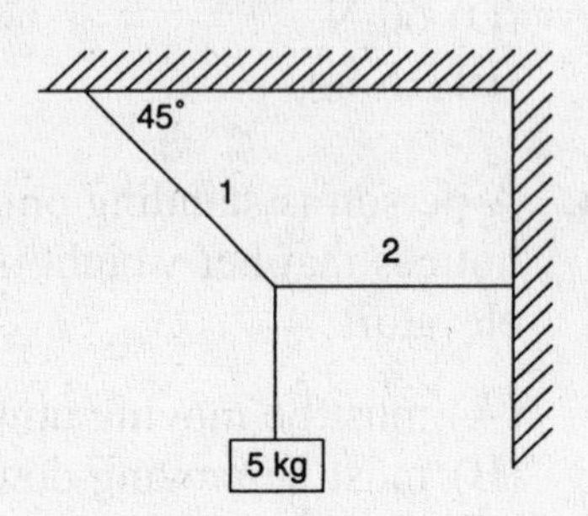

8. Two masses, M and m, are hung over a massless, frictionless pulley as shown below. If $M > m$, what is the downward acceleration of mass M?

(A) $\mathbf{g}$

(B) $\dfrac{(M-m)\mathbf{g}}{M+m}$

(C) $\left(\dfrac{M}{m}\right)\mathbf{g}$

(D) $\dfrac{Mm\mathbf{g}}{M+m}$

(E) $Mm\mathbf{g}$

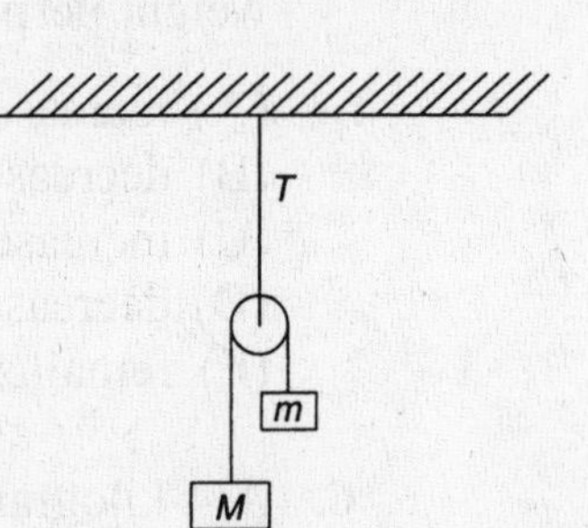

9. A 20-newton force is pushing two blocks horizontally along a frictionless floor as shown below.

F = 20 N
2 kg
8 kg

What is the force that the 8-kilogram mass exerts on the 2-kilogram mass?

(A) 4 N
(B) 8 N
(C) 16 N
(D) 20 N
(E) 24 N

10. A force of 20 N acts horizontally on a mass of 10 kilograms being pushed up a frictionless incline that makes a 30° angle with the horizontal, as shown below.

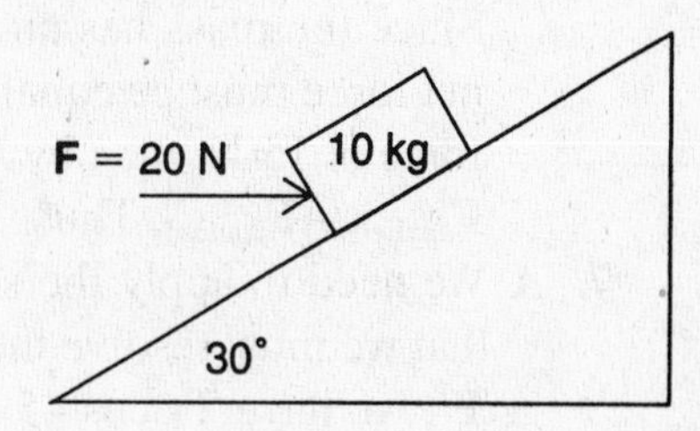

The magnitude of the acceleration of the mass up the incline is equal to

(A) 1.9 m/s^2
(B) 2.2 m/s^2
(C) 3.17 m/s^2
(D) 3.87 m/s^2
(E) 4.3 m/s^2

Explanations to Questions Chapter 6

Answers

1. (C) **4.** (E) **7.** (A) **10.** (C)
2. (D) **5.** (B) **8.** (B)
3. (A) **6.** (E) **9.** (C)

Explanations

1. C If we approximate the acceleration due to gravity as **g** = 10 m/s^2 , then the weight of a mass on the surface of Earth is given by **Fg** = *m***g** = (7.5 kg)(10 m/s^2) = 75 N.

2. D The net force acting on an accelerating mass is given by **F** = *m***a**. Using the formula, we find that the magnitude of the net force is **F** = (15 kg)(5 m/s^2) = 75 N.

3. A To find the magnitude of the force of friction that is decelerating the mass, we first need to know the value of the acceleration. Using **a** = **Δv** / Δt, we see that the acceleration (in magnitude) is (0 – 15 m/s) / 5 s = 3 m/s^2 . We can then use **F** = m**a** to get the magnitude of the net force of friction: **F** = (3kg)(3m/s^2) = 9 N.

4. E In an elevator that is accelerating, there is a net force acting on a person present. This net force is the resultant of the weight of the person and the normal (scale) force. If we write **F** = *m***a** = $\mathbf{F}_n$ – *m***g**, we see that, if the person "weighs" less in an elevator, it is because the elevator is experiencing negative acceleration. Notice that this fact does not tell you in which direction the elevator is moving! It maybe moving downward and speeding up (negative acceleration), or it may be slowing down while moving up (also with negative acceleration). Thus the correct answer is **E**.

5. B We have learned that the component of the weight perpendicular to the incline is proportional to the cosine of the angle of elevation. As this angle increases (making the incline more and more vertical), the value of the cosine decreases.

6. E The net force acting on an accelerating mass is given by $\mathbf{F} = m\mathbf{a}$. Since the mass has an acceleration of 3 m/s^2, the magnitude of the net force must be equal to $\mathbf{F}_{net} = (4 \text{ kg})(3 \text{ m/s}^2) = 12$ N. If the applied force is 15 N, then by vectors, the net force must be given by $\mathbf{F}_{net} = \mathbf{F}_{applied} - \mathbf{F}_{friction}$. Thus, $\mathbf{F}_{friction} = 3$ N.

7. A We need to apply the second law for static equilibrium. This means that we must resolve the tensions into their x- and y-components. For $\mathbf{T}_1$, we have $\mathbf{T}_1 \cos 45$ and $\mathbf{T}_1 \sin 45$. In equilibrium, the sum of all the x forces must equal zero. This means that $\mathbf{T}_1 \cos 45 = \mathbf{T}_2$ (since $\mathbf{T}_2$ is entirely horizontal). The y-component of $\mathbf{T}_1$ must balance the weight $= m\mathbf{g} = 49$ N. Thus, $\mathbf{T}_1 \sin 45 = 49$ N and $\mathbf{T}_1 = 69.3$ N.

8. B The free-body diagrams for both masses, M and m, look as shown below:

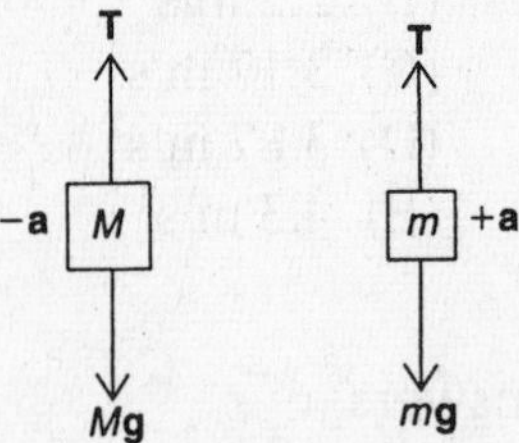

The large mass is accelerating downward, while the small mass is accelerating upward. The tension in the string is directed upward, while gravity, given by the weight, is directed downward. Using the second law for accelerated motion, we must show that $\Sigma\mathbf{F}_x = m\mathbf{a}$ and $\Sigma\mathbf{F}_y = m\mathbf{a}$ separately. Thus we have

$$\mathbf{T} - M\mathbf{g} = -M\mathbf{a} \quad \text{and} \quad \mathbf{T} - m\mathbf{g} = m\mathbf{a}$$

Eliminating tension $\mathbf{T}$ and solving for $\mathbf{a}$ gives $(M - m)\mathbf{g}/(M + m)$.

9. C The 20-N force is pushing on a total mass of 10 kg. Thus, using $\mathbf{F} = m\mathbf{a}$, we have the acceleration of both blocks equal to 2 m/s^2. We draw a free-body diagram for the 2-kg mass as shown below.

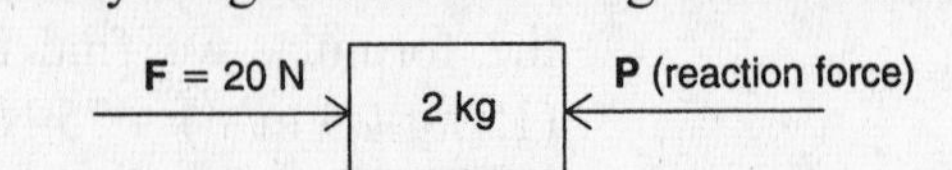

Let $\mathbf{P}$ represent the force that the 8-kg mass exerts on the 2-kg mass. Writing the second law of motion, we get $\mathbf{F} - \mathbf{P} = m\mathbf{a}$. To find $\mathbf{P}$, we substitute the known numbers; $20 - \mathbf{P} = (2)(2) = 4$. Thus, $\mathbf{P} = 16$ N.

10. C From the given diagram, we see that the force necessary to move the mass up the incline must be in excess of the component force of gravity trying to push the mass down the incline. The component of gravity down the incline is always given by $m\mathbf{g} \sin \theta$. Resolving the given force into a component parallel to the incline and a component perpendicular to the incline, we find that the force up the incline is, at the same angle, $\mathbf{F} \cos \theta$. Thus, in general we would write:

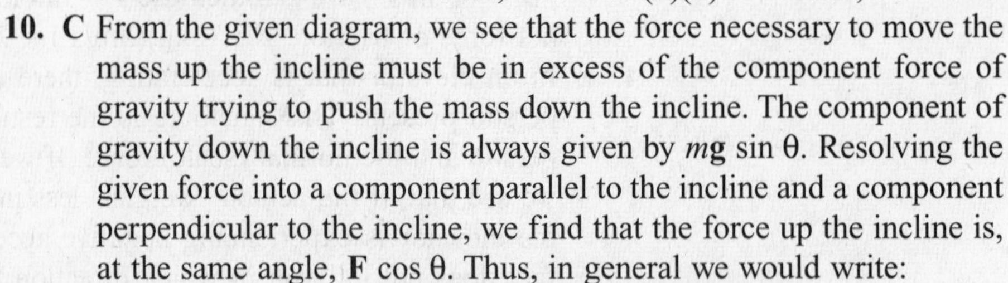

$$\mathbf{F} \cos \theta - m\mathbf{g} \sin \theta = m\mathbf{a}$$

Substituting the known numbers gives $\mathbf{a} = 3.17$ m/s^2.

CHAPTER 7

Work, Energy, Simple Machines

Work and Energy

In Figure 7.1 below, a pulley system is being used to lift an object with less effort than the object weighs. To understand how this is possible, we need to understand the concept of *work*, as it applies to physics.

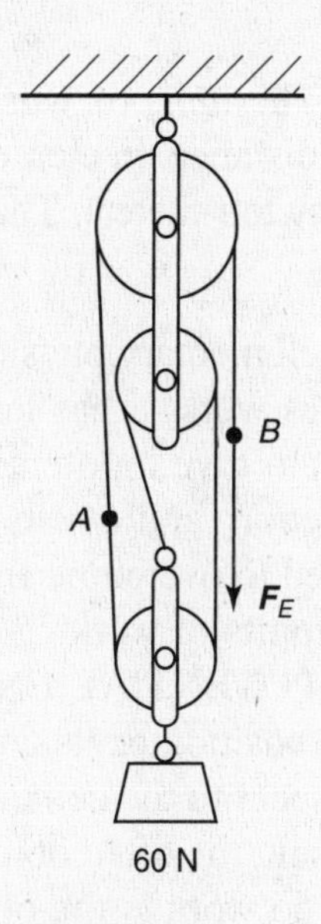

FIGURE 7.1

In physics we talk about work done on an object or work done by an object or by a force. When a force *moves* an object, the force does work on the object. If the 20-newton force in Figure 7.2 moves the 50-newton object 3 meters, 60 joules of work has been done on the object.

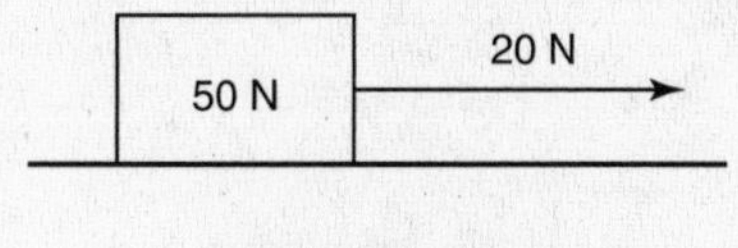

FIGURE 7.2

Work

Work is equal to the product of the force and the distance the object moves *in the direction of the force.* If the object moves in a direction other than the direction of the force, we must take the component of the force in the direction of motion of the object. For example, if a rope is used to pull a sled, the rope is flexible, and any force, $\boldsymbol{F}_1$, exerted on the upper end is to be thought of as a pull that is transmitted to the sled and acts in the direction of the rope (see Figure 7.3). This usually results in moving the sled along the road, and to calculate the work done by $\boldsymbol{F}_1$ on the sled we find the component $\boldsymbol{F}$ in the direction of motion and multiply it by the distance the sled moves while the force is applied.

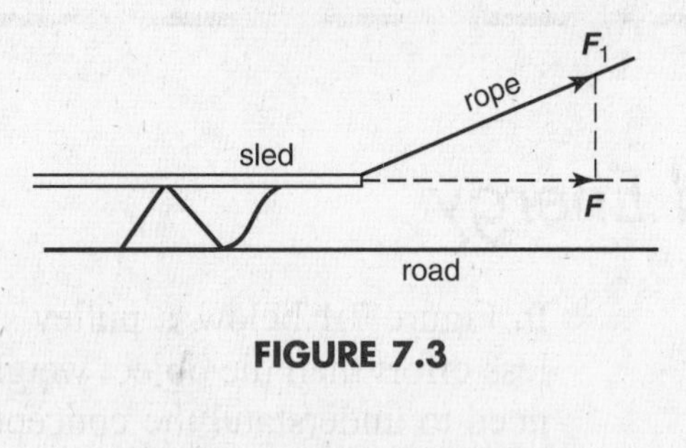

FIGURE 7.3

work = force × distance

The *units* of work are obtained by multiplying a unit of force by a unit of distance. In elementary physics courses the most common unit is the *joule* (newton-meter). The abbreviation for joule is J.

Energy

In elementary physics *energy* is defined as the ability to do work. If an object does work, it has less energy left. In mechanics, work is done on an object (1) to give it potential energy, (2) to give it kinetic energy, (3) to overcome friction, and (4) to accomplish a combination of the above three. Energy used to overcome friction is converted to heat. *Units* of energy are the same as units of work.

POTENTIAL ENERGY is the energy possessed by an object because of its position or its condition. If we lift an object we change its position with respect to the center of Earth; we do work against the force of gravity. As a result, the lifted object has a greater ability to do work. This increased ability to do work is the object's potential energy with respect to its original position, as can be illustrated by referring to Figure 7.4. Boys *A* and *B* have lifted the object of weight $\boldsymbol{w}$ a vertical distance *h*. The object now can do work: If *A* lets go, the object may be able to lift *B*, thus doing work on *B*. If $\boldsymbol{w} = 200$ newtons and $h = 15$ meters, the object gained a potential energy of 3,000 newton · meters.

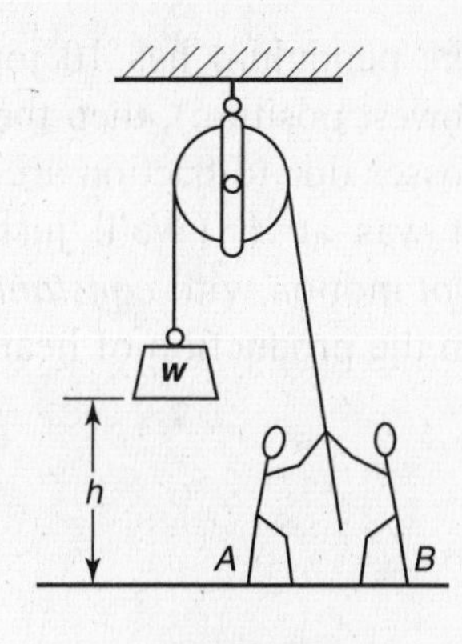

FIGURE 7.4

$$\text{gravitational potential energy} = \mathbf{w}h = mgh$$

where **w** is the weight of the object and *h* is the vertical height through which it has been lifted. Remember: the potential energy is with respect to the original position. The object has a greater potential energy with respect to a level lower than the one shown in the diagram.

If a spring is compressed or stretched, it also gains potential energy. This potential energy is equal to the work that was done in compressing or stretching the spring if frictional losses are neglected and is equal to the work that the spring can now do.

KINETIC ENERGY is the energy possessed by an object because of its motion. Winds (moving air) can do more work that stationary air, as in turning windmills or lifting roofs. The kinetic energy of an object is proportional to the square of its speed. If a car's speed is changed from 30 meters per second to 60 meters per second, the car's kinetic energy has become four times as high. A 6,000-kilogram truck moving at 30 meters per second has twice as much kinetic energy as a 3,000-kilogram car moving with the same speed. Kinetic energy is proportional to the mass of the object.

$$\text{kinetic energy} = \tfrac{1}{2} m\mathbf{v}^2$$

PRINCIPLE OF CONSERVATION OF ENERGY Energy cannot be created or destroyed but may be changed from one form into another. Mass can be considered a form of energy as a consequence of Einstein's theory of relativity. When mass is converted to a form of energy, such as heat,

$$\text{energy produced} = mc^2$$

where m is the mass converted and c is the speed of light. When m is expressed in kilograms and c in meters per second (3×10^8 m/s), the energy is expressed in joules (J).

In a swinging pendulum, Figure 7.5, kinetic energy is changed to potential energy and potential energy back to kinetic. If *A* is the highest position reached by the pendulum, the energy is all potential (=**w***h*). At *B* it is all kinetic; at *C* the energy is partly potential and partly kinetic; and at *D,* the highest point reached on the other side, the energy is all potential again. If the pendulum has 10 joules of potential energy at A (with respect to its lowest position), then the kinetic energy at B is 10 joules, provided energy losses due to friction are negligible. The height reached at D is the same as it was at A. (We'll just note in passing that the motion of the bob is *not* motion with *constant* acceleration.) Energy losses due to friction result in the production of heat.

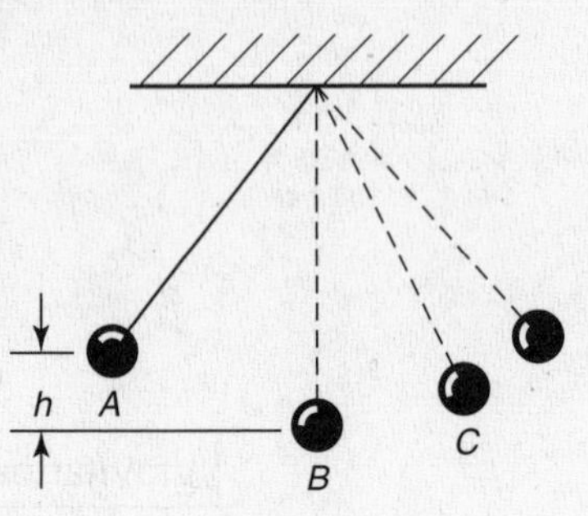

FIGURE 7.5

FRICTION AND WORK *Friction* is a force that always opposes motion or a tendency for motion. For example, in Figure 7.6, imagine a block resting on a horizontal surface. If a 1-newton pull is applied toward the left and the object doesn't move, a 1-newton force must be acting toward the right on the block. If the pull is increased to 2 newtons and the block doesn't move, the force to the right must have increased to 2 newtons. In each case the force to the right is friction, acting between the block and the surface on which it rests. Once sliding has been produced, the force of friction is practically independent of the speed with which the block moves. Friction is also practically independent of the area of contact (unless it is pointlike). Friction does depend on the nature of the surfaces in contact and on the force pushing the surfaces together (the so-called *normal force*). The coefficient of friction is used to describe the surfaces:

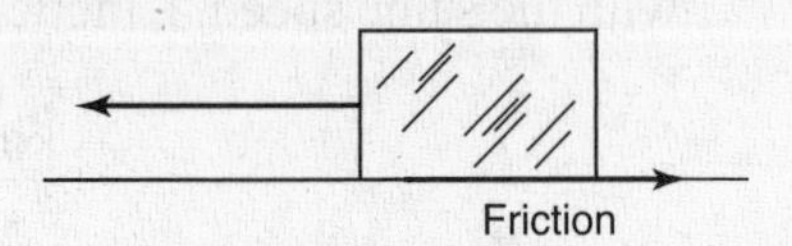

FIGURE 7.6

$$\text{coefficient of sliding friction} = \frac{\text{force of friction during motion}}{\text{normal force}}$$

Example: A 30-newton object is dragged along a horizontal surface for a distance of 10 meters. If the coefficient of friction is 0.2, what is the minimum amount of work that must be done? In this case, the normal force is the weight of 30 newtons, so, during motion, friction is 0.2×30 N = 6 N. Therefore, a pull of 6 newtons will maintain motion at constant speed and the work against friction is 6 newtons × 10 meters = 60 *joules.* A larger force could be used (requiring more work); this would result in accelerated motion.

$$\text{work against friction} = \text{friction} \times \text{distance object moves}$$

Friction during rolling motion is usually less than during sliding motion if the normal force is the same. In reference tables, therefore, we may find coefficients of rolling friction as well as coefficients of sliding friction.

When an object moves through a fluid (gas or liquid), the force of friction acting on the object does depend on the speed with which it moves through the fluid. For example, as a raindrop falls through the atmosphere, the frictional force acting on the raindrop increases as its speed increases, until the friction equals the weight of the raindrop. Maximum velocity is then reached and is known as the *limiting* or *terminal velocity*.

INCLINED PLANE Sometimes an inclined plane (see Figure 7.7) is used to slide heavy objects up to a platform or other elevated area. This is represented by the diagram, in which the height to which the object must be raised is indicated by h and the length of the plane by L. The weight of the object is **w**, but a smaller effort, force **F**, is required if we apply this

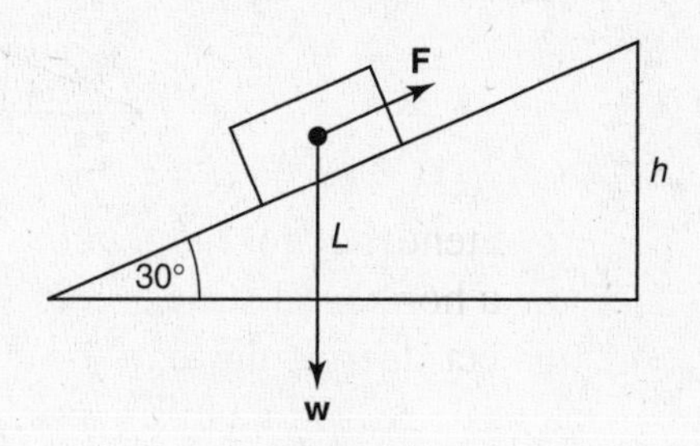

FIGURE 7.7

effort parallel to the plane. It is shown a few pages later, in the section headed "Simple Machines," that the following proportion is true when friction is negligible:

$$\frac{\mathbf{F}}{\mathbf{w}} = \frac{h}{L}$$

Since h is less than L, the length of the plane, the force **F** needed to push the object up along the plane at constant velocity is less than the weight of the object. However, we do not save work. The distance L along which the object must be pushed is proportionately larger than h, the vertical distance we want to raise the object.

Suppose the object weighs 1,000 newtons. What is the minimum effort needed to slide the object up along the plane?

We could make a scale drawing and get a graphical solution. We could also substitute in the above proportion since we have all the information needed. A third method is useful if we know the angle of the plane with the horizontal, but some of the other information is missing. We notice that the ratio h/L is equal to the sine of angle A:

$$\frac{\text{force parallel to plane}}{\text{weight}} = \sin A$$

$$\frac{\mathbf{F}}{\mathbf{w}} = \sin A$$

$$\mathbf{F} = \mathbf{w} \sin A$$

In this example, angle A is 30 degrees and weight $\mathbf{w}$ is 1,000 newtons. Therefore,

$$\mathbf{F} = (1{,}000 \text{ N})(0.5) = 500 \text{ N}$$

If we apply an effort of 500 newtons parallel to the plane, we can get the 1,000-newton object to the top of the plane.

Suppose we get the object near the top of the incline and let go. We know that the object is going to slide down the plane with increasing speed. Why? Gravity, the weight of the object, pulls vertically. We can think of a component of the weight acting parallel to the plane (and another one acting perpendicular to the plane), $\mathbf{F}_1$ (see Figure 7.8).

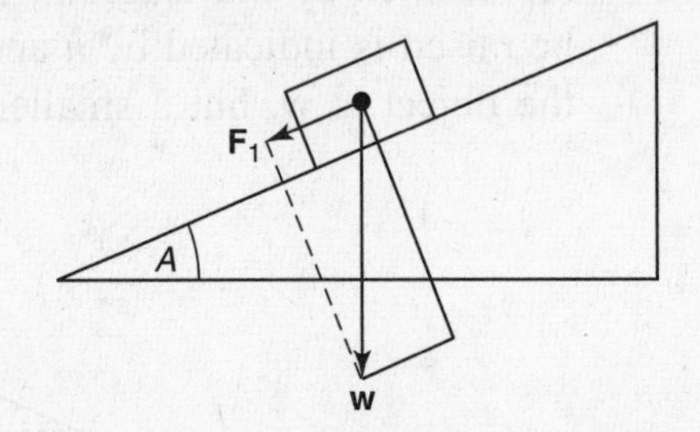

FIGURE 7.8

This component must be equal and opposite to **F**, which we just showed is equal to **w** sin A. The reason for this is Newton's first law of motion. The **F** was the force we needed to slide the object up with constant velocity. The object, therefore, was in equilibrium. Thus **F** was opposed by the equal and opposite component of the weight.

Figure 7.9 summarizes forces most often encountered in inclined plane problems. The object shown is in equilibrium, as all forces are balanced. With inclined planes, the perpendicular components of interest are not horizontal and vertical. Note also that the scale of the force vectors is independent of the inclined plane. (In the case shown, the weight vector drops below the horizontal line.)

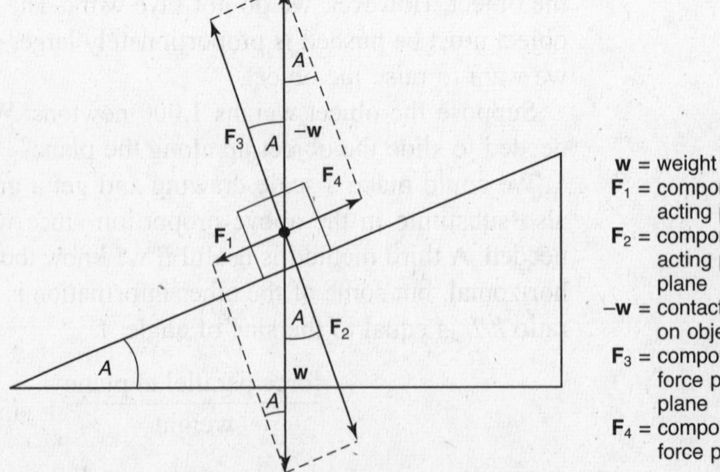

w = weight of object = mg
$\mathbf{F}_1$ = component of weight acting parallel to plane
$\mathbf{F}_2$ = component of weight acting perpendicular to plane
–**w** = contact force of plane on object
$\mathbf{F}_3$ = component of contact force perpendicular to plane
$\mathbf{F}_4$ = component of contact force parallel to plane

FIGURE 7.9 Forces on an Object in Equilibrium on an Inclined Plane

Elasticity and Hooke's Law

Work is sometimes done by forces that do not remain constant. The most common such forces arise from *springs.* The stretch or compression of a spring is proportional to the applied force ($\mathbf{F} = kx$). The constant of proportionality, k, is called the *spring constant.* It tells how much force is required to stretch a spring one unit length. For example, if a 1-newton object suspended from a spring stretches it 3 centimeters, a 2-newton object will stretch

it 6 centimeters. The spring constant = = or $k = 33$, in SI units.

The work done in stretching or compressing a spring is proportional to the square of the elongation. This work is stored as elastic potential energy.

$$\text{elastic potential energy} = \frac{1}{2}kx^2$$

Power

Power is the rate of doing work.

$$\text{power} = \frac{\text{work}}{\text{time}}$$

As the equation shows, we divide the work done by the time required to do the work. Since work is calculated by multiplying the force used (to do the work) by the distance the force moves,

$$\text{power} = \frac{\text{force} \times \text{distance}}{\text{time}}$$

The unit for power is the watt or joules per second.

Simple Machines

Probably the most direct way of doing useful work on an object is to take hold of it and lift it. This method, however, is often difficult or inconvenient, and we turn to machines to help us. A *machine* is a device that will transfer a force from one point of application to another for some practical advantage. For example, we may want to lift a flag to the top of a pole (Figure 7.10). It is inconvenient to climb to the top of the pole each day and pull the flag up. Instead, we keep a pulley at the top of the pole and with the aid of a rope apply a downward force at A and produce an upward force on the flag at B. We have six *simple machines*: pulley, lever, wheel and axle, inclined plane, screw, and wedge. The force that we apply to the machine in order to do the work (such as at A in the above case) is known as the *effort,* $\mathbf{F}_E$. The force that we have to overcome (such as the weight at B) is known as the *resistance,* $\mathbf{F}_R$.

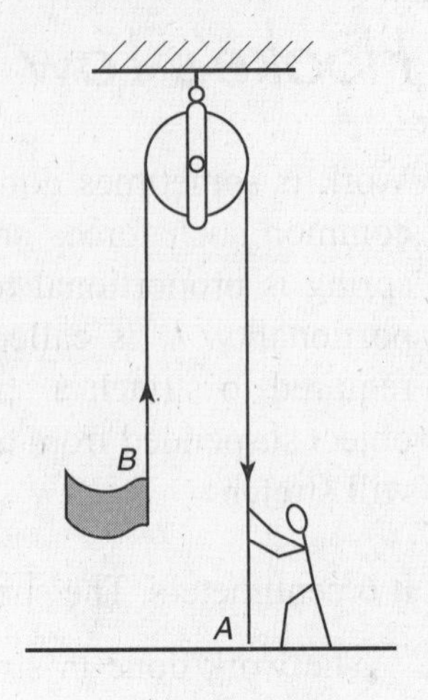

FIGURE 7.10

Mechanical Advantage

The *mechanical advantage* (MA) of a machine is the ratio of the resistance to the effort. If we want to lift a weight of 400 newtons and need to exert an effort of only 100 newtons, the MA is 4.

$$\text{mechanical advantage} = \frac{\text{resistance}}{\text{actual effort}}$$

$$MA = \frac{\mathbf{F}_R}{\mathbf{F}_E}$$

If we are interested in lifting an object, the *useful work* done with the machine, or the *work output,* or work accomplished, is equal to the weight of the object multiplied by the distance the object is lifted. In general,

$$\text{work output} = \text{resistance} \times \text{distance resistance moves}$$
$$\text{work output} = \mathbf{F}_R d_R$$

In order to accomplish this work the effort must move a certain distance, d_E. The *work input* is the work done by the effort.

$$\text{work input} = \text{effort} \times \text{distance effort moves}$$
$$\text{work input} = \mathbf{F}_E d_E$$

Under ideal conditions there is no useless work, such as work done in overcoming friction. Then,

$$\text{work output} = \text{work input}$$
$$\frac{\mathbf{F}_R}{\mathbf{F}_E} = \frac{d_E}{d_R} = \text{IMA}$$

where IMA is ideal mechanical advantage.

Note that for a given machine the ratio of the two distances is not affected by anything we do to change friction in the machine; it gives us the ideal mechanical advantage even if there is friction. The ratio of the two forces is affected by friction, since for a given situation the required effort is decreased by decreasing friction. The ratio of the two forces gives us the actual mechanical advantage, but if friction is negligible, the actual mechanical advantage equals the ideal mechanical advantage.

Efficiency

With an actual machine some useless work must be done, such as work to overcome friction. The input work equals work output plus useless work. For a machine

$$\text{efficiency} = \frac{\text{work output}}{\text{work input}}$$

Sometimes these derived formulas are useful in problems with machines:

$$\text{efficiency} = \frac{\text{MA}}{\text{IMA}}; \ \text{efficiency} = \frac{\text{ideal effort}}{\text{actual effort}}$$

The Lever

When we use crowbars, bottle openers, or oars, we are using the simple machine known as the lever. A *lever* is a rigid bar free to turn about a fixed point known as the *fulcrum* or pivot. An oar can rotate around P (Figure 7.11). We pull on the oar handle at A and as a result a push is exerted on the

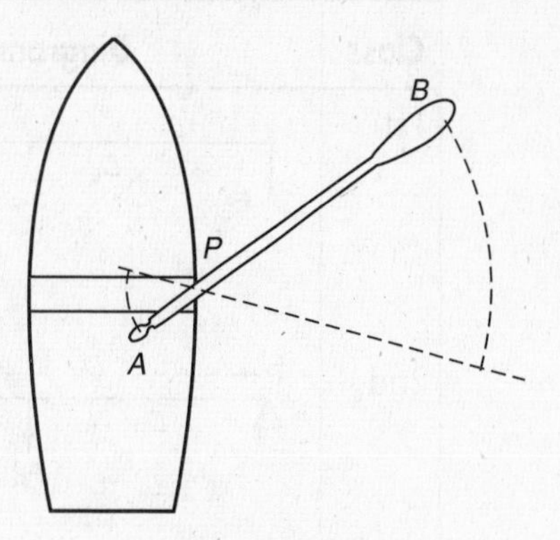

FIGURE 7.11

water at B by the oar blade. We represent this schematically in Figure 7.12. The fulcrum P is represented by a triangle; the pull we apply is the effort, $\mathbf{F}_E$; the push *by the water* on the oar is our resistance, $\mathbf{F}_R$. The moment arms are D_E and D_R, respectively. They are also called *lever arms*. From the principle of moments, $\mathbf{F}_E D_E = \mathbf{F}_R D_R$. From this we see that $\dfrac{F_R}{F_E} = \dfrac{D_E}{D_R}$, which gives the mechanical advantage of the lever. If the effort lever arm is greater than the resistance lever arm, the mechanical advantage is greater than one. In the case of the lever we often want a mechanical advantage of less than one. This, for example, is true in the illustration of rowing a boat. The effort is greater than the resistance of the water, but as a result the handle at A moves a shorter distance than the blade does during each stroke, as indicated by the dotted lines in the diagram. This gives us a speed advantage when the ideal mechanical advantage is less than one. (The discussion of the boat is somewhat simplified, because in an actual boat the pivot moves with respect to the water.)

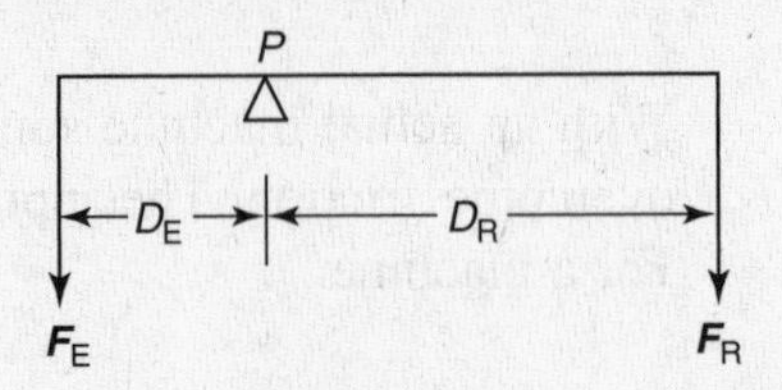

FIGURE 7.12

Notice, therefore, that the ideal mechanical advantage of a lever depends on where the fulcrum is located with respect to the points where the effort and resistance, respectively, are applied. The IMA may be less than one, greater than one, or equal to one. In the table below are shown three different arrangements of fulcrum (P), effort ($\mathbf{F_E}$), and resistance ($\mathbf{F_R}$). Look at the diagrams and, thinking in terms of the possible lengths of the lever arms, see if you can figure out the possible values of the IMA. Then compare your results with those in the table. It is not necessary to remember what class of lever each case represents.

LEVERS

Class	Diagram	Characteristics	IMA
1st	P, D_E, D_R, F_E, F_R	P between $\mathbf{F_E}$ & $\mathbf{F_R}$	MA $\geq$ 1 $\leq$
2nd	P, D_E, F_E, D_R, F_R	$\mathbf{F_R}$ between P & $\mathbf{F_E}$	MA > 1
3rd	F_E, P, D_E, D_R, F_R	$\mathbf{F_E}$ between P & $\mathbf{F_R}$	MA < 1

The Inclined Plane

When heavy objects have to be raised to a platform or put into a truck, it is often found to be convenient to slide these objects up along a board. An *inclined plane* (Figure 7.13) is a flat surface one end of which is kept at a higher level than the other. $\mathbf{F_E}$, the effort to pull or push the object up, is usually applied parallel to the plane. The input work, or work done by the effort, equals $\mathbf{F_E}$ times the length of the plane, L. The work output equals the weight of the object **w** times the height of the plane, h. Under ideal conditions the input work equals the output work, and therefore

$$\frac{\text{weight of object}}{\text{ideal effort}} = \frac{\text{length of plane}}{\text{height of plane}} = \text{IMA}$$

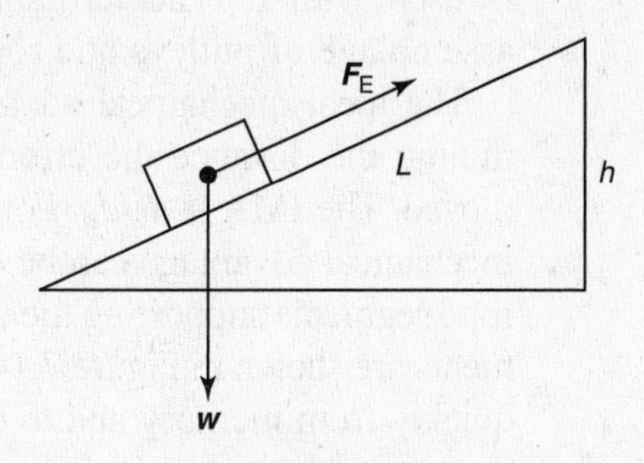

FIGURE 7.13

Under actual conditions friction opposes the motion. When the object is *moved up* the plane, the actual effort equals the ideal effort plus the force of friction. On the other hand, less force is required to keep the object from sliding down, and, when the object is allowed to *slide down* with constant speed, the effort applied up along the plane equals the ideal effort *minus* the force of friction. (Occasionally the effort may be applied parallel to the base of the plane. Then the IMA equals the base of the plane divided by the height of the plane.)

Example: An inclined plane 13 meters long and 5 meters high is used to lift a 390-newton object to a platform. The ideal MA is $\frac{13}{5}$ or 2.6; the ideal effort parallel to the plane is 390 newtons $\times \frac{5}{13}$ or 150 newtons. Actually 200 newtons may be required; then the force of friction is 50 newtons. The MA $= \frac{\text{weight}}{\text{effort}}$, or $\frac{390\text{ N}}{200\text{ N}}$. Therefore, the MA = 1.95. The useful work = $\mathbf{w} \times h = 390\text{ N} \times 5\text{ m} = 1{,}950\text{ J}$. The input work = $\mathbf{F}_E \times L = 200\text{ N} \times 13\text{ m} = 2{,}600\text{ J}$. The efficiency equals

$$\frac{\text{output work}}{\text{input work}} = \frac{1{,}950\text{ J}}{2{,}600\text{ J}} = 0.75 \text{ or } 75\%$$

Note that efficiency also equals

$$\frac{\text{ideal effort}}{\text{actual effort}} = \frac{150\text{ N}}{200\text{ N}} = 0.75$$

The Pulley

Especially when heavy objects have to be lifted through a considerable distance, the pulley is a convenient simple machine for the job. The *pulley* is a wheel so mounted in a frame that it may turn readily around the axis through the center of the wheel. If the frame and wheel move through space as the pulley is used, we have a *movable pulley.* Otherwise we say we have a fixed pulley. The wheel is usually grooved to guide the string or rope that is used with it. The term *block and tackle* is used for the commercial assemblage of pulleys and rope.

The ideal mechanical advantage of pulleys can be figured out by determining the distance the effort has to move for every foot the resistance is moved. The IMA = d_E/d_R. For many common pulley arrangements this ideal mechanical advantage can be determined visually by counting the number of rope segments supporting the movable pulley(s). Some typical pulley arrangements are shown in Figure 7.14. You should be able to reproduce the diagrams quickly from memory and to determine the IMAs of the arrangements.

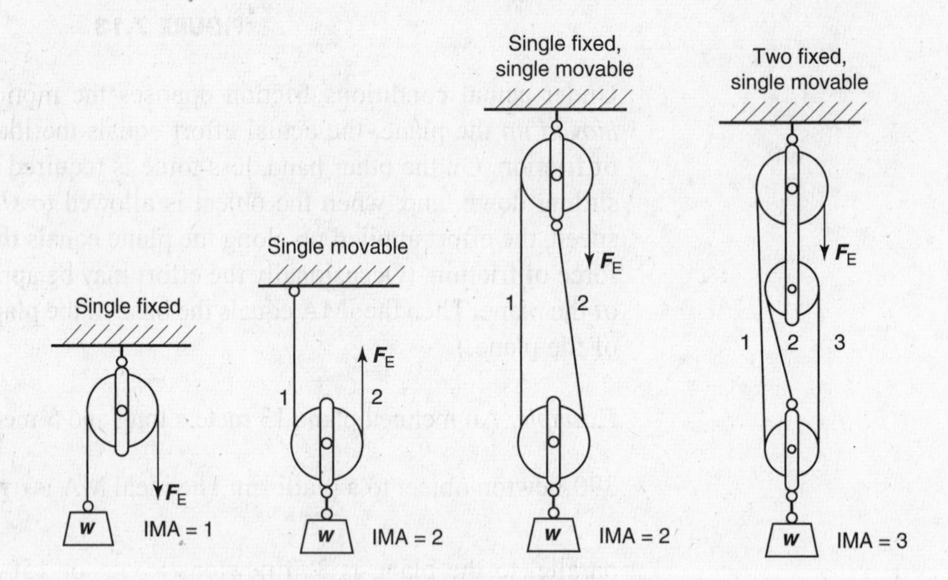

FIGURE 7.14

Note that in these diagrams the rope segments are numbered to show how the ideal mechanical advantage can be determined visually from the diagram; the effort is applied to one end of the string while the other end of the string is attached to the weight to be lifted, to the pulley frame, or to a rigid support such as the ceiling. The weight to be lifted (**w**) represents the resistance if we neglect the weight of the movable pulleys and friction.

Example: A system of pulleys having an IMA of 3 is used to lift an object weighing 600 newtons. The effort required is 300 newtons. What is the efficiency? The ideal effort = $\mathbf{F}_R$/IMA = weight/IMA = 600 N/3 = 200 N. The efficiency = ideal effort/actual effort = 200 N/300 N = 2/3 or 66.7%.

Questions Chapter 7

In each case, select the choice that best completes the statement.

1. If speed of an object is tripled, its kinetic energy is

(A) 1/9
(B) 1/3
(C) 3
(D) 6
(E) 9

times its original speed.

Questions 2–4

A person having a mass of 60 kilograms exerts a horizontal force of 200 newtons in pushing a 90-kilogram object a distance of 6 meters along a horizontal floor. He does this at constant velocity in 3 seconds.

2. The weight of this person is, in newtons, approximately,

(A) 40
(B) 90
(C) 200
(D) 400
(E) 600

3. The work done by this person is, in joules,

(A) 540
(B) 1,080
(C) 1,200
(D) 3,600
(E) 5,400

4. The force of friction is

(A) exactly 60 N
(B) between 60 and 90 N
(C) exactly 90 N
(D) exactly 200 N
(E) greater than 200 N

Questions 5 and 6

The 1,000-newton weight of a pile driver falls freely from rest; it drops 25 meters, strikes a steel beam, and drives it 3 centimeters into the ground.

5. The kinetic energy of the weight of the pile driver just before it hits the beam is, in newton · meters,

(A) 3,000
(B) 25,000
(C) 75,000
(D) 200,000
(E) 800,000

6. The speed of the weight just before hitting the beam is, in meters per second, approximately

(A) 0.25
(B) 3
(C) 25
(D) 22
(E) 64

Questions 7–10

An inclined plane 5 meters long has one end on the ground and the other end on a platform 3 meters high. A man weighing 650 newtons wishes to push a 900-newton object up this plane. The force of friction is 100 newtons.

7. The minimum force the man must exert is, in newtons, approximately

(A) 100
(B) 540
(C) 640
(D) 900
(E) 1,000

8. In order to hold the object on the plane without letting it slide, the minimum force required is, in newtons, approximately

(A) 0
(B) 100
(C) 440
(D) 540
(E) 640

9. The potential energy gained by the object when it is at the top of the plane is, in joules,

(A) 100
(B) 324
(C) 2,700
(D) 5,400
(E) 9,000

10. The power expended by the man in pushing the object to the top of the plane in 3 seconds is

(A) 400 W
(B) 800 W
(C) 1,070 W
(D) 4,500 W
(E) 4,500 J

---

11. Within normal limits of use, if a 100-gram object stretches a spring 3 centimeters, a 200-gram object will stretch the spring, in centimeters,

(A) $3 \times \sqrt{2}$
(B) 3×2
(C) 3^2
(D) 3×2^2
(E) $3^2 \times 2^2$

12. A uniform plank 1.0 meter long weighs 50 newtons. The force needed to lift one end of the plank is, in newtons,

(A) 10
(B) 12.5
(C) 25
(D) 50
(E) 100

Questions 13–16

The pulley arrangement shown is attached to a ceiling. A weight of 240 newtons is to be lifted. The weight of the pulleys is negligible.

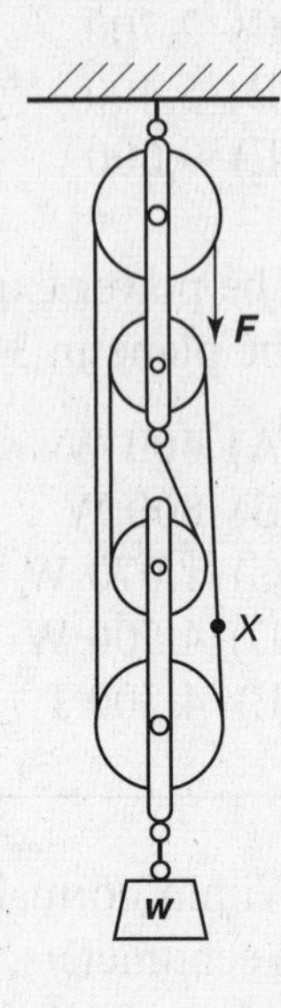

13. If frictional losses are negligible, the force **F** required to lift the weight at constant speed is

(A) 48 N
(B) 60 N
(C) 80 N
(D) 120 N
(E) 240 N

14. For the conditions in problem 13, the pull on the ceiling will be

(A) 60 N
(B) 120 N
(C) 240 N
(D) 288 N
(E) 300 N

15. For the conditions in problem 13, the tension in the rope at *X* will be

(A) 48 N
(B) 60 N
(C) 80 N
(D) 120 N
(E) 240 N

16. If the actual force **F** is 180 newtons, the percent efficiency of the system is

(A) 25
(B) 33
(C) 50
(D) 75
(E) 90

Explanations to Questions Chapter 7

Answers

1. (E)	**5.** (B)	**9.** (C)	**13.** (B)
2. (E)	**6.** (D)	**10.** (C)	**14.** (E)
3. (C)	**7.** (C)	**11.** (B)	**15.** (B)
4. (D)	**8.** (C)	**12.** (C)	**16.** (B)

Explanations

1. E Kinetic energy = $\frac{1}{2}m\mathbf{v}^2$. For a given object the kinetic energy is proportional to the square of its speed. When the speed is multiplied by 3, the kinetic energy is multiplied by 3^2, or 9.

2. E The weight of an object on Earth is the gravitational pull of Earth on it and is equal to the product of the object's mass and the value of the acceleration of a freely falling object:

$$\begin{aligned} \text{weight} &= m\mathbf{g} \\ &= (60\text{ kg})(9.8\text{ m/s}^2) \\ &= 600\text{ N} \end{aligned}$$

3. C Work done on an object is equal to the product of the force exerted on the object and the distance moved in the direction of the force:

$$\begin{aligned} \text{work} &= \text{force} \times \text{distance} \\ &= (200\text{ N})(6\text{ m}) \\ &= 1{,}200\text{ joules} \end{aligned}$$

4. D Since the force exerted is horizontal, none of it is used to overcome gravity. Also, since the velocity remains constant, none of the force (and work done) is used to give the object kinetic energy. Therefore all of the 200-N force used is required to overcome friction.

5. B The kinetic energy that the weight has just before it hits is equal to the potential energy it had at the top just before falling. The potential energy at the top = weight × height:

$$\text{PE} = 1{,}000\text{ N} \times 25\text{ m} = 25{,}000\text{ N} \cdot \text{m}$$

6. D To calculate the speed of an object falling freely from rest if we know the distance of fall:

$$\mathbf{v}^2 = 2\mathbf{g}d = 2 \times 9.8 \text{ m/s}^2 \times 25 \text{ m}$$

$$\mathbf{v} = 22 \text{ m/s}$$

7. C The man has to overcome two forces: the force of friction and the component of the weight parallel to the plane.

$$\frac{\text{parallel component}}{\text{weight}} = \frac{\text{height of plane}}{\text{length of plane}}$$

$$\frac{\text{parallel component}}{900 \text{ N}} = \frac{3 \text{ m}}{5 \text{ m}}$$

and parallel component = 540 N
Force required = 100 N + 540 N = 640 N.

8. C Friction will help in keeping the object from sliding down. The required force, therefore, is equal to the parallel component minus the force of friction: 540 N – 100 N = 440 N.

9. C PE = weight × height = 900 N × 3 m = 2,700 joules.

10. C Power is the rate of doing work: Power = work/time. The work done by the man is equal to the product of the force required and the distance the object is moved in the direction of the force (which in this case is parallel to the plane).

$$\text{work} = \text{force} \times \text{distance}$$

$$= (640 \text{ N})(5 \text{ m})$$

$$= 3{,}200 \text{ J}$$

$$\text{power} = \frac{\text{work}}{\text{time}} = \frac{3{,}200 \text{ J}}{3 \text{ s}} = 1{,}070 \text{ W}$$

(Note that the work required is greater than the potential energy gained because work had to be done against friction as well as to raise the object against gravity.)

11. B According to Hooke's law, the distortion is proportional to the distorting force. Since the distorting force is doubled, the distortion is also doubled.

12. C The fulcrum is at one end, at the left in the diagram. The center of gravity of a uniform plank is in the middle; the weight of 50 N acts downward 0.5 m from either end. Counterclockwise moment = clockwise moment; 10 **F** = 50 N × 0.5 m; **F** = 25 N.

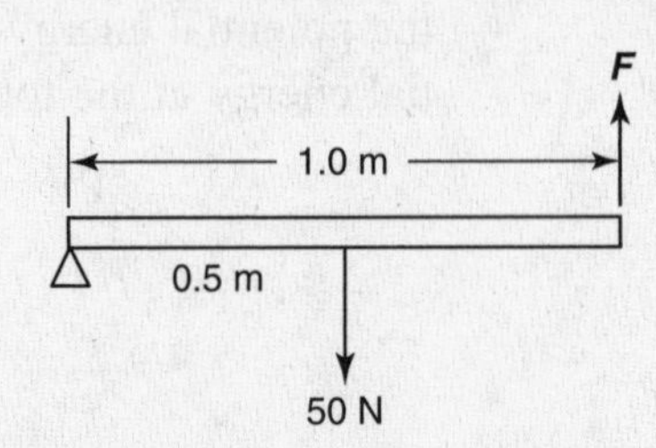

13. B There are 4 strands supporting the movable pulleys; therefore, the ideal mechanical advantage is 4.

$$\text{Then IMA} = \frac{\text{weight of object}}{\text{ideal effort}}, \text{ and ideal effort} = \frac{\text{weight of object}}{\text{IMA}}$$

$$= \frac{240 \text{ N}}{4} = 60 \text{ N}$$

14. E If the weight of the pulleys is negligible, there are two basic forces pulling on the ceiling: force **F** and the 240-N object. Pull on ceiling = 60 N + 240 N = 300 N.

15. B Each of the 4 strands supporting the movable pulleys supplies ¼ of the upward pull on the weight or ¼ of 240 N (i.e., 60 N). Another way is to think of the force **F** as transmitted throughout the rope and therefore being equal to the tension.

16. B $\text{Efficiency} = \frac{\text{ideal effort}}{\text{actual effort}} = \frac{60 \text{ N}}{180 \text{ N}} = 33\%$

CHAPTER 8

Impacts and Linear Momentum

Internal and External Forces

Consider a system of two blocks with masses m and M ($M > m$). If the blocks were to collide, the forces of impact would be equal and opposite. However, because of the different masses, the response to these forces (i.e., the changes in velocity) would not be equal. In the absence of any outside or external forces acting on the objects (such as friction or gravity), we say that the impact forces are internal.

To Newton, the "quantity of motion" discussed in his book *Principia* was the product of an object's mass and velocity. This quantity, called *linear momentum* or just *momentum*, is a vector quantity having units of kilogram · meters per second (kg · m/s). Algebraically, momentum is designated by the letter **p**: $\mathbf{p} = m\mathbf{v}$.

To understand Newton's rationale, consider the action of trying to change the motion of a moving object. Do not confuse this with the inertia of the object; here, both the mass and the velocity are important. Consider, for example, that a truck moving at a slow 1 meter per second can still inflict a large amount of damage because of its mass. Also, a small bullet, having a mass of perhaps 1 gram or less, does incredible damage because of its high velocity. In each case (see Figure 8.1) the damage is the result of a force of impact when the object is intercepted by something else. Let's now consider the nature of impact forces.

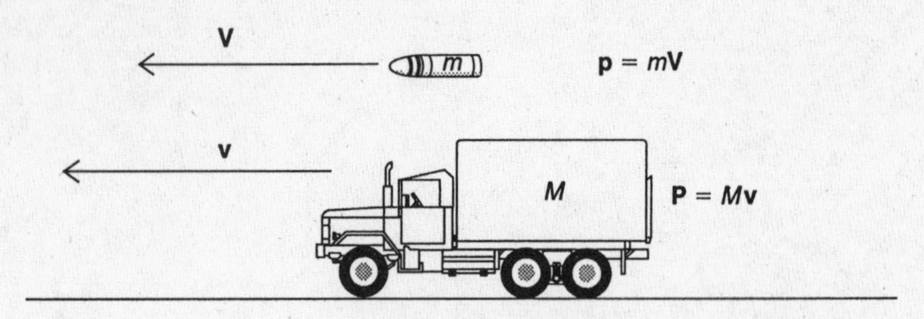

FIGURE 8.1 Comparison of the Momentum of a Truck with That of a Bullet

Impact Forces and Momentum Changes

Consider a mass m moving with a velocity $\mathbf{v}$ in some frame of reference. If the mass is subjected to some external forces, then, by Newton's second law of motion, we can write that $\Sigma\mathbf{F} = m\mathbf{a}$. The vector sum of all the forces, referred to as the *net force*, is responsible for changing the velocity of the motion (in magnitude and/or direction).

If we recall the definition of acceleration as the rate of change or velocity, we can rewrite the second law of motion as

$$\mathbf{F}_{\text{net}} = m\left(\frac{\Delta \mathbf{v}}{\Delta t}\right)$$

This expression is also a vector equation and is equivalent to the second law of motion. If we make the assumption that the mass of the object is not changing, we can again rewrite the second law in the form

$$\mathbf{F}_{\text{net}} = \frac{\Delta m\mathbf{v}}{\Delta t} = \frac{\Delta \mathbf{p}}{\Delta t}$$

This expression means that the net external force acting on an object is equal to the rate of change of momentum of the object and is another alternative form of Newton's second law of motion. This *change in momentum* is a vector quantity in the same direction as the net force applied. Since the time interval is just a scalar quantity, we can multiply both sides by Δt to get

$$\mathbf{F}_{\text{net}}\,\Delta t = \Delta \mathbf{p} = m\,\Delta \mathbf{v} = m\mathbf{v}_f - m\mathbf{v}_i$$

The quantity, $\mathbf{F}_{\text{net}}\,\Delta t$, called the *impulse*, represents the effect of a force acting on a mass during a time interval Δt, and is likewise a vector quantity. From this expression, it can be stated that the impulse applied to an object is equal to the change in momentum of the object.

Another way to consider impulse is to look at a graph of force versus time for a continuously varying force (see Figure 8.2).

The area under this curve is a measure of the impulse in units of newton · seconds (N · s). Another way to view this concept is to identify the average force, $\overline{\mathbf{F}}$, so that the rectangle formed by the average force is equal in area to the entire curve. This approach is more manageable algebraically, and we can write

$$\overline{\mathbf{F}}\,\Delta t = \Delta \mathbf{p}$$

As an example, consider the following problem: How long will an average braking force of 1500 newtons take to stop a 2500-kilogram car traveling with an initial velocity of 10 meters per second? We substitute the given numbers into the formula for impulse and momentum change:

$$(1500)\Delta t = (2500)(10)$$

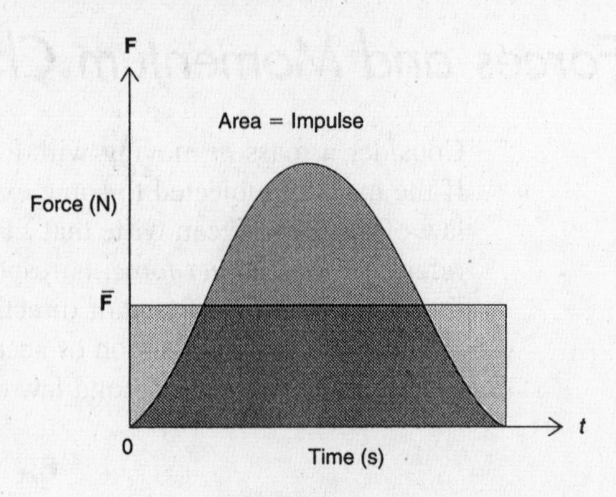

FIGURE 8.2

The change in velocity is equal to 10 meters per second because the object is stopping. Solving for the time, we find that the braking force will take about 16.7 seconds.

Law of Conservation of Linear Momentum

Newton's third law of motion states that for every action there is an equal but opposite reaction. This reaction force is present whenever we have an interaction between two objects in the universe. Suppose we have two masses, m_1 and m_2, that are approaching each other along a horizontal frictionless surface. Let $\mathbf{F}_{12}$ be the force that m_1 exerts on m_2, and let $\mathbf{F}_{21}$ be the force that m_2 exerts on m_1. According to Newton's law, these forces must be equal and opposite; that is, $\mathbf{F}_{12} = -\mathbf{F}_{21}$.

Rewriting this expression as $\mathbf{F}_{12} + \mathbf{F}_{21} = 0$ leads to an interesting implication. Since each force is a measure of the rate of change of momentum for that object, we can write

$$\mathbf{F}_{12} = \frac{\Delta \mathbf{p}_1}{\Delta t} \quad \text{and} \quad \mathbf{F}_{21} = \frac{\Delta \mathbf{p}_2}{\Delta t}$$

Therefore:

$$\frac{\Delta \mathbf{p}_1}{\Delta t} + \frac{\Delta \mathbf{p}_2}{\Delta t} = 0 \quad \text{and} \quad \frac{\Delta(\mathbf{p}_1 + \mathbf{p}_2)}{\Delta t} = 0$$

The change in the sum of the momenta is therefore zero, implying that the sum of the total momentum for the system ($\mathbf{p}_1 + \mathbf{p}_2$) is a constant all the time. This conclusion is called the *law of conservation of linear momentum*, and we say simply that the momentum is conserved.

Here is another way of writing this conservation statement in a general form for any two masses (after separating all initial and final terms):

$$m_1\mathbf{v}_{1i} + m_2\mathbf{v}_{2i} = m_1\mathbf{v}_{1f} + m_2\mathbf{v}_{2f}$$

Extension of the law of conservation of momentum to two or three dimensions involves the recognition that momentum is a vector quantity. Given two masses moving in a plane relative to a coordinate system, conservation of momentum must hold simultaneously in both the horizontal and vertical directions. These vector components of momentum can be calculated using the standard techniques of vector analysis used to resolve forces in components.

Elastic and Inelastic Collisions

During any collision between two pieces of matter, momentum is always conserved. This statement is not, however, necessarily true about kinetic energy. If two masses stick together after a collision, it is observed that the kinetic energy before the collision is not equal to the kinetic energy after it. If the kinetic energy is conserved as well as the momentum, the collision is described as *elastic*. If the kinetic energy is not conserved after the collision (e.g., energy being lost to heat or friction), the collision is described as *inelastic*.

Questions Chapter 8

In each case, select the choice that best answers the question or completes the statement.

1. Which of the following expressions, where **p** represents the linear momentum of the particle, is equivalent to the kinetic energy of a moving particle?

(A) $m\mathbf{p}^2$
(B) $m^2/2\mathbf{p}$
(C) $2\mathbf{p}/m$
(D) $\mathbf{p}/2m$
(E) $\mathbf{p}^2/2m$

2. Two carts having masses 1.5 kilograms and 0.7 kilogram, respectively, are initially at rest and are held together by a compressed massless spring. When released, the 1.5-kilogram cart moves to the left with a velocity of 7 meters per second. What are the velocity and direction of the 0.7-kilogram cart?

(A) 15 m/s right
(B) 15 m/s left
(C) 7 m/s left
(D) 7 m/s right
(E) 0 m/s

3. The product of an object's instantaneous momentum and its acceleration is equal to its

 (A) applied force
 (B) kinetic energy
 (C) power output
 (D) net force
 (E) displacement

4. A ball with a mass of 0.15 kilogram has a velocity of 5 meters per second. It strikes a wall perpendicularly and bounces off straight back with a velocity of 3 meters per second. The ball underwent a change in momentum equal to

 (A) 0.30 kg · m/s
 (B) 1.20 kg · m/s
 (C) 0.15 kg · m/s
 (D) 5 kg · m/s
 (E) 7.5 kg · m/s

5. What braking force is supplied to a 3000-kilogram car traveling with a velocity of 35 meters per second that is stopped in 12 seconds?

 (A) 29,400 N
 (B) 3000 N
 (C) 8750 N
 (D) 105,000 N
 (E) 150 N

6. A 0.1-kilogram baseball is thrown with a velocity of 35 meters per second. The batter hits it straight back with a velocity of 60 meters per second. What is the magnitude of the average impulse exerted on the ball by the bat?

 (A) 3.5 N · s
 (B) 2.5 N · s
 (C) 7.5 N · s
 (D) 9.5 N · s
 (E) 12.2 N · s

7. A 1-kilogram object is moving with a velocity of 6 meters per second to the right. It collides and sticks to a 2-kilogram object moving with a velocity of 3 meters per second in the same direction. How much kinetic energy was lost in the collision?

 (A) 1.5 J
 (B) 2 J
 (C) 2.5 J
 (D) 3 J
 (E) 0 J

8. A 2-kilogram mass moving with a velocity of 7 meters per second collides elastically with a 4-kilogram mass moving in the opposite direction at 4 meters per second. The 2-kilogram mass reverses direction after the collision and has a new velocity of 3 meters per second. What is the new velocity of the 4-kilogram mass?

(A) −1 m/s
(B) 1 m/s
(C) 6 m/s
(D) 4 m/s
(E) 5 m/s

9. A mass m is attached to a massless spring with a force constant k. The mass rests on a horizontal frictionless surface. The system is compressed a distance x from the spring's initial position and then released. The momentum of the mass when the spring passes its equilibrium position is given by

(A) $x\sqrt{mk}$
(B) $x\sqrt{k/m}$
(C) $x\sqrt{m/k}$
(D) $x\sqrt{k^2m}$
(E) xmk

10. During an inelastic collision between two balls, which of the following statements is correct?

(A) Both momentum and kinetic energy are conserved.
(B) Momentum is conserved, but kinetic energy is not conserved.
(C) Momentum is not conserved, but kinetic energy is conserved.
(D) Neither momentum nor kinetic energy is conserved.
(E) Momentum is sometimes conserved, but kinetic energy is always conserved.

Explanations to Questions Chapter 8

Answers

1. (E)	5. (C)	9. (A)
2. (A)	6. (D)	10. (B)
3. (C)	7. (D)	
4. (B)	8. (B)	

Explanations

1. **E** If we multiply the formula for kinetic energy by the ratio m/m, we see that the formula for kinetic energy becomes

$$\text{KE} = \left(\frac{1}{2m}\right)(m^2)(\mathbf{v}^2) = \frac{\mathbf{p}^2}{2m}$$

2. A Momentum is conserved, so $(1.5)(7) = 0.7\mathbf{v}$. Thus $\mathbf{v} = 15$ m/s. The direction is to the right since in a recoil the masses go in opposite directions.

3. C An object's instantaneous momentum times its acceleration will equal its power output in units of joules per second or watts.

4. B The change in momentum is a vector quantity. The rebound velocity is in the opposite direction, so $\Delta\mathbf{v} = 5 - (-3) = 8$ m/s. The change in momentum is

$$\Delta\mathbf{p} = (0.15)(8) = 1.20 \text{ kg} \cdot \text{m/s}$$

5. C The formula is $\mathbf{F}\,\Delta t = m\,\Delta\mathbf{v}$. Solving for the force, we get

$$\mathbf{F} = \frac{(3000)(35)}{12} = 8750 \text{ N}$$

6. D Impulse is equal to the change in momentum, which is

$$\Delta\mathbf{p} = |\,(0.1[-60 - 35]\,| = 9.5 \text{ N} \cdot \text{s}$$

because of the change in direction of the ball.

7. D First, we find the final velocity of this inelastic collision. Momentum is conserved, so we can write $(1)(6) + (2)(3) = (3)\mathbf{v}'$ since both objects are moving in the same direction. Thus $\mathbf{v}' = 4$ m/s. The initial kinetic energy of the 1-kg object is 18 J, while the initial kinetic energy of the 2-kg mass is 9 J. Thus the total initial kinetic energy is 27 J. After the collision, the combined 3-kg object has a velocity of 4 m/s and a final kinetic energy of 24 J. Thus, 3 J of kinetic energy has been lost.

8. B Momentum is conserved in this elastic collsion but the directions are opposite, so we must be careful with negative signs. We therefore write $(2)(7) - (4)(4) = -(2)(3) + 4\mathbf{v}'$ and get $\mathbf{v}' = 1$ m/s.

9. A We set the two energy equations equal to solve for the velocity at the equilibrium position. Thus

$$\frac{1}{2}kx^2 = \frac{1}{2}m\mathbf{v}^2$$

since no gravitational potential energy is involved, and we write that the velocity is $\mathbf{v} = x\sqrt{k/m}$. Now momentum $\mathbf{p} = m\mathbf{v}$, so we multiply by m and factor the "mass" back under the radical sign, where it is squared, so $\mathbf{p} = x\sqrt{mk}$.

10. B Momentum is always conserved in an inelastic collision, but kinetic energy is not conserved because the objects stick together.

CHAPTER 9

Torque and Angular Momentum

Introduction

In Chapter 8, we discussed the fact that, if a single force is directed toward the center of mass of an extended object, the result is a linear or translational acceleration in the same direction as the force. If, however, the force is not directed through the center of mass, then the result is a rotation about the center of mass. In addition, if the object is constrained by a fixed pivot point, the rotation will take place about that pivot (e.g., a hinge). These observations are summarized in Figure 9.1.

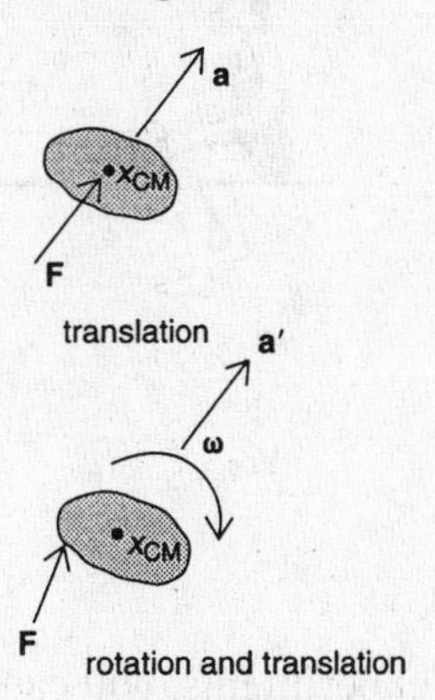

FIGURE 9.1

Parallel Forces and Moments

If two forces are used, it is possible to prevent the rotation if these forces are parallel and are of suitable magnitudes. If the object is free to move in space, translational motion may result (see Figure 9.2).

As an example of parallel forces, consider two people on a seesaw, as shown in Figure 9.3. The two people have weights $\mathbf{W}_1$ and $\mathbf{W}_2$, respectively, and are sitting distances d_1 and d_2 from the fixed pivot point (called the *fulcrum*). From our discussion about forces, we see that the tendency of each force (pro-

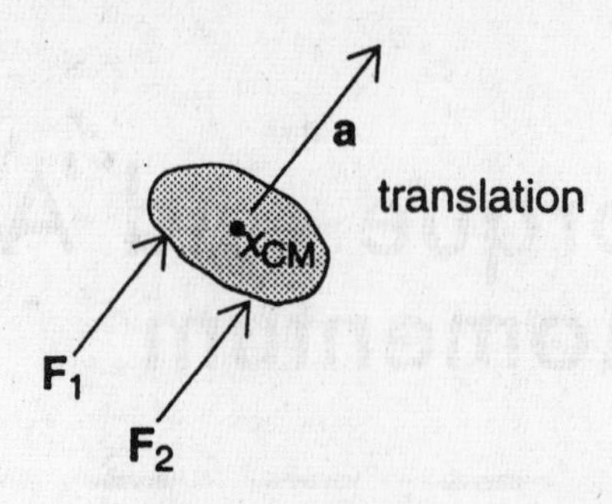

FIGURE 9.2

vided by gravity), is to cause the seesaw to rotate about the fulcrum. Force W_1 will tend to cause a counterclockwise rotation; force W_2, a clockwise rotation. Arbitrarily, we state that a clockwise rotation is taken as being a negative, while counterclockwise is taken as positive. What factors will influence the ability of the two people to remain in "balance"; that is, what conditions must be met so that the system remains in rotational equilibrium (it is already in a state of translational equilibrium and constrained to remain that way)?

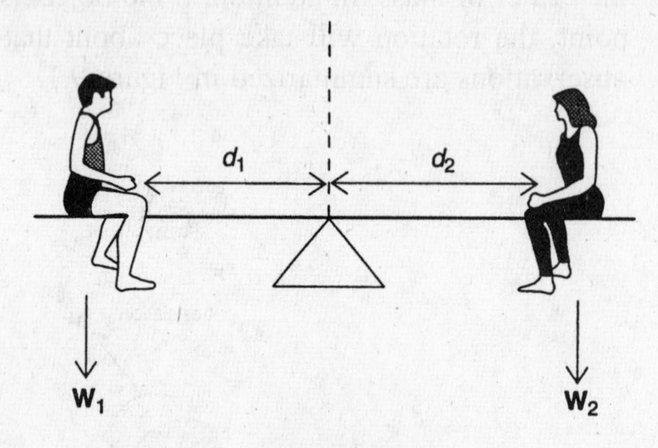

FIGURE 9.2

Through experiments, you can show that distances d_1 and d_2 (called *moment arm distances*) play a crucial role since we take the weights as being constant. If the two weights were equal, it should not be a surprise to learn that $d_1 = d_2$. In these examples, the weight of the seesaw is taken to be negligible (this is not a realistic scenario).

It turns out that, if W_2 is greater, that person must sit closer to the fulcrum. If equilibrium is to be maintained, the following condition must hold:

$$(W_1)(d_1) = (W_2)(d_2)$$

These are force and distance products, but they do not represent work in the translational sense because the two vectors are not parallel or resolvable into parallel components. In rotational motion, the product of a force and a perpendicular moment arm distance (relative to a fulcrum) is called *torque*. Let us now consider some more examples of torques and equilibrium.

Torque

When you tighten a bolt with a wrench, you apply a force to create a rotation or twist. This twisting action, called torque in physics, is represented by the Greek letter $\boldsymbol{\tau}$. Even though its units is the newton · meter (N · m), you must not confuse it with translational work. In static equilibrium, the two conditions met are that the vector sum of all forces acting on the object equals zero and that the vector sum of all torques equals zero:

$$\Sigma \mathbf{F} = 0 \quad \text{and} \quad \Sigma \tau = 0$$

When we say that $\boldsymbol{\tau} = \mathbf{F}d$, d is the moment arm distance from the center of mass, or the pivot point, and $\mathbf{F}$ is the force perpendicular to the vector displacement. If the force is applied at some angle θ to the object, as seen in Figure 9.4, the component of the force, perpendicular to the vector displacement out from the pivot, is taken as the force used. Algebraically this is stated as

$$\tau = rF \sin \theta$$

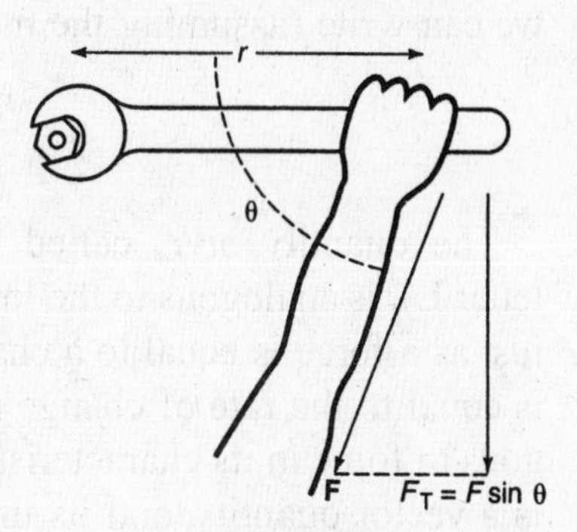

FIGURE 9.4

Remember: torque is a vector quantity, but work is not! The direction of the torque is taken as either positive or negative, depending on whether the rotation is counterclockwise or clockwise.

As an example, consider a light string wound around a frictionless and massless wheel, as shown in Figure 9.5. The free end of the string is attached to a 1.2-kilogram mass that is allowed to fall freely. The wheel has a radius of 0.25 meter. What torque is produced?

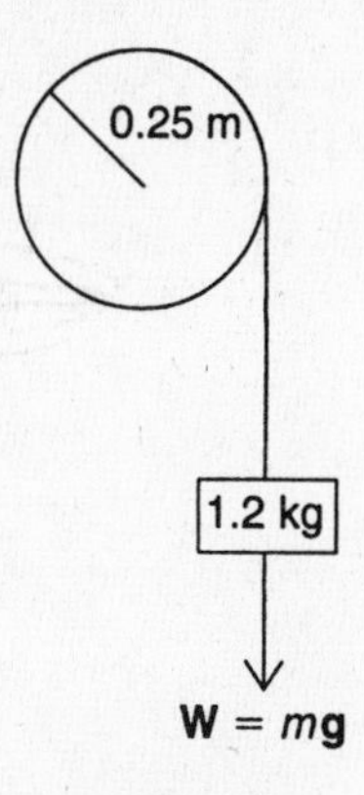

FIGURE 9.5

The force acting at right angles to the center of the wheel is the weight of the mass, given by

$$\mathbf{W} = m\mathbf{g} = (1.2)(9.8) = 11.76 \text{ N}$$

The radius serves as the moment arm distance, $d = 0.25$ m, so we can write

$$\tau = -\mathbf{F}d = (11.76)(0.25) = -2.94 \text{ N} \cdot \text{m}$$

The torque is negative since the falling weight induces a clockwise rotation in this example.

Angular Momentum and Its Conservation

If we have a force producing a torque a distance **r** from a pivot point, we can write that $\tau = \mathbf{Fr}$. If we recall that $\mathbf{F} = m\mathbf{a}$, we can write $\tau = m\mathbf{ar}$. Now, the acceleration is the rate of change of the velocity; that is, $\mathbf{a} = \Delta\mathbf{v}/\Delta t$, so we can write (assuming the mass and radius are constant)

$$\tau = \frac{m\Delta\mathbf{vr}}{\Delta t} = \frac{\Delta m\mathbf{vr}}{\Delta t} = \frac{\Delta\mathbf{L}}{\Delta t}$$

The quantity $m\mathbf{vr}$, called the *angular momentum* (designated by the letter **L**), is analogous to the linear momentum discussed in Chapter 8. Also, just as a force is equal to a change in linear momentum over time, a torque is equal to the rate of change of angular momentum. Thus torque is analogous to force in its characteristics to produce rotations. Angular momentum is a vector quantity, and its unit is the kilogram · square meter per second ($\text{kg} \cdot \text{m}^2/\text{s}$).

If the net torque acting on a system is zero, then the rate of change of angular momentum is zero, and we say that the angular momentum has been conserved. An example of this concept occurs when an ice skater starts to spin and draws his arms inward. Since angular momentum is conserved, the response to a decrease in radius is an increase in angular velocity, ω, so he spins faster. (See Figure 9.6.). As a ratio we can state that, for two given times t_1 and t_2,

$$\frac{\mathbf{v}_1}{\mathbf{v}_2} = \frac{r_2}{r_1}$$

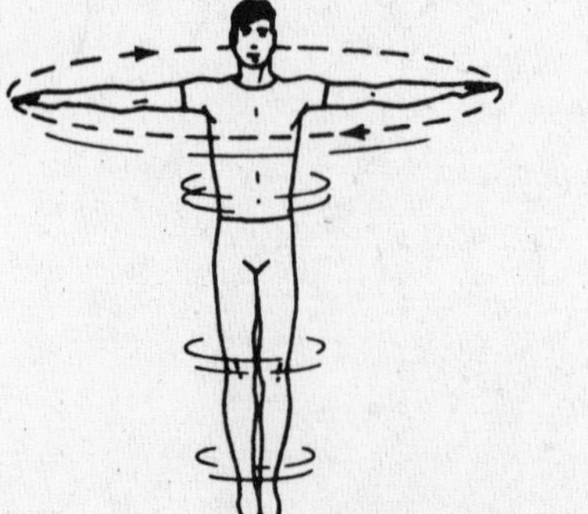

FIGURE 9.6

We know that angular velocity can be expressed as $\boldsymbol{\omega} = \mathbf{v}/r$, so we can also write

$$\frac{\omega_1}{\omega_2} = \frac{r_2^2}{r_1^2}$$

Another example of a system in which no net torque is acting is the solar system. Since the force of gravity is radial, it acts parallel to the radius vector swept out from the Sun, thus producing zero torque.

Questions Chapter 9

In each case, select the choice that best answers the question or completes the statement.

1. A point mass m is undergoing uniform circular motion with an angular frequency ω in a horizontal circle of radius r. Which of the following is a representation of the angular momentum of the mass?

(A) $mr^2\omega$
(B) mr^2/ω
(C) $r\omega^2/m$
(D) $mr\omega$
(E) $m\omega$

2. A skater extends her arms, holding a 2-kilogram mass in each hand. She is rotating about a vertical axis at a given rate. She brings her arms inward toward her body in such a way that the distance of each mass from the axis changes from 1 meter to 0.50 meter. Her rate of rotation (neglecting her own mass) will

(A) be doubled
(B) be halved
(C) be quadrupled
(D) be quartered
(E) remain the same

3. A 45-kilogram girl is sitting on a seesaw 0.6 meter from the balance point, as shown below. How far, on the other side, should a 60-kilogram boy sit so that the seesaw will remain in balance?

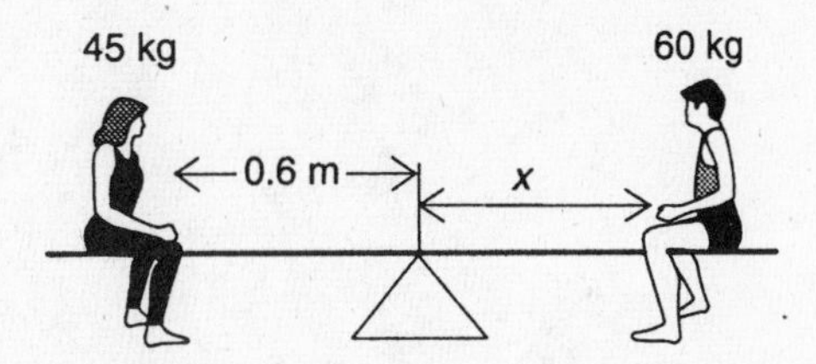

(A) 0.30 m
(B) 0.35 m
(C) 0.40 m
(D) 0.45 m
(E) 0.50 m

4. A balanced meter stick is shown below. The distance from the fulcrum is shown for each mass except the 10-gram mass. What is the approximate position of the 10-gram mass, based on the diagram?

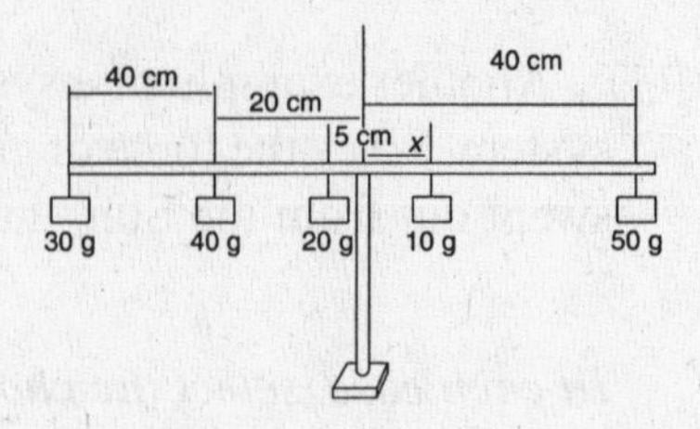

(A) 7 cm
(B) 9 cm
(C) 10 cm
(D) 15 cm
(E) 21 cm

5. A 1-kilogram mass swings in a vertical circle after having been released from a horizontal position with zero initial velocity. The mass is attached to a massless rigid rod of length 1.5 meter. The angular momentum of the mass, in kilogram · square meters per second, when it is in its lowest position is approximately

(A) 4
(B) 5
(C) 8
(D) 10
(E) 12

6. A rock with a mass of 0.05 kilogram is swung overhead in a horizontal circle of radius 0.3 meter at a constant rate of 5 revolutions per second. The angular momentum of the rock is

(A) 0.14 kg · m^2/s
(B) 0.0056 kg · m^2/s
(C) 0.32 kg · m^2/s
(D) 1.32 kg · m^2/s
(E) 2.45 kg · m^2/s

7. A solid cylinder consisting of an outer radius R_1 and an inner radius R_2 is pivoted on a frictionless axle as shown below. A string is wound around the outer radius and is pulled to the right with a force $\mathbf{F}_1 = 3$ N. A second string is wound around the inner radius and is pulled down with a force $\mathbf{F}_2 = 5$ N. If $R_1 = 0.75$ meter and $R_2 = 0.35$ meter, what is the net torque acting on the cylinder?

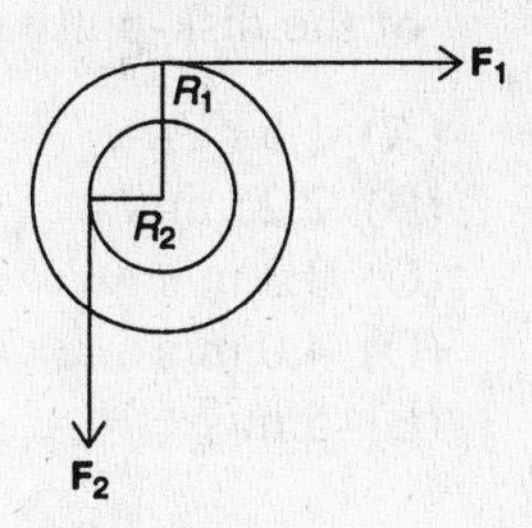

(A) 2.25 N · m
(B) −2.25 N · m
(C) 0.5 N · m
(D) −0.5 N · m
(E) 0 N · m

<u>*Questions 8 and 9*</u> are based on the following diagram. The rod is considered massless.

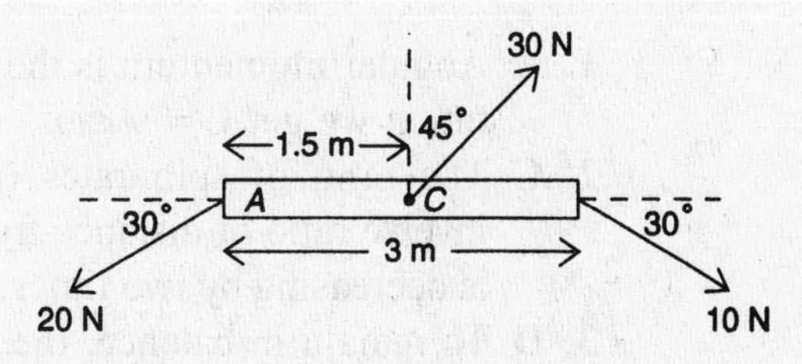

8. What is the net torque about an axis through point A?

(A) 16.5 N · m
(B) 15.2 N · m
(C) −5.5 N · m
(D) −7.8 N · m
(E) 6 N · m

9. What is the net torque about an axis through point C?

(A) 3.5 N · m
(B) 7.5 N · m
(C) −15.2 N · m
(D) 5.9 N · m
(E) 7 N · m

10. A small disk of mass 2 kilograms slides on a frictionless horizontal surface, and is constrained to move in a circular path by a light, rigid rod of length 0.5 meter. The disk is undergoing uniform circular motion at a velocity of 4 meters per second at any instant. A piece of putty, with a mass of 0.4 kilogram, is dropped onto the disk. If the radius of the circular path remains constant, what is the new velocity of the disk-putty system?

(A) 1.5 m/s
(B) 2.25 m/s
(C) 3.3 m/s
(D) 4.0 m/s
(E) 5 m/s

Explanations to Questions Chapter 9

Answers

1. (A)	**4.** (C)	**7.** (D)	**10.** (C)
2. (C)	**5.** (C)	**8.** (A)	
3. (D)	**6.** (A)	**9.** (B)	

Explanations

1. A Angular momentum is the quantity $m\mathbf{v}r$. Since $\mathbf{v} = r\boldsymbol{\omega}$, upon substitution we get $\mathbf{L} = mr^2\boldsymbol{\omega}$.

2. C The ratio of spin rates (angular velocity) is proportional to the inverse ratio of distance from the axis of rotation. Since the distance is decreasing by two times, the spin rate must increase by four times.

3. D To remain in balance, the two torques must be equal. The force on each side is given by the weight, $m\mathbf{g}$. The moment arm distances are 0.6 m and x. Since the factor $\mathbf{g}$ will appear on both sides of the torque balance equation, we can eliminate it and write

$$(45)(0.6) = (60)x \quad \text{implies} \quad x = 0.45 \text{ m}$$

4. C In this problem, again, the sum of the torques on the left must equal the sum of the torques on the right. We could convert all masses to kilograms and all distances to meters, but in the balance equation the same factors appear on both sides. Therefore, for simplicity and time efficiency, we can simply write

$$(30)(40) + (40)(20) + (20)(5) = (10)x + (50)(40)$$

$$x = 10 \text{ cm}$$

5. C The formula for angular momentum is $m\mathbf{v}r$; $m = 1$ kg and $r = 1.5$ m in this problem. The velocity at the bottom of the swing can be determined from conservation of energy. The kinetic energy at the bottom, $(1/2)m\mathbf{v}^2$, equals the loss of potential energy during the swing and is equal to mgh, where $h = r = 1.5$ m. Solving for velocity, we get $\mathbf{v} = 5.42$ m/s. Substituting for the angular momentum gives us 8.1 or approximately 8 kg $\cdot$ m^2/s.

6. A Angular momentum is expressed as $m\mathbf{v}r$. The velocity is given by the circumference, $2\pi r$, multiplied by the frequency, 5 rev/s. Making the necessary substitutions gives us the answer: 0.14 kg · m²/s.

7. D The net torque is given by the vector sum of all torques. $\mathbf{F}_2$ provides a counterclockwise positive torque, while $\mathbf{F}_1$ provides a clockwise negative torque. Each radius is the necessary moment arm distance. Thus we have

$$\tau_{net} = (5)(0.35) - (3)(0.75) = -0.5 \text{ N} \cdot \text{m}$$

8. A In this problem, the net torque "through point A" implies that the force passing through point A does not contribute to the net torque. Also, we need the components of the remaining forces perpendicular to the beam. From the diagram, we see that the 30-N force acts counterclockwise (positive), while the 10-N force acts clockwise (negative). Thus:

$$\tau_{net} = (30)(\cos 45)(1.5) - (10)(\sin 30)(3) = 16.5 \text{ N} \cdot \text{m}$$

9. B In this problem, since the pivot is now set at point C, we can eliminate the 30-N force passing through point C as a contributor to the torque. Again, we see that the 20-N force will act in a counterclockwise direction, while the 10-N force will act clockwise. Each force is 1.5 m from the pivot. We also need the component of each force perpendicular to the beam. Thus:

$$\tau_{net} = (20)(\sin 30)(1.5) - (10)(\sin 30)(1.5) = 7.5 \text{ N} \cdot \text{m}$$

10. C Angular momentum is conserved since the dropping of the putty does not add a net torque to the system. Thus we can state that $m_1\mathbf{v}_1r_1 = m_2\mathbf{v}_2r_2$. The radius is remaining the same, and the mass is increasing by 0.4 kg. Therefore

$$(2)(4)(0.5) = (2.4)(\mathbf{v}_2)(0.5) \quad \text{implies} \quad \mathbf{v}_2 = 3.3 \text{ m/s}$$

CHAPTER 10

Gravitation

Kepler's Laws of Planetary Motion

In the early seventeenth century, in an effort to support the belief by Copernicus that the Sun is the center of the universe, Johannes Kepler developed three laws of planetary motion. Briefly, these laws state that:

1. All the planets orbit the Sun in elliptical paths.
2. A planet sweeps out an equal area of space in an equal amount of time in its orbit.
3. The ratio of the cube of a planet's mean distance from the Sun and the square of its period is a constant.

We can understand Kepler's first law by looking at Figure 10.1. An ellipse consists of two fixed points, called *foci* (one point is a *focus*). An *ellipse* is defined to be the set of all points such that the sum of the distances to the two fixed points, from anywhere, equals a constant. Thus, if a point P is a distance P_{f1} from the first focal point and a distance P_{f2} from the second, then $P_{f1} + P_{f2}$ is always a constant. In the solar system, the Sun is located at one of the foci. The extent to which an ellipse differs from a circle is referred to as its *eccentricity*.

The second law can be understood from the law of conservation of angular momentum. In Figure 10.1, if it takes 1 month to go from point A to point B on the ellipse and 1 month to go from point C to point D, then Kepler's second law asserts that the areas swept out (as outlined) are equal. From our knowledge of forces, we suspect that the force acting on the planet increases as distance decreases since, as the planet gets closer, it begins to accelerate in its orbit and the centripetal acceleration directed toward the center also increases.

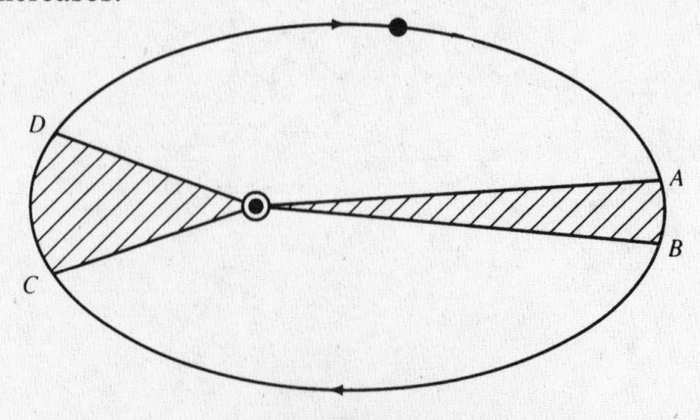

FIGURE 10.1

The eccentricity of the planetary orbits is so small that they can almost be considered circles. Therefore, since the force acting on the planets is radial, no net torque acts on the system. In that case, there is no change in the angular momentum of the system (it is conserved). Therefore, as the planet gets closer to the Sun, it must orbit faster in order to maintain the same amount of angular momentum.

Kepler's third law of motion can be written mathematically in the form

$$\frac{R^3}{T^2} = K_S = 3.35 \times 10^{18}\ \text{m}^3/\text{s}^2$$

This constant is the same for all planets in the solar system for which R is the mean radius and T is the period. The universality of Kepler's third law lies in the fact that for any system (such as the Earth-Moon system) a suitable "Kepler's constant" can be found. The Moon obeys Kepler's laws, as do also the satellites of Jupiter and stars in distant galaxies!

Newton's Law of Universal Gravitation

We already know that all objects falling near the surface of Earth have the same acceleration, given by $\mathbf{g}$ = 9.8 meters per second squared. The value of this constant can be determined from the independence of the period of a pendulum on the mass of the bob. This "empirical" verification is independent of any "theory" of gravity.

The weight of an object on Earth is given by $\mathbf{W} = m\mathbf{g}$, which represents the magnitude of the force of gravity due to Earth that is acting on the object. We know that gravity causes a projectile to assume a parabolic path. Isaac Newton, in his book *Principia*, extended the idea of projectile motion to an imaginary situation in which the velocity of the projectile was so great that the object would fall and fall, but the curvature of Earth would bend away and leave the projectile in "orbit." Newton conjectured that this might be the reason why the Moon orbits the Earth, as shown in Figure 10.2.

To answer this question, Newton first had to determine the centripetal acceleration of the Moon based on observations from astronomy. He knew the relationship between centripetal acceleration and period. We have written that relationship as

$$\mathbf{a}_c = \frac{4\pi^2 R}{T^2}$$

where R is the distance to the Moon (in meters) and T is the orbital period (in seconds). From astronomy we know that $R = 3.8 \times 10^8$ meters, and since the Moon orbits Earth in 27.3 days, we have $T = 2.3 \times 10^6$ seconds. Using these values, we find that the magnitude of $\mathbf{a}_c = 2.8 \times 10^{-3}$ meters per second squared.

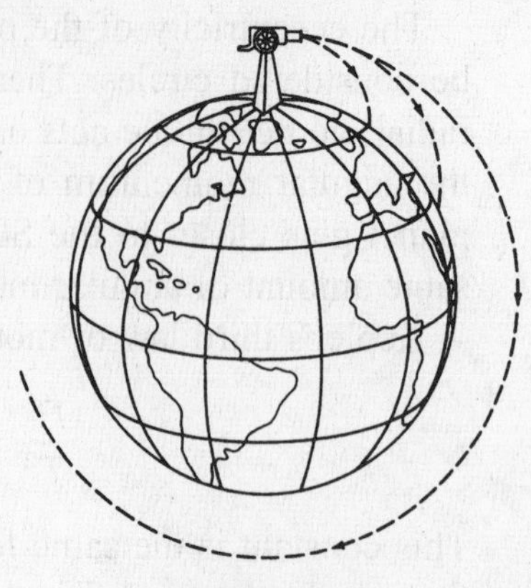

FIGURE 10.2

Using the formula for centripetal acceleration, Newton was able to observe that, since $\mathbf{F} = m\mathbf{a}$ by his second law, the formula for centripetal force is given by

$$\mathbf{F}_c = \frac{M4\pi^2 R}{T^2}$$

Using Kepler's third law of planetary motion, and setting M equal to the mass of the Moon, Newton could express the force of gravity in the form

$$\mathbf{F}_g = \frac{M_{\text{Moon}} 4\pi^2 K_{\text{Earth}}}{R^2}$$

Then Kepler's constant could be evaluated for the Earth-Moon system by using Kepler's third law.

Using his own third law of motion, Newton realized that the force that Earth exerts on the Moon should be exactly equal, but opposite, to the force that the Moon exerts on Earth (observed as tides). This relationship implies that the constant K should be dependent on the mass of Earth, and this mutual interaction implies that the force of gravity should be proportional to the product of both masses. In other words, Newton's law of gravity could be expressed as

$$\mathbf{F}_g = \frac{GM_1M_2}{R^2}$$

This is a vector equation in which the force of gravity is directed inward toward the center of mass for the system (in this case, near the center of Earth). Since Earth is many times more massive than the Moon, the Moon orbits Earth, and not vice versa.

The value of G, called the *universal gravitational constant*, was experimentally determined by Henry Cavendish in 1795. In modern units $G = 6.67 \times 10^{-11}$ newton · square meter per square kilogram. If we recognize that $\mathbf{F} = m\mathbf{a}$, and then if we consider M_1 equal to the mass of an object of mass m, and M_2 to equal the mass of Earth, M_E, the acceleration of a mass m near the surface of Earth is given by

$$\mathbf{a} = \frac{GM_E}{{R_E}^2}$$

where R_E is the radius of Earth in meters. Using known values for these quantities, we discover that **a** = 9.8 meters per second squared = **g**!

Thus we have a theory that accounts for the value of the known acceleration due to gravity. In fact, if we replace the mass of Earth by the mass of any other planet, and the radius of Earth by the corresponding radius of the other planet, the above formula allows us to determine the value of **g** on any planet or astronomical object in the universe! For example, the value of **g** on the Moon is approximately 1.6 meters per second squared, or about one-sixth the value on Earth. Thus, objects on the Moon weigh one-sixth as much as they do on Earth.

Now, Newton's prediction for the acceleration of the Moon toward Earth is that the value of **g** decreases as the square of the distance from Earth. The ancient Greek astronomers discovered that the mean distance to the Moon is approximately equal to 60 times the radius of Earth. Thus, the acceleration of the Moon should be 1/3600 of the acceleration of an object near Earth's surface: **a** = 9.8/3600 = 0.0028 meters per second squared! Newton's theory of gravity was confirmed in one simple, triumphant demonstration. Further proof came when his friend Edmund Halley used Newton's law to predict that a certain comet (now known as Halley's comet) would return every 76 years, appearing next in 1758.

Newton's Interpretation of Kepler's Laws

We are now in a position to discuss Kepler's third law of planetary motion in light of Newton's law of gravitation. First, we should observe that Newton's "theory" merely describes the relationship between two masses and their displacement from each other. It is a vector law, so the familiar rules for vector mechanics will apply. However, as Newton himself wrote, nowhere is the "cause" of gravity discussed. In *Principia*, his seminal work, Newton himself boldly states that he "frames no hypotheses" regarding the origin of the force of gravity.

Nevertheless, the theory allows us to understand the meaning behind Kepler's laws. For example, we have noted that Kepler's second law can be thought of as a consequence of the law of conservation of angular momentum. The reason why the shapes of the orbits of the planets are elliptical was determined by Newton and depends somewhat on the amount of total energy available to a particular planet when it formed. We will not go into more detail on this topic.

Kepler's third law relates the distance of a planet from the Sun and the velocity of its orbit about the Sun:

$$\frac{R^3}{T^2} = K_S$$

To begin, let us write the equation for the force of gravity between a planet and the Sun:

$$\mathbf{F} = \frac{GM_PM_S}{R^2}$$

Recall now that, if the eccentricities of ellipses are small, the ellipses can be treated approximately as though they were circles. In this case, the mean radius of the orbit can be thought of as the radius of a circular path in which a centripetal force (provided by gravity) acts on the planet:

$$\frac{M_P 4\pi^2 R}{T^2} = \frac{GM_PM_S}{R^2}$$

The mass of the planet can be eliminated from both sides, and we are left with an equation that depends simply on the mass of the Sun and the distance to the Sun (which is Kepler's law!):

$$\frac{R^3}{T^2} = \frac{GM_S}{4\pi^2} = k_S$$

If we want to know the value of the constant for an Earth-centered system, we simply use the mass of Earth instead of the Sun's mass in the equation above. The mass of the Sun is approximately 2×10^{30} kilograms, and the mass of Earth is approximately 6×10^{24} kilograms.

Kepler's third law implies that, for any system, the ratio of distance cubed to time squared is a constant for any two intervals of time:

$$\frac{R_1^3}{T_1^2} = \frac{R_2^3}{T_2^2}$$

We can use this relationship to find any corresponding period for any given distance from the center of Earth (or the Sun) if we know any "standard." For example, if we take the Earth system and use the Moon's motion as our standard, we can compute the orbital period of an artificial satellite or spacecraft (recognizing that R is actually the sum of the height above the surface of Earth and Earth's radius).

This approach leads directly to the concept of geostationary or geosynchronous orbits. At an altitude of about 22,000 miles, the orbital period of an orbiting spacecraft is exactly equal to the rotational period of Earth (about 24 hours). The spacecraft (or satellite) appears to remain stationary above a specified point on Earth. These satellites are used for communication relays and weather forcasting since a series of them, strategically placed, can always monitor a given region of Earth's surface.

Gravitational Energy

In Chapter 7 we saw that the amount of work done (by or against gravity) when vertically displacing a mass is given by the change in the gravitational potential energy:

$$\Delta PE = \Delta m\mathbf{g}h$$

We now know that the value of **g** is not constant but varies inversely with the square of the distance from the center of Earth. Also, since we want the potential energy to become weaker the closer we get to Earth, we can use the results from the section on Newton's law of universal gravitation to rewrite the potential energy formula as

$$PE = \frac{-GM_0M_E}{R_E}$$

This equation is valid if an object is located near the surface of Earth. Actually, we should have used a distance variable $R = R_E + h$ (where h is the height above the surface). If h is very small compared with the radius of Earth, it can effectively be eliminated from the equation.

The "escape velocity" from the gravitational force of Earth can be determined by considering the situation where an object has just reached infinity, with zero final velocity, given some initial velocity at any direction away from the surface of Earth. We designate that escape velocity as $\mathbf{v}_{esc}$, and state that, when the final velocity is zero (at infinity), the total energy must be zero, which implies

$$\frac{1}{2}M_0\mathbf{v}_{esc}^2 = \frac{GM_0M_E}{R_E}$$

The mass of the object can be eliminated from the relationship, leaving

$$\mathbf{v}_{esc} = \sqrt{\frac{2GM_E}{R_E}}$$

An object can leave the surface of Earth at any velocity. There is, however, one minimum velocity at which, if the spacecraft coasted, it would not fall back to Earth because of gravity.

The orbital velocity can be determined by assuming that we have an approximately circular orbit. In this case, we can set the centripetal force equal to the gravitational force:

$$\frac{GM_0M_E}{R_E^2} = \frac{M_0\mathbf{v}_0^2}{R_E}$$

Eliminating the mass of the object from the equation leaves

$$\mathbf{v}_{orbit} = \sqrt{\frac{GM_E}{R_E}}$$

Questions
Chapter 10

In each case, select the choice that best answers the question or completes the statement.

1. What is the value of **g** at a height above Earth's surface that is equal to the radius of Earth?

(A) 9.8 N/kg
(B) 4.9 N/kg
(C) 6.93 N/kg
(D) 2.45 N/kg
(E) 1.6 N/kg

2. Another planet has half the mass of Earth and half the radius. Compared to the acceleration due to gravity near the surface of Earth, the acceleration of gravity near the surface of this other planet is

(A) twice as much
(B) one-fourth as much
(C) half as much
(D) the same
(E) zero

3. Which of the following is an expression for the acceleration of gravity with uniform density ρ and radius R?

(A) $G(4\pi\rho/3R^2)$
(B) $G(4\pi\rho R^2/3)$
(C) $G(4\pi\rho/3R)$
(D) $G(4\pi R\rho/3)$
(E) None of these is correct.

4. Given that the mean radius of the Moon's orbit is 3.84×10^8 meters and its period is 2.36×10^6 seconds, at what altitude above the surface of Earth does a geostationary satellite orbit? (The radius of Earth is 6.0×10^{24} m.)

(A) 4.23×10^7 m
(B) 3.59×10^7 m
(C) 6.2×10^8 m
(D) 2.2×10^7 m
(E) 5.8×10^7 m

5. What is the orbital velocity of a satellite at a height of 300 kilometers above the surface of Earth? (The mass of Earth is approximately 6×10^{24} kg, and its radius is 6.0×10^{24} m.)

(A) 5.42×10^1 m/s
(B) 1.15×10^6 m/s
(C) 7.7×10^3 m/s
(D) 6×10^6 m/s
(E) 3×10^8 m/s

6. What is the escape velocity from the Moon, given that the mass of the Moon is 7.2×10^{22} kg and its radius is 1.778×10^6 m?

(A) 1.64×10^3 m/s
(B) 2.32×10^3 m/s
(C) 2.69×10^6 m/s
(D) 5.38×10^6 m/s
(E) 3×10^8 m/s

7. A "black hole" theoretically has an escape velocity that is greater than or equal to the velocity of light (3×10^8 m/s). If the effective mass of the black hole is equal to the mass of the Sun (2×10^{30} kg), what is the effective "radius" (called the "Schwarzchild radius") of the black hole?

(A) 3×10^3 m
(B) 1.5×10^3 m
(C) 8.9×10^6 m
(D) 4.45×10^6 m
(E) 0 m

8. Kepler's second law of planetary motion is a consequence of the law of conservation of

(A) linear momentum
(B) charge
(C) energy
(D) mass
(E) angular momentum

9. What is the gravitational force of attraction between two trucks, each of mass 20,000 kilograms, separated by a distance of 2 meters?

(A) 0.057 N
(B) 0.013 N
(C) 0.0067 N
(D) 1.20 N
(E) 0 N

10. The gravitational force between two masses is 36 newtons. If the distance between the masses is tripled, the force of gravity will be

(A) the same
(B) 18 N
(C) 9 N
(D) 4 N
(E) 27 N

Explanations to Questions Chapter 10

Answers

1. (D)	**4.** (B)	**7.** (A)	**10.** (D)
2. (A)	**5.** (C)	**8.** (E)	
3. (D)	**6.** (B)	**9.** (C)	

Explanations

1. D The value of **g** varies inversely with the square of the distance from the center of Earth; therefore, if we double the distance from the center (as in this case), the value of **g** decreases by one-fourth: $1/4(9.8) = 2.45$. Since $\mathbf{g} = \mathbf{F}/m$, alternative units are newtons per kilogram.

2. A If, using the formula for **g**, we take half the mass, the value decreases by one-half. If we decrease the radius by half, the value will increase by four times. Combining both effects results in an overall increase of two times.

3. D The formula for **g** is $\mathbf{g} = GM/R^2$. The planet is essentially a sphere of mass M, radius R, and density ρ (with $M = V\rho$, where V is the volume). The volume of the planet is given by $V = 4/3\pi R^3$. Making the substitutions yields:

$$g = G\frac{4\pi R\rho}{3}$$

4. B From Kepler's third law, we can determine the value of the constant K for the Earth system by using the data for the Moon:

$$\frac{(3.84 \times 10^8)^3}{(2.36 \times 10^6)^2} = 1.01664 \times 10^{13}\ \text{m}^3/\text{s}^2$$

Now, we use Kepler's third law again, but this time with a period of 86,400 s (which equals the 24-h period of a geostationary orbit) and the above constant for K:

$$\frac{R^3}{T^2} = K = 1.01664 \times 10^{13} = \frac{R^3}{(86{,}400)^2}$$

Solving for R (don't forget to take the cube root!) yields R = 42,338,119 m as the distance of the satellite from the center of Earth. To get the answer, we subtract the radius of Earth, which is given in the question, leaving $h = 3.59 \times 10^7$ m.

5. C The formula for orbital velocity is

$$\mathbf{v}_{\text{orbit}} = \sqrt{\frac{GM}{R}}$$

where R is the distance from the center of Earth. In this case we must add 300 km = 300,000 m to the radius of Earth. Thus, $R = 6.7 \times 10^6$ m. Substituting the given values yields $\mathbf{v}_{\text{orbit}} = 7728$ or 7.7×10^3 m/s.

6. B The formula for escape velocity is

$$\mathbf{v}_{esc} = \sqrt{\frac{2GM_{Moon}}{R_{Moon}}}$$

Substituting the given values yields $\mathbf{v}_{esc} = 2324$ or 2.32×10^3 m/s.

7. A From question 6, we know that the escape velocity is given by $\mathbf{v}_{esc} = \sqrt{2GM/R}$. To find R, we need to square both sides, and then solve for the radius. This yields $R = 2GM/\mathbf{v}^2$. Substituting the given values yields $R = 3000$ or 3×10^3 m. (This is only a "theoretical" size for the black hole. As an interesting exercise, try calculating the value of **g** on such an object!)

8. E Kepler's second law can be explained in terms of the conservation of angular momentum. Since the force of gravity is radial, the net torque acting on a planet in the solar system is zero. Therefore, the change in angular momentum is zero and so this quantity is conserved.

9. C We use the formula for gravitational force:

$$\mathbf{F} = \frac{GM_1M_2}{R^2}$$

Substituting the given values (don't forget to square the distance!) yields $\mathbf{F} = 0.0067$ N.

10. D The force of gravity is an inverse-square-law relationship. This means that, as the distance is tripled, the force is decreased by one-ninth. One-ninth of 36 N is 4 N.

CHAPTER 11

Heat and Temperature

The Kinetic Theory

The kinetic theory of matter states that all matter is composed of molecules that are relatively far apart and exert force on each other only when they collide. These molecules move freely and quickly between collisions. The pressure of a gas on its container is due to the collisions of the gas molecules with the walls of the container. The molecules of liquids are comparatively close to each other, but can move past one another readily. Even when a fluid is "still," its molecules continue to vibrate, as indicated by Brownian movements in which large particles are seen to move because of bombardment by invisible molecules. In solids the molecules are much closer together. There are considerable bonding forces between the molecules and little freedom of motion. The molecules of solids vibrate, however, and therefore possess kinetic energy.

Temperature and Heat

One way of defining *temperature* is to say it is the degree of hotness or coldness of an object. If we think in terms of the *kinetic theory* we get another definition. The molecules of a substance are in constant random motion. If we heat a gas, its molecules move faster; that is, when the temperature of a gas rises, the average speed of its molecules also rises. When the speed of motion of a molecule goes up, its kinetic energy goes up. The molecules in a substance are not all moving with the same speed and therefore do not all have the same kinetic energy. We can speak, however, of the average kinetic energy of these molecules. When we heat a gas, the average kinetic energy of its molecules rises. This fact leads to thinking of *temperature* as a measure of the average kinetic energy per molecule of a substance. If we take a cupful of water from a bathtub full of water, the *average* kinetic energy per molecule, and therefore the temperature, are the same in the cup as in the tub.

If we have two different gases at the same temperature, perhaps a mixture of oxygen and hydrogen at 0°C, the average kinetic energy of the oxygen molecules is the same as that of the hydrogen molecules. However, on the average, the oxygen molecules will move more slowly than the hydrogen molecules because the oxygen molecule has the greater mass. (Remember: $E_k = \frac{1}{2}m\mathbf{v}^2$).

We have seen that the molecules of a substance have kinetic energy. In the ideal gases we assume that this is kinetic energy of translation only. This assumption is accurate enough for a monatomic gas such as helium and for diatomic gases such as oxygen and nitrogen at moderate temperatures and pressures. In actual gases the molecules may also have kinetic energy of rotation, as well as potential energy with respect to each other. For example, when a solid expands, work is done to pull the molecules away from each other against the cohesive force. Until recently *heat* was thought of as the sum total of the energy of the molecules in a substance. We now say that a substance has *internal energy* as a result of the kinetic and potential energy of its molecules. In Chapter 20 we shall see that some of the internal energy is also due to the potential energy of the atoms. In crystalline solids much of the internal energy is due to the vibration of the nuclei of the atoms. At the present time physicists' knowledge of internal energy is incomplete.

When a hot piece of copper is brought into contact with a cold piece, the hot piece gets colder and the cold piece gets hotter—both pieces end up at the same temperature. The hot piece lost some internal energy and the cold piece gained some. We define *heat* as the form of energy that flows between two bodies because they are at different temperatures. It is energy in transit.

Thermometers

When an object is heated, many of its characteristics may change. For example, when a solid gets hotter there may be a change in its color, electrical resistance, and/or dimensions. We can use these changes as means for measuring temperature. Many thermometers (Figure 11.1) depend on the fact that substances tend to expand when heated.

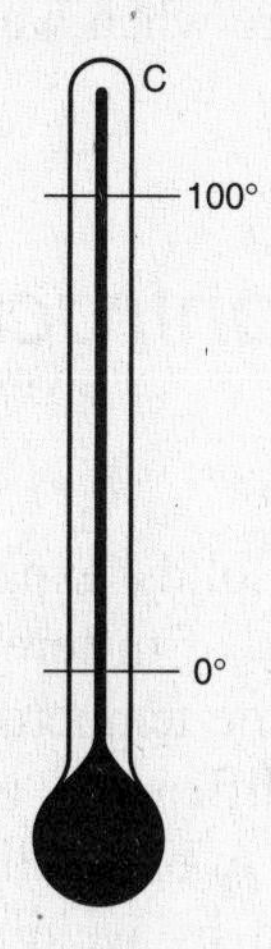

FIGURE 11.1

Let us describe briefly the making and calibration of a mercury thermometer. A uniform, thick-walled capillary tube with a glass bulb at one end is partly filled with mercury. The bulb is heated, the mercury expands and fills the tube, driving out all the air. Then the top of the tube is sealed. The selection of a temperature scale requires the choice of a reference point and the size of a unit. In practice, two reference points are usually used for the mercury thermometer. If we insert the bulb of a mercury thermometer in water and heat the water steadily, the mercury level in the thermometer keeps rising until the water starts boiling. As long as the water keeps boiling, the level of the mercury in the thermometer remains constant, no matter how vigorously the water is boiled, and, therefore a reference point has been readily obtained. As the water is cooled, the level of the mercury keeps dropping until ice starts forming. As the water is cooled still further, more ice keeps forming, but the level of the mercury remains the same as long as both ice and water are present and are mixed. A second reference point, known as the *ice point*—the temperature at which ice melts or water freezes,—has been obtained. On the *Celsius* scale the ice point is marked 0°.

The temperature at which water boils is affected significantly by variation in atmospheric pressure. Therefore in specifying the first reference point we emphasize the pressure to be used: the *standard pressure* is the pressure that would be produced by a column of mercury 76 centimeters high. For short, we speak of a pressure of 76 centimeters of mercury, or 760 millimeters of mercury, or of 1 (standard) atmosphere. The first reference point is known as the *steam point:* the temperature at which water under standard pressure boils. On the Celsius scale the steam point is marked 100°.

The *Kelvin scale* is used as an *absolute temperature* scale. This scale has no negative temperatures. *Absolute zero* = –273°C = 0 K. Therefore, 273 K = 0°C. Each Kelvin unit on the Kelvin scale has the same size as each degree on the Celsius scale. We shall say more about absolute temperature when we describe the behavior of gases.

Expansion and Contraction

Solids

Most solids expand when heated and contract when cooled. The same lengths of different solids expand different amounts when heated through the same temperature change. For example, brass expands more than iron. This difference is made use of in thermostats, which feature bimetallic strips. A bimetallic strip (sometimes known as a *compound bar*) may be formed by welding or riveting together a brass and an iron strip. When heated, the bimetallic strip bends, with the brass forming the outside of the curve (see Figure 11.2). When the strip bends, it can make or break an electric contact. In cooling, the reverse happens; making a circuit can result in the operation of a heater.

When a solid is heated, it expands not only in length but also in width and thickness. Therefore, thermal expansion results in an increase in length, area, and volume. If a hollow solid, such as a glass flask, is heated, the empty space increases in volume just as though it were made of the same material as the walls.

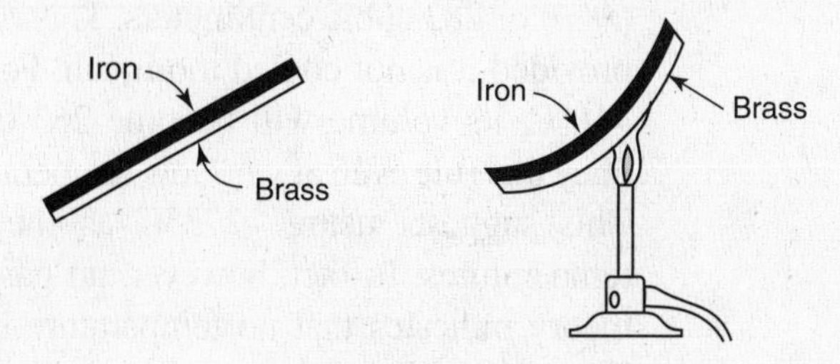

FIGURE 11.2

Liquids

Most liquids expand when heated and contract when cooled. Different liquids expand different amounts when heated through the same temperature change. The coefficient of volume expansion tells what change in volume takes place per unit volume when the temperature goes up 1 degree. This difference is greater for most liquids than for solids. For example, when the mercury thermometer is heated, the mercury expands more than the glass bulb. Therefore the level of the mercury rises to compensate for the difference in volume expansion.

Water behaves peculiarly in this respect. As water is cooled from 100°C it contracts until its temperature reaches 4°C. If it is cooled further the water will expand; that is, water is densest at 4°C. Therefore, as a lake is cooled at the surface by the atmosphere, the bottom of the lake tends to reach and maintain a temperature of 4°C. If the surface freezes, the resulting ice stays at the top since the specific gravity of ice is only about 0.9 that of water. Water molecules form hydrogen bonds with each other. These are abundant in ice and decrease in relative abundance as the temperature of water goes up (Figure 11.3).

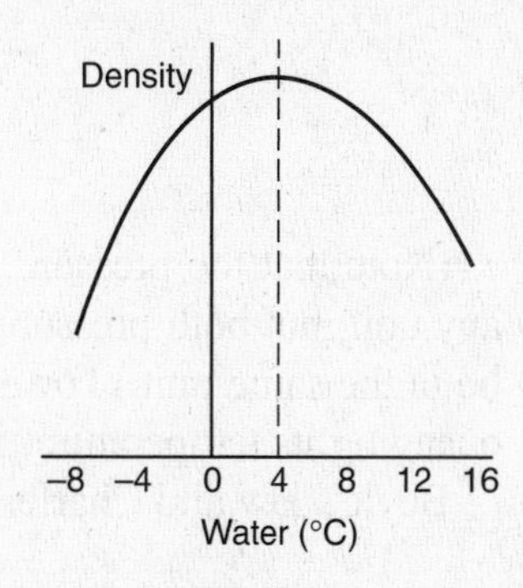

FIGURE 11.3

Gases

If the pressure on a gas is kept constant, heating the gas will result in an increase in its volume. For all gases the coefficient of volume expansion is nearly the same. For example, if we start with a gas at 0°C, for each Celsius degree rise in temperature the volume increases by $\frac{1}{273}$ of whatever volume the gas occupied at 0°C, provided the pressure of the gas is not allowed to change. Thus, a gas occupying a volume of 273 cubic centimeters at a pressure of 700 millimeters is heated from 0°C to 10°C, and the pressure is kept at 700 millimeters, the volume of the gas will become 273 cubic centimeters + 273 (10/273) or 283 cubic centimeters. This rule also operates when the gas is cooled, provided it is not cooled too much. For example, if the above gas is cooled to –10°C, its volume will become 263 cubic centimeters. *If* gases continued to obey this rule even at very low temperatures, the gas would disappear at –273C. This suggests using –273°C as the zero temperature on a gas scale of temperatures. In fact, however, no gas behaves in this way. Fortunately, other theory indicates that no temperature below approximately –273°C can exist. Therefore, this is taken as the zero temperature on the absolute scales; 0 K = –273°C. Each Kelvin unit measures the same temperature change as each Celsius degree.

Actual gases become liquefied at least a few degrees above absolute zero. The kinetic theory states that the speed of motion of the molecules in a substance decreases as the substance's temperature decreases. This does not mean, however, that all motion within the atom ceases at absolute zero.

A more convenient way than the one above to calculate the volume of a gas as the result of a temperature change is to use *Charles' law:* If the pressure on a gas is kept constant, its volume is directly proportional to its *absolute* temperature:

$$\frac{V_1}{V_2} = \frac{T_1}{T_2}$$

where V_1 is the volume of the gas at absolute temperature T_1. The volume may be expressed in any unit used for this property; of course, the same unit must be used for both volumes.

Another important gas law is *Boyle's law:* If the temperature of a gas is kept constant, the volume of the gas varies inversely with its pressure ($V = k/\mathbf{P}$):

$$\mathbf{P}_1 V_1 = \mathbf{P}_2 V_2$$

(Note that the pressure may be in any unit and that the volume may be in any unit, but both pressures must be in the same unit and both volumes must be in the same unit.) For example, if the pressure on a gas is doubled without changing its temperature, its volume is reduced to half of its original volume.

Boyle's law and Charles' law may be combined to give the general gas law:

$$\frac{\mathbf{P}_1 V_1}{T_1} = \frac{\mathbf{P}_2 V_2}{T_2}$$

Note: Gauges used to measure air pressure in automobile tires do not give the true pressure that is required for the gas laws stated in this chapter. To get the true or absolute pressure, we must add to the gauge pressure the standard pressure of the atmosphere, about 101,000 newtons per square meter = 101,000 pascals.

The SI unit for pressure is the pascal: 1 Pa = 1 N/m^2.

Questions Chapter 11

In each case, select the choice that best answers the question or completes the statement.

1. A temperature of 450 K is equivalent to a Celsius temperature of

(A) 723°
(B) 10°
(C) 273°
(D) 450°
(E) 177°

2. A temperature of 100°C is equivalent to a temperature on the Kelvin scale of

(A) 0 K
(B) 173 K
(C) 212 K
(D) 273 K
(E) 373 K

3. A change in temperature of 20°C is equivalent to a change in temperature on the Kelvin scale of

(A) 0.9 K
(B) 11 K
(C) 6.7 K
(D) 20 K
(E) 68 K

4. According to the kinetic theory of gases, what is the relationship between the average kinetic energy of the molecules and the absolute temperature of an ideal confined gas?

(A) The KE is inversely proportional to the absolute temperature.
(B) The KE is linearly proportional to the absolute temperature.
(C) The KE is not related to the absolute temperature.
(D) The KE is parabolically related to the absolute temperature.
(E) The KE is proportional to the square root of the absolute temperature.

5. If the volume of a gas is doubled without changing its temperature, the pressure of the gas is

(A) reduced to one-half of the original value
(B) reduced to one-fourth of the original value
(C) not changed
(D) doubled
(E) quadrupled

6. The volume of a given mass of gas is doubled without changing its temperature. As a result the density of the gas

(A) is reduced to one-fourth of the original value
(B) is halved
(C) remains unchanged
(D) is doubled
(E) is quadrupled

7. The volume of a given mass of gas will be doubled at atmospheric pressure if the temperature of the gas is changed from 150°C to

(A) 300°C
(B) 423°C
(C) 573°C
(D) 600°C
(E) 743°C

8. All of the following express approximately standard atmospheric pressure EXCEPT

(A) 1 atm
(B) 101 kPa
(C) 760 cm of mercury
(D) 76 cm of mercury
(E) 760 mm of mercury

9. If 500 cubic centimeters of gas, having a pressure of 760 millimeters of mercury, is compressed into a volume of 300 cubic centimeters, the temperature remaining constant, the pressure of the gas will be, in millimeters of mercury, approximately

(A) 500
(B) 900
(C) 1,100
(D) 1,270
(E) 1,500

10. How many cubic meters of air at an atmospheric pressure of 10^5 pascals must be pumped into a 2-cubic meter tank containing air at atmospheric pressure in order to raise the absolute pressure to 8×10^5 pascals? Assume that the temperature remains constant.

(A) 8
(B) 10
(C) 14
(D) 20
(E) 24

Explanations to Questions Chapter 11

Answers

1. (E)	**4.** (B)	**7.** (C)	**9.** (D)
2. (E)	**5.** (A)	**8.** (C)	**10.** (C)
3. (D)	**6.** (B)		

Explanations

1. E 450 K – 273° = 177°C

2. E T = °C + 273° = (100 + 273) K = 373 K

3. D Change in °C = change in K = 20

4. B The average kinetic energy of the molecules is linearly proportional to the absolute temperature of the ideal gas.

5. A Boyle's law applies. The pressure of the gas varies inversely with its volume. Since the volume is doubled, the pressure must be reduced to one-half of its original value.

6. B Density = mass/volume. The mass of the gas does not change in the problem; the volume is doubled. Therefore the density must go down to one-half of its original value.

7. C Charles' law applies: At constant pressure, the volume of a gas is proportional to its absolute temperature. Therefore, to double the volume of the gas we must double the absolute temperature. The starting temperature: T = °C + 273° = (150 + 273) K = 423 K. The final absolute temperature is 2 × 423 K = 846 K. The final Celsius temperature is (846 – 273)° = 573°.

8. C 760 cm of mercury is not equivalent to standard atmospheric pressure. Note that 1 kPa = 1,000 Pa.

9. D Boyle's law applies.

$$\mathbf{P}_1 V_1 = \mathbf{P}_2 V_2$$

$$\mathbf{P}_2\ (300\ \text{cm}^3) = 760\ \text{mm} \times 500\ \text{cm}^3$$

$$\mathbf{P}_2 = \frac{760 \times 500}{300}\ \text{mm} = 1{,}270\ \text{mm}$$

10. C The total volume of air at atmospheric pressure is the volume to be added (V) plus the 2 m^3 already in the tank. Therefore

$$10^5\ \text{Pa}\ (V + 2\ \text{m}^3) = 8 \times 10^5\ \text{Pa}\ (2\ \text{m}^3)$$

$$V + 2\ \text{m}^3 = 16\ \text{m}^3$$

$$V = 14\ \text{m}^3$$

CHAPTER 12

Measurement of Heat

Internal energy, work, and heat being transferred are all measured in joules. In order to raise the temperature of 1 kilogram of liquid water 1°C, 4.19 kilojoules of heat are needed.

From the definition of the calorie we know that, if the temperature of 1 gram of water goes up from 10°C to 11°C, 1 calorie has been added to the water. However, we find that less heat is required to raise the temperature of 1 gram of most other substances 1 Celsius degree. We often define the *specific heat capacity* (sp. ht.) of a substance as the amount of heat required to raise the temperature of 1 gram of the substance 1°C. From the above we can see that the specific heat of water is 4.19 kilojoules per kilogram • °C. The specific heat capacity of copper is 0.39 kilojoules per kilogram • °C.

The specific heat capacity $\left(\frac{\text{kJ}}{\text{kg}\cdot{}^\circ\text{C}}\right)$ of a substance is often obtained in the laboratory by using the method of mixtures. For example, a 100-gram block of iron is moved from boiling water (100°C) to a beaker containing 110 grams of water at 18°C. Assume that this change raises the temperature of the water in the beaker to 25°C. We can now calculate the specific heat of iron:

$$\text{heat lost by hot object} = \text{heat gained by cold object}$$

Also, for any object that loses or gains heat *without changing phase,*

$$\text{heat lost (or gained)} = \text{mass} \times \text{specific heat} \times \text{temperature change}$$

Using SI units, we can state that

$$\text{heat lost lost by the iron} = 0.100\text{ kg} \times \text{sp. ht.} \times (100^\circ\text{C} - 25^\circ\text{C})$$

$$\text{heat gained by the water} = 0.110\text{ kg} \times 4.19\text{ kJ/kg}\cdot{}^\circ\text{C} \times (25^\circ\text{C} - 18^\circ\text{C})$$

If we assume that no other substances are involved in the heat exchange, then

$$0.100\text{ kg} \times \text{sp. ht.} \times 75^\circ\text{C} = 0.110\text{ kg} \times 4.19\text{ kJ/kg}\cdot{}^\circ\text{C} \times 7^\circ\text{C}$$

$$c = \text{sp. ht.} = 0.43\ \frac{\text{kJ}}{\text{kg}\cdot{}^\circ\text{C}}$$

Change of Phase

Earlier we defined solid, liquid, and gas. These are three phases of matter. If we heat a solid, its temperature goes up until it starts melting. Further transfer of heat to the solid results in melting without a change of temperature. (Removing the same amount of heat from the liquid will then result in *fusion* without a change of temperature.) Continued heating after all the solid has melted again leads to a temperature change—this time of the liquid. In the above equation the specific heat capacity of the liquid must be used; this is usually different from the specific heat capacity of the solid. For example, the specific heat capacity of ice is 2.05 kilojoules per kilogram • °C. Recall that for liquid water the value is 4.19 kilojoules per kilogram • °C.

When a substance is changed from solid to liquid, or from liquid to gas, heat must be supplied to pull the molecules away from each other against cohesive forces. This heat represents an increase in potential rather than kinetic energy; therefore the temperature remains the same. When the reverse change of phase, for example, solidification, takes place, this heat is released. For a given crystalline substance (a metal or, ice, e.g.), melting and freezing occur at the same temperature, at constant pressure. (See Figure 12.1.)

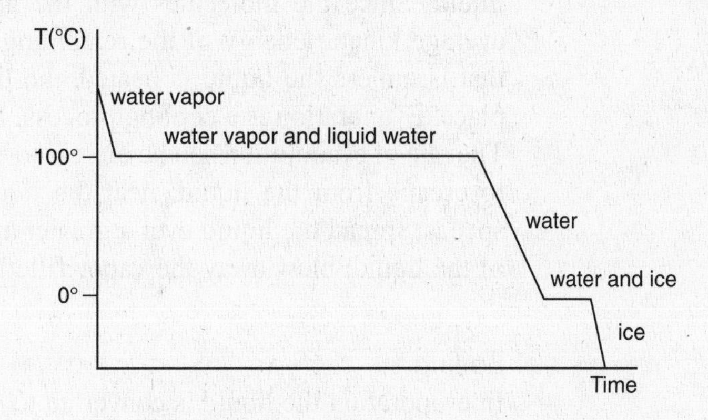

FIGURE 12.1 Cooling Curve for Water

Heat of Fusion

Heat of fusion of a substance is the amount of heat required to melt a unit mass of the substance without change of temperature. For ice this is 335 kilojoules per kilogram.

$$\text{heat required for melting} = \text{mass} \times H_F$$

where H_F is the heat of fusion of the substance.

When we defined the ice point in Chapter 11, we made use of this fact that the temperature does not change while melting takes place. It was also indicated there that this temperature is practically independent of ordinary changes in atmospheric pressure. If the pressure on ice is twice ordinary atmospheric pressure, its melting point becomes –0.0075°C. However, in ice skating a person's weight is supported by a small area—the area of the bottom of the blades. This results in a high pressure (F/A) on the ice—enough to melt it; when the skater passes, the water freezes again. *Regelation* is this melting under increased pressure and refreezing when the pressure is reduced. Regelation happens with a substance such as ice that expands on freezing. On the other hand, a substance that contracts on solidifying has a higher melting point at higher pressures. A lower melting point also results from dissolving solids in liquids and explains why salt sprinkled on an icy sidewalk tends to melt the ice.

Vaporization

Vaporization is the process of changing a substance to a vapor. It includes evaporation, boiling, and sublimation.

Evaporation

As explained above, the molecules of a liquid are in constant motion. The speeds of the different molecules are different. Some of the fastest moving molecules at the surface can escape from the liquid in spite of gravity and the attractive force between molecules. The molecules that escape are relatively far apart from each other, as in a gas, and form the vapor of the liquid. Since the molecules with the greatest kinetic energy escaped, the average kinetic energy of the remaining molecules in the liquid decreased; that is, unless the liquid is heated, the liquid cools when evaporation takes place. Evaporation is a cooling process. It occurs at the surface of the liquid. The rate of evaporation can be increased by making it easier for the molecules to escape from the liquid: heat the liquid to give more molecules higher speeds; spread the liquid over a greater area; decrease the air pressure on top of the liquid; blow away the vapor-filled air above the liquid.

Boiling

In evaporation the liquid is converted to vapor at the surface of the liquid. In boiling the liquid is converted to vapor within the body of the liquid. We observe the vapor in the form of bubbles. Another way to define boiling is in terms of pressure. The molecules of the vapor move around freely like the molecules of a gas and exert pressure. We can speak of the pressure of a vapor as we speak of the pressure of the atmosphere. *Boiling* occurs when the pressure of the vapor forming inside the liquid equals the external pressure on the liquid; this is usually atmospheric pressure. During boiling the temperature of the liquid doesn't change. The temperature at which change occurs is the *boiling point* of the liquid. Although the temperature of the liquid does not change during boiling, energy must be supplied for the vaporization (to do work against the cohesive force between molecules). *Heat of vaporization* (H_v) of a substance is the amount of heat needed to vaporize a unit mass of the substance without changing its temperature. For water at 100°C this is 2,260 kilojoules per kilogram.

$$\text{heat required for vaporizaton} = \text{mass} \times H_V$$

When the pressure on a liquid is increased, its boiling point goes up; when the pressure on a liquid is decreased, its boiling point goes down. For example, at standard atmospheric pressure (760 millimeters) the boiling point of water is 100°C. In mountainous areas the atmospheric pressure is usually considerably below 760 millimeters and water boils at temperatures significantly below 100°C. Cooking at these lower temperatures proceeds rather slowly. To hasten the cooking, pressure cookers may be used; in these the pressure is allowed to build up by using special covers that do not let the vapor escape. The buildup of the pressure raises the boiling point.

Sublimation

Sublimation is the direct change from solid to vapor. For example, at ordinary room temperatures, solid carbon dioxide ("dry ice") changes directly to carbon dioxide gas. *Condensation* is the changing of a vapor to a liquid. This is accompanied by the release of the same amount of heat that would be required to vaporize the same quantity of the liquid. The term *condensation* is also used occasionally to refer to the reverse of sublimation, that is, to the direct conversion of a vapor to a solid, as in the formation of frost on a cold night.

When there is a change of phase, the method of mixtures described above has to be modified:

$$\begin{aligned}\text{heat gained (or lost)} &= \text{mass} \times \text{specific heat} \times \text{temperature change}\\ &\quad + \text{mass melted} \times \text{heat of fusion}\\ &\quad + \text{mass vaporized} \times \text{heat of vaporization}\end{aligned}$$

In this formula any term may be omitted that does not fit the situation. For example, to calculate the heat required to boil away completely 10 grams of water at 100°C if the starting temperature of the water is 60°C, we omit the "mass melted" term since no melting is involved.
Heat gained = $(10 \times 4.19 \times 40) + (10 \times 2{,}260) \doteq 24{,}000$ J.

Table of Constants for Liquid Water

System	sp. gr.	density	sp. ht.	H_f	H_v
SI	1	10^3 kg/m^3*	4.19 kJ/kg • °C	335 kJ/kg	2,260 kJ/kg

*This can also be expressed as 1 kg/l.

Questions Chapter 12

In each case, select the choice that best answers the question or completes the statement. Use Appendix IV for reference.

1. Four thousand joules of heat are added to 100 grams of water when its temperature is 40°C. The new temperature of the water is approximately

(A) 30°C
(B) 45°C
(C) 50°C
(D) 60°C
(E) 70°C

2. If 2 kilojoules of heat are added to 500 grams of water at 100°C and standard pressure, after 2 minutes the temperature of the water will be

(A) 100°C
(B) 110°C
(C) 115°C
(D) 120°C
(E) 125°C

3. A girl has 240 grams of water at 50°C. How many grams of ice, at 0°C, are needed to cool the water to 0°C?

(A) 35
(B) 150
(C) 250
(D) 500
(E) 750

4. A certain alloy has a melting point of 1,000°C. The specific heat of the solid is 1.257 joules per gram • C, and its heat of fusion is 84 joules per gram. Approximately how many kilojoules of heat are required to change 10 grams of the material from a solid at 20°C to a liquid at 1,000°C?

(A) 5.6
(B) 8.2
(C) 13.2
(D) 15.7
(E) 20

5. As the temperature of a solid increases, its specific heat

(A) increases
(B) decreases
(C) increases, then decreases
(D) decreases, then increases
(E) remains the same

6. When 1 gram of atmospheric water vapor condenses in the air, it results in

(A) always cooling the surrounding air
(B) always heating the surrounding air
(C) always leaving the surrounding air temperature unchanged
(D) heating the surrounding air only if the change occurs below 100°C
(E) cooling the surrounding air only if the change occurs below 100°C

7. Approximately how much heat energy (in kilojoules) is required to raise the temperature of 200 grams of lead from its melting point to its boiling point?

(A) 28
(B) 45
(C) 31
(D) 37
(E) 53

8. How much heat energy is required to completely vaporize 2 kilograms of alcohol at its boiling point?

(A) 855 kJ
(B) 109 kJ
(C) 1,710 kJ
(D) 218 kJ
(E) 2.43 kJ

9. Which of the following substances is a liquid at 150°C and standard air pressure?

(A) Mercury
(B) Alcohol
(C) Zinc
(D) Ammonia
(E) Water

Explanations to Questions Chapter 12

Answers

1. (C)	**4.** (C)	**7.** (D)
2. (A)	**5.** (E)	**8.** (C)
3. (B)	**6.** (B)	**9.** (A)

Explanations

1. C Heat gained = mass × sp. ht. × temp. change.

4,000 J = 100 g × 4.19 kJ/kg • °C × temp. change. Temp. change = 9.5°C.

Therefore, new temperature = 49.5°C or approximately 50°C.

2. A At its boiling point, the temperature of the water remains the same, 100°C.

3. B The heat lost by the water is transferred to the melting ice at 0°C:

$$(240\text{ g}) \times (4.19\text{ kJ/kg}\cdot{}^\circ\text{C}) \times (50^\circ\text{C}) = \text{mass} \times (335\text{ J/g})$$
$$\text{mass of ice} = 150\text{ g}$$

4. C Heat gained = (mass × sp. ht. × temp. change) + (mass × heat of fusion)

Heat gained = (10 g) × (1.257 J/g • °C) × (980°C) + (10 g) × (84 J/g)
Heat gained = 13,158.6 J, which is approximately equal to 13.2 kJ

5. E The specific heat of a substance remains constant in the solid state.

6. B Heat is required to vaporize a liquid. Whenever the vapor condenses to a liquid, heat has to be given off again to the surroundings. In this case the condensation results in heating the surrounding air.

7. D Heat gained = mass (in kg) × sp. ht. × change in temperature.

$$\text{Heat gained (in kJ)} = (0.2\text{ kg}) \times (0.13\text{ kJ/kg} \cdot {}^\circ\text{C}) \times (1{,}412^\circ\text{C})$$
$$= 36.7\text{ kJ or } 37\text{ kJ.}$$

8. C From Appendix IV, we see that the heat of vaporization of alcohol is 855 kJ/kg. The total amount of heat needed to vaporize the alcohol completely is the product of its mass and heat of vaporization:

$$\text{heat (in kJ)} = \text{mass (in kg)} \times \text{(heat of vaporization)}$$
$$\text{heat} = (2\text{ kg}) \times (855\text{ kJ/kg}) = 1{,}710\text{ kJ}$$

9. A From the list of boiling points in Appendix IV, we see that only mercury remains a liquid at 150°C.

CHAPTER 13

Heat and Work

It was the English scientist James Prescott Joule who, probably more than anyone else, was responsible for our belief that heat is a form of energy. He created the device shown in Figure 13.1. The masses are raised by turning the drum at the top. The container is filled with water. As the masses fall, they give up their potential energy to the water, which in turn exhibits a rise in average kinetic energy (temperature).

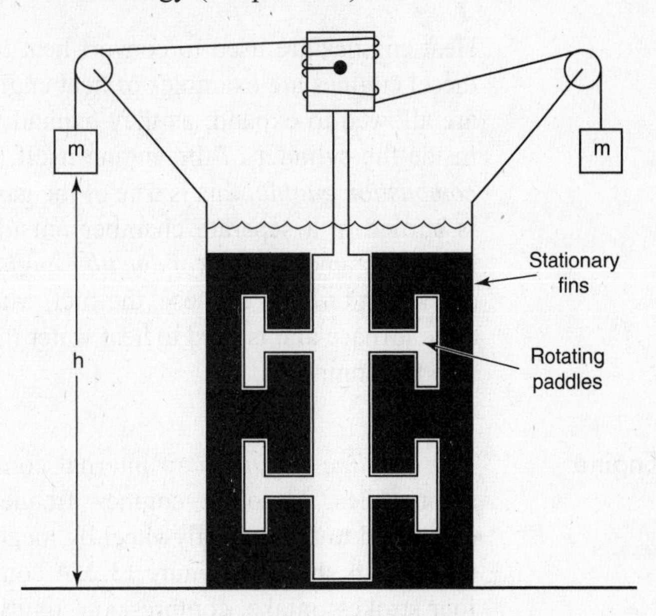

FIGURE 13.1 Joule's Apparatus

Since energy is the ability to do work, we should not be surprised that we can use hot objects to do work for us; we have already studied the reverse—when work is done with a machine, heat is produced since some energy is used to do work against friction. How can a hot object be used to do work for us?

Imagine a gas in a cylinder with a movable piston. The molecules of the gas bombard the piston and produce a pressure on it. If we want to compress the gas, we must exert a force against the piston. We do work in compressing the gas. As we push the piston, we exert a force on the molecules of the gas thus giving them greater speed. The temperature of the gas rises; if the cylinder is insulated, the work done in compressing the gas results in an equivalent increase in the internal energy of the gas. If the cylinder is not insulated, the temperature of the gas can be kept constant by letting this heat equivalent go instead to the material outside the cylinder. Mechanical energy can be converted completely to heat.

If the gas in the cylinder is allowed to expand, the gas does work as it moves the piston. If the cylinder is insulated, as the gas expands its temperature goes down; the energy required for the work comes from the internal energy of the gas. If the cylinder is not insulated, the temperature of the gas can be kept constant. Then, as the gas expands, heat flows into it from outside the cylinder; this heat equals the work done by the gas in expanding.

Recall that an isothermal process is one in which the temperature remains constant. Boyle's law ($pV = k$) applies to a gas that expands or contracts while its temperature is kept constant. An *adiabatic* process is one in which no heat is allowed to enter or leave the system, as in the above process in which the cylinder is kept insulated. Boyle's law does not apply to an adiabatic process.

Heat Engines

Heat engines are used to convert heat to mechanical energy. Gasoline and diesel engines are examples of heat engines. In these heat engines hot gases are allowed to expand; as they expand they do work. If the fuel is burned inside the cylinder of the engine itself, the engine is known as an *internal combustion engine;* this is true of the gasoline and diesel engines. If the fuel is burned in a separate chamber outside the engine proper, the engine is known as an *external combustion engine;* this is true of the steam engine and steam turbine. In these, the fuel, which may be coal, is burned in a separate furnace and is used to heat water in a boiler. The steam is then directed into the engine.

Gasoline Engine

The *gasoline engine* is an internal combustion engine commonly used in automobiles. Gasoline engines frequently have six or eight cylinders connected to the same flywheel by means of a common shaft. One of these cylinders is shown in Figure 13.2. A complete cycle of operation consists of four strokes: intake, compression, ignition or power, exhaust. The cylinder is shown at the beginning of the intake stroke. The exhaust valve E is closed. Gasoline and air mixture enters through the open intake valve I, and the piston moves down. Then both valves are closed and the mixture is compressed as the piston moves up during the second stroke. At the proper instant, high voltage is applied to the spark plug S; the resulting spark ignites the mixture, producing an explosion that provides the power for pushing the piston and turning the crankshaft, flywheel, and ultimately the wheels of the car. During the last stroke the piston is pushed up again; this time the intake valve is closed and the exhaust valve is open. The waste gases are pushed out and the cycle can start again. Notice that there are two complete up and down motions during each cycle. The *carburetor* provides the proper mixture of air and gasoline vapor, which is then burned in the cylinders. The engine is kept running between power strokes by the flywheel, F. The whole process is usually started by briefly switching a battery to a starting motor.

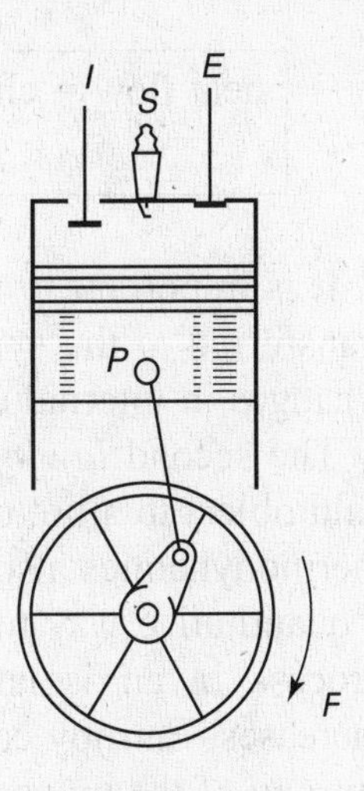

FIGURE 13.2

Diesel Engine

The *diesel engine* is also an internal combustion engine. Neither carburetor nor spark plug is needed. The air in the cylinder is compressed so much that its temperature rises sufficiently to ignite the fuel, which is then sprayed in. The fuel is denser and cheaper than gasoline. The efficiency of a diesel engine (about 40 percent) is higher than that of a steam engine or gasoline engine.

Jet and Rocket Engines

Jet and rocket engines operate on the same principle of Newton's third law. Gases escaping under pressure in one direction exert a push on the engine in the opposite direction. The jet engine takes from the atmosphere the oxygen needed for burning its fuel, while the rocket engine carries oxygen as well as its own fuel.

Thermodynamics

Thermodynamics is the branch of science that deals with the relationship among changes in energy, the flow of heat, and the performance of work. For example, in the preceding section we mentioned heat engines. Let us consider one in more detail. Heat is provided to a cylinder with a movable piston as shown for the gasoline engine. The heated gas pushes the piston, which then does useful work in turning the wheel before returning to its starting position. During the cycle some of the heat supplied also goes into increasing the internal energy of the system, such as increasing the kinetic energy of translation and rotation of the molecules of the cylinder, piston, and waste gases.

These ideas can be quantified for any isolated system. The *first law of thermodynamics* is a form of the law of conservation of energy: the heat supplied to a system is equal to the increase in internal energy of the system plus the work done by the system.

heat flow = change in internal energy + work done by system
$Q = \Delta U + W$

It is important to use signs correctly. *Heat flow* is *positive* when heat is transferred *to* the system. *Work* is *positive* when it is done *by* the system. Changes in internal energy will be accordingly positive or negative.

The second *law of thermodynamics* states that heat cannot flow from a cold object to a hot object unless external work is done. The second law of thermodynamics also gives rise to the concept of *entropy.* Entropy provides a quantitative way to measure the disorder of a system. For any complete process in an isolated system, the entropy either remains constant or increases. Entropy can remain constant only in reversible processes, so the entropy of the universe is increasing. For reversible processes,

entropy change = heat added/Kelvin temperature

In thermodynamics, it is important to understand certain terms. If no heat is added or subtracted from an isolated system, the resulting processes are said to occur *adiabatically.* If a change occurs without a change in temperature, the process is called *isothermal*. If a change occurs without a change in pressure, the process is *isobaric*. Finally, if there is no change in volume, the process is *isochoric*.

In the nineteenth century, the physicists Sadi Carnot, in France, and Rudolph Clausius, in Prussia, developed the modern concepts of thermodynamics. We can quickly summarize some of these:

1. Heat gained is equal to heat lost.
2. Heat never flows from a cooler body to a hotter body of its own accord.
3. Heat can never be taken from a reservoir without something else happening to the system.
4. Entropy is a measure of the random disorder in a system.
5. The entropy of the universe is increasing.

Methods of Heat Transfer

Sometimes we want heat to escape; sometimes we want to keep it from escaping. In the case of heat engines, such as the gasoline engine, only some of the heat can be used to do work. It is necessary to get rid of the excess heat; otherwise, the engine will rapidly overheat. In the case of heating a home, we want the heat to get from the furnace to the apartment, but we may also want to keep the heat from getting out of the apartment. The three methods of heat transfer are conduction, convection, and radiation.

Conduction

Conduction is the process of transferring heat through a medium that does not involve appreciable motion of the medium. For example, if we heat one end of a copper rod, the other end gets hot, too. The energy transfer in metallic conductors is chiefly by means of “free” electrons, but also involves bombardment of one molecule by the next. In general, metals are good conductors; good conductors of heat are also good conductors of electricity. Silver is the best metallic conductor of heat and electricity. Copper and aluminum are also very good. Liquids and gases, as well as nonmetallic solids, are poor conductors of heat. Poor conductors are known as *insulators*.

Convection

Convection is the process of transferring heat in a fluid that involves motion of the heated portion to the cooler portion of the fluid. The heated portion expands, rises, and is replaced by cooler fluid, thus producing so-called convection currents. This process may be observed over a hot radiator: The heated air expands; since its density is different from that of the surrounding colder air it reflects some light and can be "seen" to move upward. Radiators heat rooms chiefly by convection and, therefore, are more effective when placed near the floor than when placed near the ceiling. Also note that, when the heated fluid rises, it mixes with the cooler fluid.

Radiation

Radiation is the process of transferring heat that can occur in a vacuum; it takes place by a wave motion similar to light. The wave is known as an *electromagnetic wave* and will be discussed further in Chapter 18. The higher the temperature of an object, the greater the amount of heat it radiates. Black objects radiate more heat from each square centimeter of surface than light-colored objects at the same temperature.

The *vacuum bottle* (thermos bottle) is designed with the three methods of heat transfer in mind (Figure 13.3). The bottle is made of glass—a good insulator. The stopper is made of cork or plastic; these good insulators minimize heat transfer by conduction. The space between the double walls is evacuated, also minimizing convective heat transfer. The inside surfaces (facing the vacuum) are silvered; shiny surfaces reflect electromagnetic waves, thus minimizing heat transfer by radiation. As a result, hot liquids inside the bottle stay hot and cold liquids stay cold for many hours.

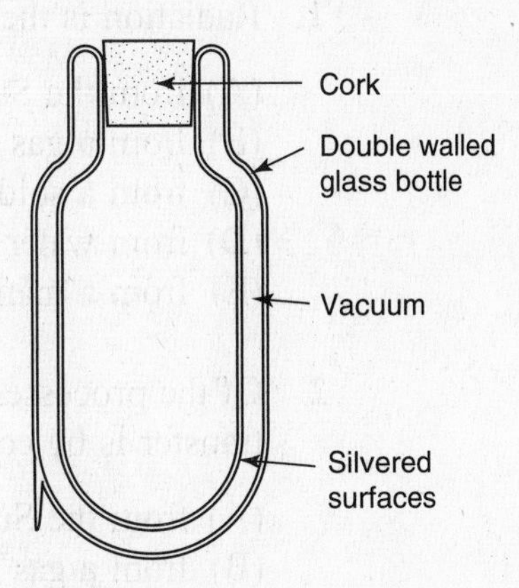

FIGURE 13.3

Energy Sources

For heating buildings and operating machines, human beings have depended almost solely on fossil fuels such as coal and oil. Although new deposits keep being located and tapped, many people think that these fuels will be found inadequate in very few years. Waterfalls, of course, can be used in locations close enough to exploit economically this source of energy. Waterfalls and fossil fuels ultimately owe their energy to the Sun. Energy keeps coming to Earth from the Sun by radiation (about 8 joules per minute on each square centimeter in full sunlight). In some parts of the world large, curved mirrors are arranged to permit the use of sunlight for cooking and heating. A tremendous amount of energy is available also in the constant motion of the tides. It is believed, however, that, as the human population grows and the standard of living rises, all these sources of energy will prove inadequate. Some think the hope of the future lies in nuclear energy—energy released when certain changes take place in the nucleus of atoms. Scientists have already obtained nuclear energy by the splitting or fission of the nuclei of a heavy element, such as uranium, in a controlled manner in a nuclear reactor (or atomic pile). Only relatively small quantities of such fissionable materials are available. Nuclear energy can also be obtained by the combining or fusion of the nuclei of a light atom, such as hydrogen. If this procedure can be done in a controllable manner, a practically endless supply of energy will be available, because hydrogen, the necessary "fuel," is obtainable from the oceans in almost unlimited quantity. Nuclear energy will be discussed further in Chapter 20. Some houses have roof-top installations with circulating fluids to absorb solar energy for producing hot water and heat.

Questions Chapter 13

In each case, select the choice that best answers the question or completes the statement.

1. Radiation is the chief method of energy transfer

(A) from the Sun to an Earth satellite
(B) from a gas flame to water in a teakettle
(C) from a soldering iron to metals being soldered
(D) from water to an ice cube floating in it
(E) from a mammal to the surrounding air

2. Of the processes below, the one in which practically all the heat transfer is by conduction is

(A) from the Sun to an Earth satellite
(B) from a gas flame to the top layer of water in a teakettle
(C) from a soldering iron to metals being soldered
(D) from the bottom of a glass of water to an ice cube floating in it
(E) from a mammal to the surrounding air

3. Two kilograms of water are heated by stirring. If this raises the temperature of the water from 15°C to 25°C, how much work, in joules, was done to the water by the stirring?

(A) 20,000
(B) 40,000
(C) 60,000
(D) 80,000
(E) 100,000

4. In a certain steam engine, the average pressure on the piston during a stroke is 50 newtons per square meters. The length of each stroke is 12 centimeters, the area of the piston is 120 square centimeters, and the diameter of the flywheel is 5 meters. The amount of work done on the piston during each stroke is, in newton meters, approximately

(A) 250
(B) 0.072
(C) 0.54
(D) 1.63
(E) 12.62

5. The work done *to* a system is characterized as

(A) positive
(B) negative
(C) either positive or negative
(D) indeterminate
(E) of no consequence

6. Which of the following can actually lower the internal energy of ("cool") a room?

(A) a fan
(B) a refrigerator with its door open
(C) a refrigerator with its door closed
(D) an air conditioner in the middle of a room
(E) an air conditioner partially exposed to the outside

7. Thirty joules of heat flow into a system. The system in turn does 50 joules of work. The internal energy of the system has

(A) increased by 80 J
(B) decreased by 80 J
(C) increased by 20 J
(D) decreased by 20 J
(E) remained constant

8. Dark, rough objects are generally good for

(A) conduction
(B) radiation
(C) convection
(D) reflection
(E) refraction

9. Black plastic handles are often used on kitchen utensils because

(A) the black material is a good radiator
(B) the plastic is a good insulator
(C) the plastic is a good conductor
(D) the plastic softens gradually with excessive heat
(E) the material is thermoplastic

10. A person seated in front of a fire in a fireplace receives heat chiefly by

(A) convection of carbon dioxide
(B) convection of carbon monoxide
(C) convection of air
(D) conduction
(E) radiation

11. If friction is negligible, when a gas is compressed rapidly

(A) its temperature remains the same
(B) its temperature goes down
(C) its temperature rises
(D) it is liquefied
(E) work is done by the gas

Explanations to Questions Chapter 13

Answers

1. (A)	**4.** (B)	**7.** (D)	**10.** (E)
2. (C)	**5.** (B)	**8.** (B)	**11.** (C)
3. (D)	**6.** (E)	**9.** (B)	

Explanations

1. A The space between the Sun and an Earth satellite is an almost perfect vacuum. Conduction and convection cannot take place through a vacuum. Although some particles are shot out of the Sun and thus carry kinetic energy with them to the satellite, this energy is small in comparison with the amount constantly conveyed by electromagnetic waves, that is, by radiation. (Also see answer to question 2.)

2. C The soldering iron is in good contact with the metal to be soldered and thus provides excellent opportunity for heat transfer by conduction. The tip of a soldering iron is often made of copper to provide better heat conduction. If a teakettle is made of metal, there is some heating, chiefly by convection, of the top layer of water in the kettle by conduction along the kettle; the bottom layer gets heated, rises toward the top, where it mixes with the colder water, and also pushes the top layer toward the bottom (B). Similar reasoning applies to choice (D). Mammals lose heat by the exhalation of warm air, evaporation of perspiration from the skin, convection, and other processes.

3. D 100% efficiency is implied.

work done = heat supplied
= mass × specific heat × temperature change
= (2 kg)(4.19 kJ/kg • °C)(25° – 15°) = 80 kJ = 80,000 J.

4. B pressure = force/area; force = ***PA***

work = force × distance = pressure × area × distance
= 50 N/m^2 × 0.012 m^2 × 0.12 m = 0.072 N • m

Note that the diameter of the flywheel is irrelevant.

5. B By convention, the work done *by* a system is positive. Therefore, the work done *to* a system is negative.

6. E To cool a room, heat must be *removed.* In fact, *all* of the other choices result in an increase in room temperature, as the electric energy used to run the devices is dissipated as heat.

7. D $Q = \Delta U + W$

$+30\text{ J} = \Delta U + 50\text{ J}$

$\underline{-50\text{ J} \qquad\qquad -50\text{ J}}$

$-20\text{ J} = \Delta U$

8. B This question involves mere recall of a fact mentioned in the chapter, that everyone is expected to know.

9. B Although the black material is a good radiator, this is irrelevant (A). The important fact is that the plastic is a good insulator of heat. The term *thermoplastic* (E) which refers to the property of softening when heat is applied, is not essential for elementary physics. Such a plastic is usually not desirable in handles of kitchen utensils; choice (E) is essentially the same as choice (D).

10. E Convection of any kind would lead to the rising of the heated gases, in this case up the chimney, not to the person. Very little conduction to the person can take place, since the intervening air is a poor conductor.

11. C When a gas is compressed, work is done on the gas and tends to heat the gas even if friction is absent. If the compression takes place rapidly, the heat produced can't escape completely and the temperature of the gas rises. Similarly, if a gas is allowed to expand rapidly, its temperature falls.

CHAPTER 14

Wave Motion and Sound

Wave Motion

Wave motion is a method of transferring energy through a medium by means of a distortion that travels away from the place where the distortion of the medium is produced. The medium itself moves only slightly. For example, a pebble dropped into quiet water disturbs it. The water near the pebble does not move far, but a disturbance travels away from the spot. This disturbance is wave motion. As a result some of the energy lost by the pebble is carried through the water by a wave, and a cork floating at some distance away can be lifted by the water; thus the cork obtains some energy. We can set up a succession of waves in the water by pushing a finger rhythmically through the surface of the water. A vibrating tuning fork produces waves in air.

Waves can be classified in different ways. The wave produced by the pebble dropped into the water is a *pulse:* a single vibratory disturbance that travels away from its source through the medium. If we push our finger regularly and rapidly down and up through the surface of the water, we produce a *periodic* wave.

Periodic Motion

Periodic motion is motion that is repeated over and over again. The motion of a pendulum is periodic. As long as its arc of swing is small, the time required for a back-and-forth swing is constant. This time is called the *period* of the motion and is given by the formula $T = 2\pi\sqrt{L/\mathbf{g}}$, where L is the length of the string from which the bob is suspended, and $\mathbf{g}$ is acceleration due to gravity.

When a mass (m) is attached to a light spring, stretched, and then released, the resulting motion is once again periodic. The reaction force on the stretched spring is given by Hooke's law:

$$\mathbf{F} = -kx$$

where k is the *force constant* and x is the *elongation* of the spring.

The period of the motion is given by the formula:

$$T = 2\pi\sqrt{\frac{m}{k}}$$

The motion of both the pendulum and the oscillating spring is referred to as *simple harmonic motion.* It should be noted that the simple harmonic motion of the pendulum is only an approximation for small angles.

The motion of tuning fork prongs is periodic. As the prongs vibrate back and forth, they push particles in the air; these push other particles, and so on. When the prong moves to the right, particles are pushed to the right; when the prong moves to the left, the particles move back into the space vacated by the prong. As the prong moves back and forth, the particles of the medium move back and forth *with the same frequency* as the prong. The energy travels away from the vibrating source, but the particles of the medium vibrate back and forth past the same spot (their equilibrium position).

The ability of a material to generate periodic waves is related to the property of *elasticity.* We have seen that Hooke's law expresses the relationship between an applied force and the resulting elongation or compression of the spring. It is the elasticity of the spring that enables it to oscillate back and forth. In fact, the force constant, *k,* can be thought of as a measure of the *modulus of elasticity.*

When solids are acted on by forces, their shapes can be distorted in a number of ways (see Figure 14.1). For example, the forces can act to stretch the solid under a state of *tension,* the forces can squeeze the solid inward under a state of *compression,* or the forces can cause layers of the solid to slide over one another. This last process is known as *shearing.*

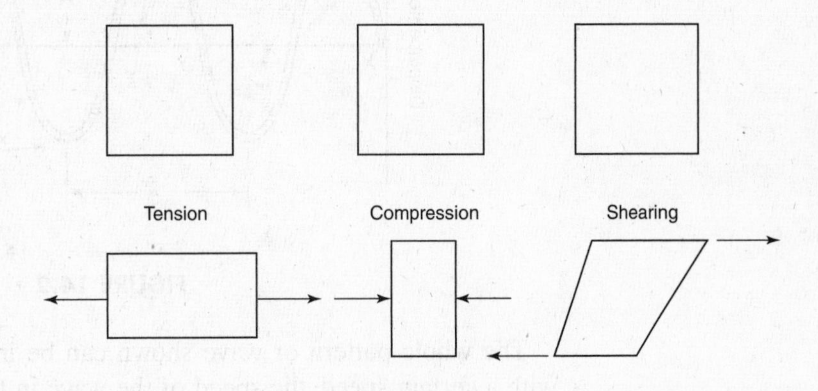

FIGURE 14.1

In physics, *stress* is a measure of the force per unit cross-sectional area. The response to stress is called *strain.* The ratio of stress to strain is a measurement of the modulus of elasticity, mentioned above. Different materials propagate waves at different speeds depending on the conditions of the medium (tension, elasticity, density, etc.). These waves are often called *mechanical waves.* Light waves are termed *electromagnetic waves* because they arise from oscillations of electromagnetic fields.

Longitudinal and Transverse Waves

Two basic waves are the longitudinal wave and the transverse wave. The wave produced in air by a vibrating tuning fork illustrates the former. A *longitudinal wave* is a wave in which the particles of the medium vibrate back and forth along the path that the wave travels. Sound waves are longitudinal waves.

A *transverse wave* in a medium is a wave in which the vibrations of the medium are at right angles to the direction in which the wave is traveling. A transverse wave is readily set up along a rope. Either a longitudinal or a transverse wave can be readily sent along a helical spring. A water wave is approximately transverse. The propagation of light can be described as a transverse wave. Since light can travel through a vacuum, an interesting question arises: What is vibrating? This topic will be discussed further in connection with electromagnetic waves in Chapter 18.

Simple longitudinal and simple transverse waves can be represented by a sine curve. Let us first consider the production of a transverse wave set up in a rope (Figure 14.2). Imagine the rope stretched horizontally with a prong of a vibrating tuning fork attached to the left end of the rope so as to produce a transverse wave in it. When the prong moves up, the rope is pulled up and the upward disturbance starts to travel to the right. As the prong moves down, the rope is pulled down and a downward displacement moves along the rope, following the upward displacement. When the prong moves up again, it produces the second peak to follow peak 1.

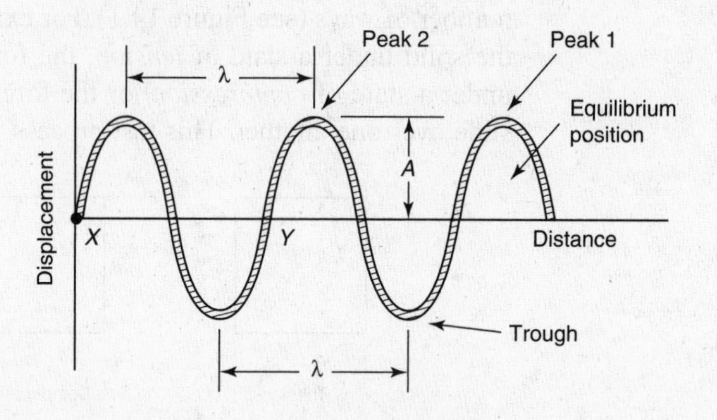

FIGURE 14.2

The whole pattern or wave shown can be imagined moving to the right with a certain speed, the speed of the wave in the medium (the rope). If we keep our eye on some specific spot on the rope, we see that part of the rope going up and down, transverse, or at right angles to the direction of motion of the wave as a whole.

The *frequency of vibration of each part of the rope is the same as the frequency of vibration of the source (the tuning fork) and the same as the number of peaks or troughs produced per second. The amplitude* of the wave is the maximum displacement or distance moved by each part of the medium away from its average or equilibrium position. In Figure 14.2, A represents the amplitude of the wave, and λ is the *wavelength*—the distance between any two successive peaks, or the distance between any two successive corresponding parts of the wave, such as X and Y.

As we watch the wave moving past a given spot in the medium, we can see peak after peak moving past the spot. The time required for two successive peaks to pass the spot is known as the *period* of the wave. It is the time required for the wave to move a distance equal to one wavelength. This time is the same as the period of vibration of the source.

Since, for motion with constant speed, distance = speed × time, the speed of the wave equals the wavelength divided by the period. The *frequency* (f) of the wave is the number of complete vibrations back and forth per second. Since the period is the time for one vibration,

$$T = \frac{1}{f}$$

Therefore, the velocity of the wave equals the wavelength divided by 1 over the frequency. Instead of dividing, we invert and multiply. The end result is this important relationship for a wave:

velocity of a wave = frequency of wave × wavelength

$$\mathbf{v} = f \times \lambda$$

The Greek letter *lambda* (λ) is frequently used for wavelength.

When something occurs over and over again, the term *cycle* is sometimes used to refer to the event that is repeated. We can speak of a wave having 100 vibrations per second or having 100 cycles per second. The metric and SI unit for frequency is the hertz (Hz). The term *wave* is sometimes used to refer to one cycle of the disturbance in the medium. In the sine curve shown in Figure 14.2, the portion of the curve from X to Y would then be one wave.

Example: A wave having a frequency of 1,000 cycles per second has a velocity of 1,200 meters per second in a certain solid. Calculate the period and the wavelength.

The period $T = (1/f) = (1/1{,}000)$ s = 0.001 s.
The wavelength $\lambda = (v/f) = (1{,}200 \text{ m/s})/(1{,}000 \text{ cycles/s}) = 1.2$ m/cycle.
Notice that, in the calculation of wavelength, *second* cancels. Often we also drop *cycle* from the answer and give the wavelength as 1.2 m.

The sine curve shown in Figure 14.2 can be used to represent a longitudinal wave as well as a transverse wave, as will be described in the next section. The above relationships and definitions apply to both types of waves. All waves also exhibit certain phenomena in common: reflection, refraction, interference, and diffraction. In addition, transverse waves exhibit the phenomenon of polarization.

Sound

In physics, when we speak about sound, we usually mean the sound wave. Sound waves are longitudinal waves in gases, liquids, or solids. They are produced by vibrating objects. Sound cannot be transmitted through a vacuum.

To describe the sound wave, let us imagine the production of sound by means of a tuning fork vibrating in air. As the prong moves to the right, air is pushed to the right; since the air at the right is pushed closer together, a region of higher pressure is produced. This region of higher pressure is known as a *compression* or *condensation.* Because of the higher pressure, some air from the compression travels on to the right, thus producing another compression. In this way a compression travels away from the tuning fork. The speed with which it travels is the speed of the wave in air. In the meantime, because of its elasticity, the prong of the tuning fork starts moving back toward the left. As it does so, air from the right rushes into the space left by the prong. The space from which the air rushed then becomes a region of lower pressure than normal. A region of reduced pressure is known as a *rarefaction.* Air at the right of the rarefaction is pushed toward the left, in turn producing another rarefaction further to the right. Thus a rarefaction travels to the right, following the compression. When the prong begins to move to the right again, it starts another condensation, and so on. The *longitudinal wave* is a succession of compressions and rarefactions, one complete vibration of the fork producing one compression and one rarefaction.

If the wave hits another medium, such as a solid, compressions and rarefactions will be set up in the new medium. The frequency of the wave in the new medium will be the same as in air, but the speeds, and therefore the wavelengths, will be different.

If a sine curve is used to represent a longitudinal wave, the *y*-axis may serve to represent the change in pressure of the medium as a result of the wave. A compression is a region of increased pressure and can be represented as a positive number, while a rarefaction, with a decrease in pressure, is represented by a negative number. The abscissa is used as for the transverse wave (Figure 14.3).

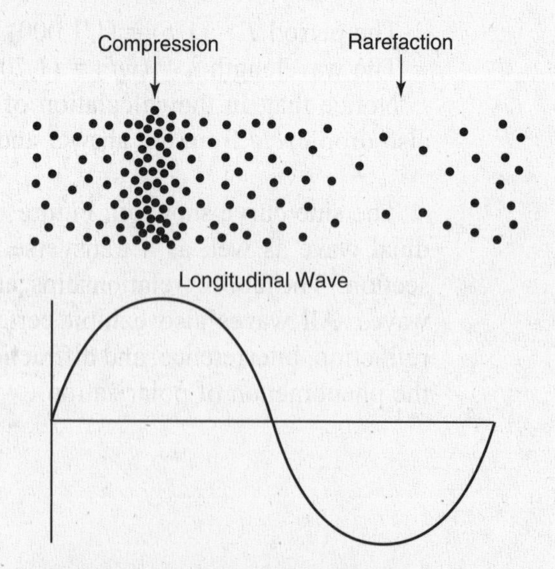

FIGURE 14.3 Transverse Wave or Transverse Representation of a Longitudinal Wave

The abscissa may be the distance along the direction of wave travel. The curve then represents the change in pressure at a given moment plotted against distance along the medium. A crest represents a compression; a trough, a rarefaction. The wavelength is the distance between successive compressions or successive rarefactions.

A sine curve can also be drawn to represent the change in pressure at a *specific point* in the medium against *time.* The distance between successive compressions on such a graph will indicate the *period* of the wave as well as the period of vibration of a particle in the medium.

The *velocity of sound* in air is approximately 331 meters per second at 0°C. The speed goes up 0.6 meters per second for each Celsius degree that the temperature goes up. Don't get mixed up in memorizing this fact—note that the change in speed is given per Celsius degree in both systems of units. The speed of sound in air is independent of the atmospheric pressure. In general, sound travels faster in liquids and solids than in air.

Musical Sounds

Sounds produced by regular vibrations of the air are musical. Irregular vibrations of the air are classified as unpleasant sounds or noise. This is often demonstrated with a *siren* consisting of a wheel with concentric sets of holes (Figure 14.4). The wheel rotates at constant speed around an axis perpendicular to the wheel and passing through its center. The holes in the two innermost circles are evenly spaced. When air is blown gently at either of these circles, a pleasant, musical sound is heard. The wheel blocks the air at a constant rate, and the blasts of air going through the holes produce compressions at a constant rate. The holes in the outermost circle, however, are irregularly spaced. When air is blown gently at this circle, air allowed through the holes produces successive compressions that follow each other at irregular intervals. What a person perceives is noise.

FIGURE 14.4 Siren

The *range in frequencies* of musical sounds is approximately 20 to 20,000 hertz or cycles per second. Some people can hear higher frequencies than others. Longitudinal waves whose frequencies are higher than those within the audible range are called *ultrasonic* frequencies. Ultrasonic frequencies are used in sonar for such purposes as submarine detection and depth finding. Ultrasonic frequencies are also being tried for sterilizing food since these frequencies kill some bacteria.

Sound waves of all frequencies in the audible range travel at the same speed in the same medium. In the audible range, the higher the frequency of the sound the higher is the *pitch.* For example, in the siren described above, a higher pitch is produced with the middle circle of holes than with the innermost circle. The term *supersonic* refers to speed greater than that of sound. An airplane traveling at supersonic speed is moving at a speed greater than the speed of sound in air at that temperature. *Mach 1* means a speed equal to that of sound; *Mach 2* means a speed twice that of sound; and so on.

Musical sounds have three basic *characteristics:* pitch, loudness, and quality or timbre. As was indicated above, *pitch* is determined largely by the frequency of the wave reaching the ear. The higher the frequency, the higher is the pitch. *Loudness* depends on the amplitude of the wave reaching the ear. For a given frequency, the greater the amplitude of the wave, the louder the sound. To discuss quality of sound, we need to clarify the concept of overtones. Sounds are produced by vibrating objects; if these objects are given a gentle push, they usually vibrate at one definite frequency, producing a pure tone. This is the way a tuning fork is usually used. When an object vibrates freely after a force is momentarily applied, it is said to produce its *natural frequency.* Some objects, such as strings and air columns, can vibrate naturally at more than one frequency at a time. The lowest frequency that an object can produce when vibrating freely is known as its *fundamental frequency;* other frequencies that the object can produce are its *overtones.* The *quality* of a sound depends on the number and relative amplitudes of the overtones present in the wave reaching the ear.

When a wave reaches another medium, part of the wave is usually reflected. When a sound wave is reflected, a distinct *echo* is heard if the reflected sound wave reaches the ear at least ⅒ second after the sound traveling directly from the vibrating source to the ear. Echoes may be used to determine the distances of reflecting objects, as in sonar. The total distance traveled by the wave is equal to the speed of the wave in the medium multiplied by the time between the start of the sound and the arrival of the echo. If the reflected wave comes back along the same path as the incident wave, the distance of the reflecting surface, such as a wall, is one-half of the total distance.

Resonance and Interference

As was stated above, objects tend to vibrate at their natural frequencies. A simple pendulum has a natural frequency. Recall that

$$\text{period of a simple pendulum} = 2\pi\sqrt{\frac{\text{length}}{\text{acceleration due to gravity}}}$$

and frequency equals 1 over the period: $f = 1/T$. Similarly, a swing has a natural frequency of vibration. If given a gentle push, the swing will vibrate at this frequency without any other push. Because of friction the amplitude of vibration decreases gradually; to keep the swing going, it is necessary to supply only a little push to the swing at its natural frequency or at a submultiple of this frequency. The swing can be forced to vibrate or oscillate at many other frequencies. All one has to do is to hold onto the swing and push it at the desired frequency. Such oscillation is known as *forced vibration.* For example, the human eardrum can be forced to vibrate at any of a wide range of frequencies by the sound reaching the ear. We can think of a distant vibrating tuning fork forcing the eardrum to oscillate by means of a longitudinal wave from the fork to the ear (see Figure 14.5)

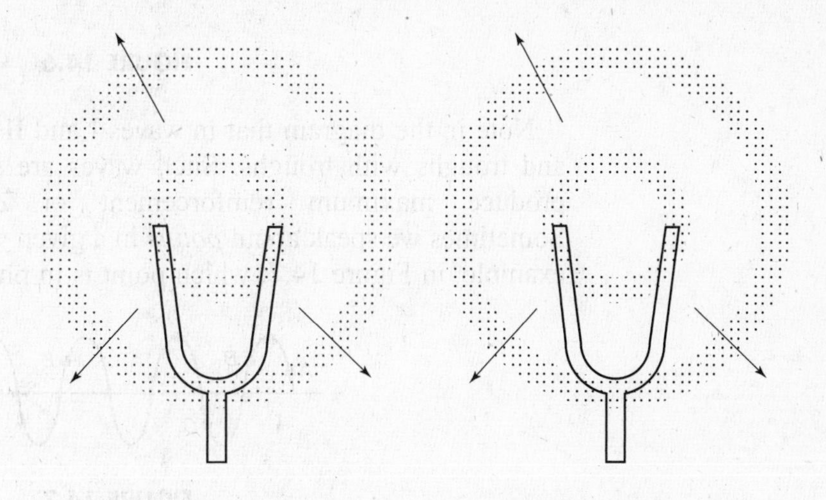

FIGURE 14.5 Resonance in Two Identical Tuning Forks

Resonance

Resonance between two systems exists if vibration of one system results in vibration or oscillation of the second system, whose natural frequency is the same as that of the first. In the case of sound the two systems may be two tuning forks of the same frequency. We mount the two forks on suitable boxes at opposite ends of a table. We strike one tuning fork, let it vibrate for a while, and then put our hand on it to stop the vibration. We will than be able to hear the second fork. It resonated to the first one. What caused the second tuning fork to vibrate? Compressions starting from the first fork gave the second fork successive pushes at the right time to build up the amplitude of its oscillations. Thus energy was transferred from the first fork to the second fork by means of the wave. When resonance occurs with sound, the term *sympathetic vibration* is also used: the second fork is said to be in sympathetic vibration with the first one. Later we shall discuss cases of resonance that do not involve sound waves.

One can think of an *air column* in a tube as having a natural frequency. The natural frequency of an air column depends on its length. If a vibrating tuning fork is held over an air column of the right length, resonance will occur between the fork and the air column; if the length of the air column is varied, the sound elicited by resonance will be loud. The length of a closed resonant air column is about one-fourth of the wavelength, as will be discussed in the next section.

Interference

Interference occurs when two waves pass through the same portion of the medium at the same time. If, at a given time and place, the two waves tend to make the medium vibrate in the same direction, *reinforcement* or constructive interference occurs. For example, in Figure 14.6 waves I and II combined to produce wave III.

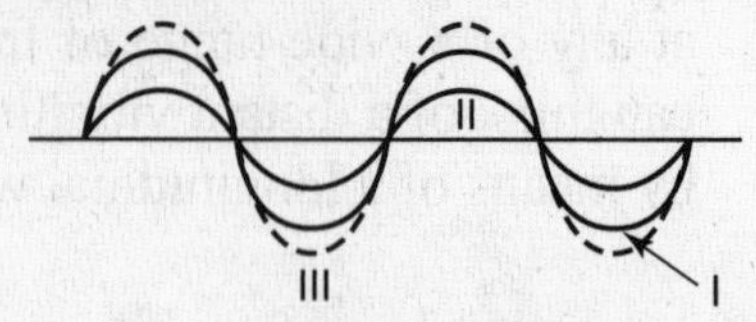

FIGURE 14.6

Note in the diagram that in waves I and II peaks coincide with peaks and troughs with troughs. Such waves are said to be in *phase.* They produce maximum reinforcement or *constructive interference.* Sometimes we speak about *points* in a given wave as being in phase. For example, in Figure 14.7, which point is in phase with point *B*?

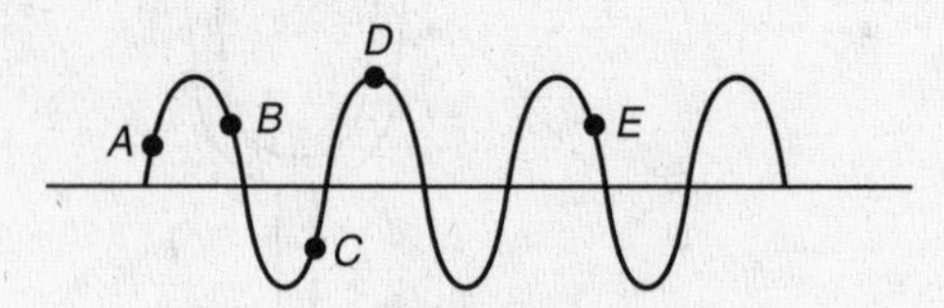

FIGURE 14.7

We can think of the diagram as representing a periodic wave moving to the right. The particles have been displaced from their equilibrium position represented by the horizontal line. Points on a wave that are a whole number of wavelengths apart are in phase; they are going through the same part of the vibration at the same time. Point *E* is two full wavelengths ahead of point *B*. They are in phase. Point *C* is about one-half of a wavelength ahead of point *B*. They are opposite in phase.

If two waves tend to make the medium vibrate in opposite directions, we have *destructive interference.* The two waves then tend to cancel each other. For example, in Figure 14.8, waves IV and V together produce wave VI.

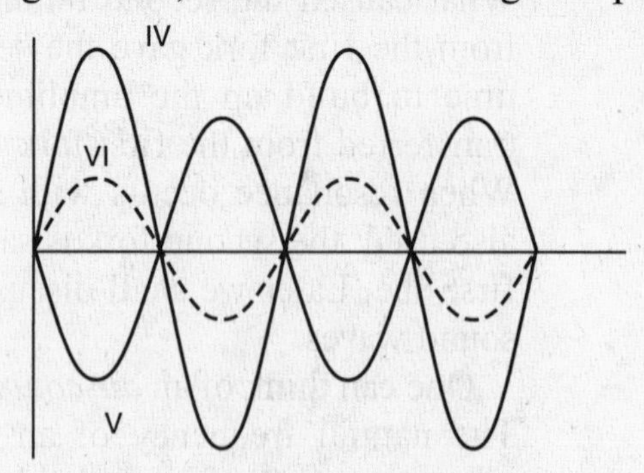

FIGURE 14.8

The peaks of one coincide with the troughs of the other. This is the condition for maximum destructive interference. Waves IV and V are said to be opposite in phase, or 180° apart, or 180° out of phase.

If two sounds reach the ear at the same time, interference takes place. *Beats* are heard when two notes of slightly different frequencies reach the ear at the same time.

the number of beats = the difference between the two frequencies

A beat is an outburst of sound followed by comparative silence. Beats result from the interference of the two waves reaching the ear. When two compressions or two rarefactions combine, the eardrum vibrates with relatively large amplitude and a loud sound is heard. When a compression combines with a rarefaction, the two tend to annul each other and relatively little sound is heard.

The resultant wave produced by two sources of slightly different frequencies sounded at the same time can be represented as shown in Figure 14.9.

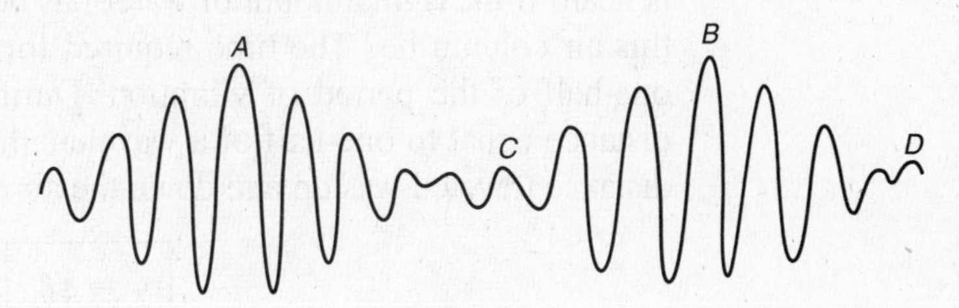

FIGURE 14.9

Notice that at *A* and *B* a relatively large amplitude corresponds to an outburst of sound and at *C* and *D* a rather small amplitude corresponds to comparative silence.

Vibrating Air Columns

Some musical instruments, such as flutes or even soda bottles, produce sound by means of vibrating air columns. We can set the air column into vibration by blowing across or into one end of the pipe.

Closed Pipes

A closed pipe has one end closed. The end we blow into is always considered open. As indicated before, one way to describe a vibrating air column is to say that it has natural frequencies of vibration. Its lowest frequency of vibration (fundamental) is one for which the wavelength equals approximately four times the length of the air column. For example, a closed air column that is 1 meter long produces a sound whose wavelength is 4 meters. If the speed of sound is 331 meters per second,

frequency of vibration = speed of sound divided by wavelength

or 331 meters per second divided by 4 meters, which gives approximately 83 vibrations per second for the frequency (f). This is the fundamental frequency. The *overtones* that a closed air column can produce are odd multiples of the fundamental frequency (3*f*, 5*f*, 7*f*, etc.).

Another way to describe a vibrating air column is in terms of resonance. Imagine a narrow cylinder with some water at the bottom, as in Figure 10.14 which gives us an air column of length ℓ_a. Since water is practically incompressible, the air in contact with the water cannot move down. This is, therefore, a closed air column. Imaging a tuning fork vibrating above the air column. Concentrate on the bottom prong of the fork. Its rest or equilibrium position is shown at *r;* its highest position at *4;* its lowest position during vibration at *2.* As the prong moves from *4* toward *2,* a compression forms at *r* and travels down the air column; when it reaches the water surface the compression is reflected. If the reflected compression reaches the prong just when it is at *r;* sending a compression in the direction away from the tube, the fork's vibration will be reinforced. The result in resonance between the vibrating tuning fork and air column—a loud sound is heard if the right amount of water has been poured in. How long should this air column be? The time required for the prong to go from *4* to *2* is one-half of the period of vibration. During this time the wave travels a distance equal to one-half of a wavelength (λ), but in this experiment this distance traveled was up and down the air column. Therefore, $2\ell_a = \lambda/2$, or

$$\boxed{\lambda = 4\ell_a}$$

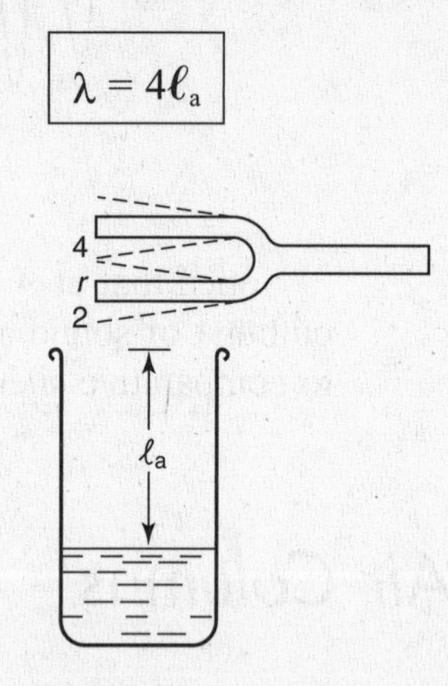

FIGURE 14.10

In words, for a resonant closed air column, the wavelength of the fundamental frequency is equal to four times the length of the air column, as was stated before. However, suppose that with the same air column we use a tuning fork whose frequency is three times higher than the one we used before. Its prong will vibrate from *4* to *2* in one-third the time, but the speed of the wave is not changed. Therefore, during this time the wave will have traveled only one-third of the total back-and-forth distance, which is two-thirds of the length of the air column. By the time the wave is reflected by the water and comes back to the tuning fork, the prong has gone back to *4* and has just reached *2* again. As before, the reflected air wave reinforces the vibration of the tuning fork. Again the air column resonated to the tuning fork. This time the closed air column vibrated at three times its fundamental frequency and produced its first overtone.

Open Pipes

An *open pipe* is open at both ends; the air can move freely at either end. When an open pipe is vibrating so as to produce its fundamental frequency, the length of the air column (ℓ_a) is equal to one-half of the wavelength (λ) produced; or the wavelength is equal to twice the length of the air column:

$$\boxed{\lambda = 2\ell_a}$$

An open pipe can produce not only odd multiples of its fundamental frequency but also even multiples. The first overtone is twice the fundamental frequency; the second overtone, three times, and so on ($2f$, $3f$, $4f$, $5f$, etc.).

Vibrating Strings

If a string is fastened at both ends and plucked or bowed at some point in between, it will vibrate transversely. Some frequencies of vibration will be natural for the string and will persist. The *fundamental* frequency of the string is the lowest of these natural or free vibrations; for this frequency, the wavelength in the string is equal to twice the length of the string (ℓ_s):

$$\boxed{\lambda = 2\ell_s}$$

Notice that this formula describes the wave in the *string*. The transverse vibration of the string sets up a longitudinal wave in the surrounding air: the sound wave. The frequency of the sound wave is the same as the frequency of vibration of the string; but since the speed of the wave in the string is different from the speed of the wave in the air, the wavelengths in the two media are also different ($v = f\lambda$). The *overtones* that a string can produce are the odd and even multiples of the fundamental frequency ($2f$, $3f$, $4f$, $5f$, etc.). Notice that in this respect a vibrating string is similar to a vibrating open air column.

Vibrating strings can also be described in terms of *standing waves* or stationary waves. When the string is plucked, the wave travels to both ends, from which it is then reflected. These reflections can take place over and over again. As a result waves go in opposite directions through the string and interference takes place. At the ends, where the string is fastened, no vibration can take place. These places of no vibration are known as *nodes.* In between there may be one or more places where the waves reinforce each other completely. These places of maximum vibration are known as *antinodes.* When overtones are produced, there will also be one or more nodes along the length of the string. The distance between successive nodes is one-half a wavelength of the wave in the string. When a string produces its fundamental frequency, it is said to vibrate as a whole; when it produces overtones, it vibrates in parts. See Figure 14.11.

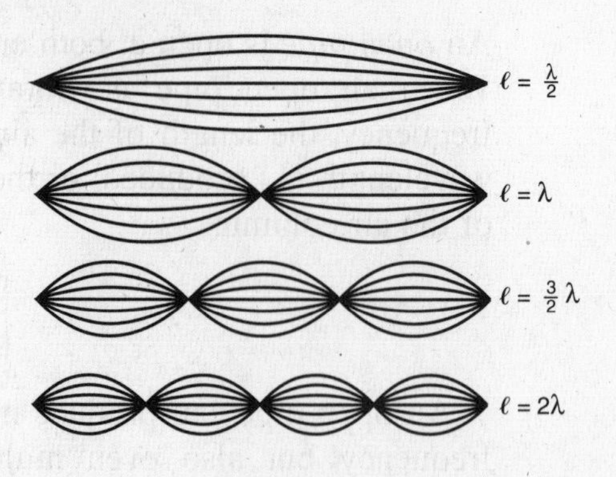

FIGURE 14.11

Laws of Vibrating Strings

The fundamental frequency of vibration of a string depends on its length, its tension, and its mass per unit length. If one of these is varied at a time, the effect is as follows:

1. The frequency varies inversely as the length; that is, making the string twice as long produces a frequency one-half as great.
2. The frequency varies directly as the square root of the tension; that is, to double the frequency the force stretching the string must be quadrupled.
3. The frequency varies inversely as the square root of the mass per unit length.

Doppler Effect

If there is relative motion between the source of a wave and the observer, the frequency of vibrations received by the observer will be different from the frequency produced by the source. In the case of sound this difference will affect the pitch of the perceived sound, since pitch depends on the number of vibrations reaching the ear per second. When the source and the observer approach each other, the pitch will be higher than if both were stationary; the pitch will be lower if source and observer are moving further apart.

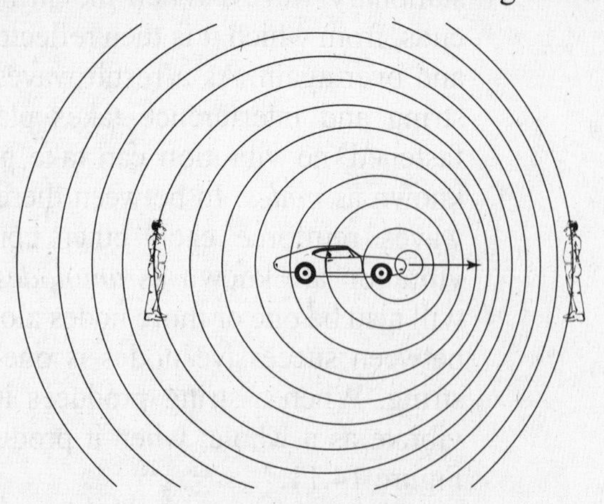

FIGURE 14.12

We can see the same phenomenon by using a car horn. In Figure 14.12 a car is moving toward one observer and away from the other. The pitch increases as the car approaches but decreases as it recedes. To an observer moving with the car, the frequency of the horn remains the same.

We can describe these relationships using the following formulas:

$$f = f_o[\mathbf{v}/(\mathbf{v} - \mathbf{v}_s)] \quad \text{toward observer}$$

$$f = f_o[\mathbf{v}/\ (\mathbf{v} + \mathbf{v}_s)] \quad \text{away from observer}$$

where

f = observed frequency f_o = transmitted frequency

$\mathbf{v}$ = wave velocity $\mathbf{v}_s$ = source velocity

If the source is moving faster than the waves in the medium, a *shock wave* is produced, as illustrated in Figure 14.13. The ratio of the speed of the source to the speed of the wave is called the *Mach number,* named for nineteenth century physicist Ernst Mach. An aircraft breaking the sound barrier at a given altitude produces a *sonic boom*, which is a shock wave in air.

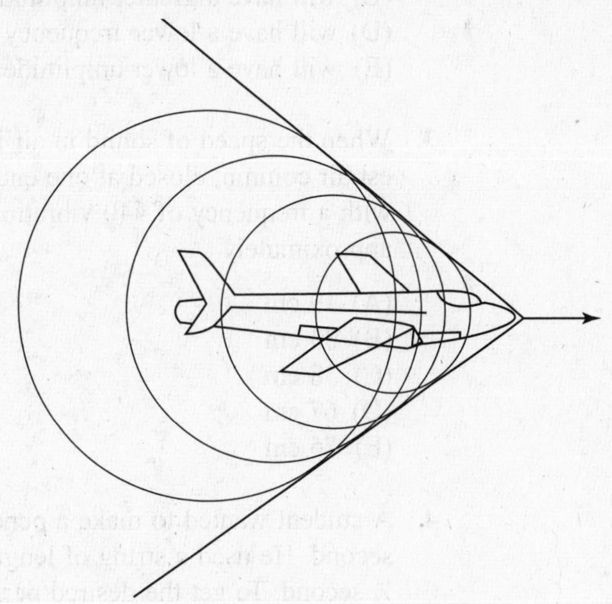

FIGURE 14.13 Shock Wave Pattern

Questions Chapter 14

In each case, select the choice that best completes the statement.

1. In order for two sound waves to produce beats, it is most important that the two waves

(A) have the same frequency
(B) have the same amplitude
(C) have the same number of overtones
(D) have slightly different frequencies
(E) have slightly different amplitudes

2. A girl produces a certain note when she blows gently into an organ pipe. If she blows harder into the pipe, the most probable change will be that the sound wave

(A) will travel faster
(B) will have a higher frequency
(C) will have a greater amplitude
(D) will have a lower frequency
(E) will have a lower amplitude

3. When the speed of sound in air is 330 meters per second, the shortest air column, closed at one end, that will respond to a tuning fork with a frequency of 440 vibrations per second has a length of approximately

(A) 19 cm
(B) 33 cm
(C) 38 cm
(D) 67 cm
(E) 75 cm

4. A student wanted to make a pendulum whose period would be 1 second. He used a string of length L and found that the period was ½ second. To get the desired period, he should use a string whose length equals

(A) $\frac{1}{4}L$
(B) $\frac{1}{2}L$
(C) $2L$
(D) $4L$
(E) L^2

5. If a vibrating body is to be in resonance with another body, it must

(A) be of the same material as the other body
(B) vibrate with the greatest possible amplitude
(C) have a natural frequency close to that of the other body
(D) vibrate faster than usually
(E) vibrate more slowly than usually

Questions 6–8

In the accompanying diagram, a narrow-diameter tube X is held with its lower end immersed in water. The level of the water is adjusted for maximum sound while a vibrating tuning fork is held over the tube. Then the length of the tube above the water is Y, and the length of the tube in the water is Z.

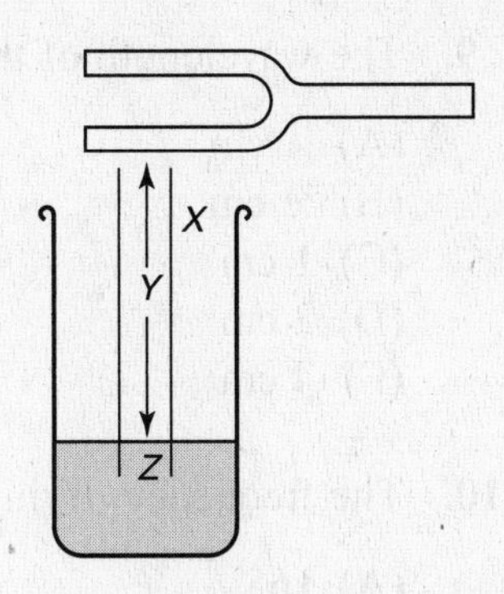

Assume that fundamental frequencies only are involved.

6. The wavelength of the sound in the air will be approximately

(A) $2Y$
(B) $2Z$
(C) $4Y$
(D) $4Z$
(E) $Y–Z$

7. The wavelength of the sound in the water will be

(A) less than $2Z$
(B) between $2Z$ and $2Y$
(C) between $2Y$ and $4Y$
(D) $4Y$
(E) greater than $4Y$

8. In the time required for the tuning fork to make one complete vibration, the wave in air will travel a distance equal to

(A) wavelength/4
(B) wavelength/2
(C) wavelength
(D) twice the wavelength
(E) four times the wavelength

Questions 9 and 10

A vibrating diaphragm sets up strong vibrations at the mouth of a horizontal tube containing air and a small amount of fine powder. The powder becomes arranged in piles 1 centimeter apart, and the speed of sound in air is 330 meters.

9. The wavelength of this sound in air is

(A) ¼ cm
(B) ½ cm
(C) 1 cm
(D) 2 cm
(E) 4 cm

10. The frequency of this sound is, in vibrations/second,

(A) 165
(B) 330
(C) 8,300
(D) 16,500
(E) 33,000

---

11. If a man moves, with a speed equal to 0.5 that of sound, away from a stationary organ producing a sound of frequency f, he will probably hear a sound of frequency

(A) less than f
(B) f
(C) $1.5f$
(D) $2.25f$
(E) $2.5f$

12. A certain stretched string produces a frequency of 1,000 vibrations per second. For the same string to produce a frequency twice as high, the tension of the string should be

(A) doubled
(B) quadrupled
(C) reduced to one-half of the original value
(D) reduced to one-fourth of the original value
(E) increased by a factor of $\sqrt{2}$

Explanations to Questions Chapter 14

Answers

1. (D)	**4.** (D)	**7.** (E)	**10.** (D)
2. (C)	**5.** (C)	**8.** (C)	**11.** (A)
3. (A)	**6.** (C)	**9.** (D)	**12.** (B)

Explanations

1. D The number of beats is equal to the difference in the frequencies of the two sounds. If the frequencies are the same, there will be no beats. Even if the amplitudes are not the same, at times the two sounds will reinforce each other and at other times they will partially annul each other, giving rise to beats.

2. C The speeds of all audible sound frequencies in air are independent of the frequency or the amplitude of vibration. However, the greater energy made available by blowing harder will lead to a greater amplitude of the wave. (It may also lead to the production of overtones, but these are higher in frequency than the fundamental.)

3. A speed of sound = frequency × wavelength
wavelength = $\mathbf{v}/f$ = (330 m/s)/(440 vib/s) = 0.75 m = 75 cm
shortest closed air column for resonance = $L/4$ = 75 cm/4 = 19 cm.

4. D For a simple pendulum, period = $2\pi\sqrt{L/g}$. The period is proportional to the square root of the length; if the length is multiplied by 4, the period is multiplied by the square root of 4, or by 2. Therefore, for the period to be doubled, the student must use a length of string four times the original length or $4L$.

5. C This is described as the condition for resonance: natural frequencies that are close to each other. The vibrations can be of rather small amplitude; the materials may be the same or different.

6. C This concept was described in the chapter as resonance for a closed air column. The resonant air column Y^1 is then one-fourth of the wavelength in air, or the wavelength in air = $4Y$.

7. E Z has nothing to do with the wavelength in water. However, the speed of sound is greater in water than in air; the frequency of the sound in water is the same as that of the original sound in air. Since speed of sound = frequency × wavelength, if the speed increases and the frequency remains the same, the wavelength has to increase also and will then be greater than $4Y$.

8. C This situation is described in the chapter as part of the discussion of the *period*.

9. D The problem describes the production of a standing wave. The powder forms piles at the nodes. The distance between successive nodes is ½ wavelength. Therefore, ½ wavelength = 1 cm; wavelength = 2 cm.

10. D Frequency = (speed of wave)/wavelength = (33,000 cm/s)/2 cm = 16,500 vibrations per second.

11. A This problem is an example of the Doppler effect. Since the man is moving away from the source of the sound, successive compressions will not reach him in as rapid succession as when he is stationary. (However, since he is moving at a speed only 0.5 that of sound, the compressions do catch up with him until the amplitude is too small to have any effect.) Therefore, the frequency he hears is less than f.

12. B The frequency of vibration of a string is proportional to the square root of its tension. The tension must be increased fourfold in order for the frequency to be doubled.

CHAPTER 15

Geometrical Optics: Reflection and Refraction

Why is it that images seen in plane mirrors seem smaller than the corresponding objects, although "the size of the image in a plane mirror is the same size as the object"? Why is it that we can take a photograph of our image in a plane mirror, although "the image is virtual and cannot be seen on a screen placed at the position of the image"? We need to understand reflection, refraction, the operation of lenses, and the eye to answer these and similar questions. Any phenomenon that can be described by thinking of light in a given medium as traveling in straight lines comes under the heading of geometrical optics. It is convenient to use the *ray* to represent the direction in which the light travels.

Reflection

When light hits a surface, usually some of the light is reflected. The *normal* for this light is the line drawn perpendicular to the surface at the point where the light ray touches the surface. The light that goes toward the surface is known as *incident* light and is represented by the incident ray. The angle of incidence (i) is the angle between the incident ray and its normal. The reflected light is represented by the reflected ray; the angle of reflection (r) is the angle between the reflected ray and its normal. See Figure 15.1.

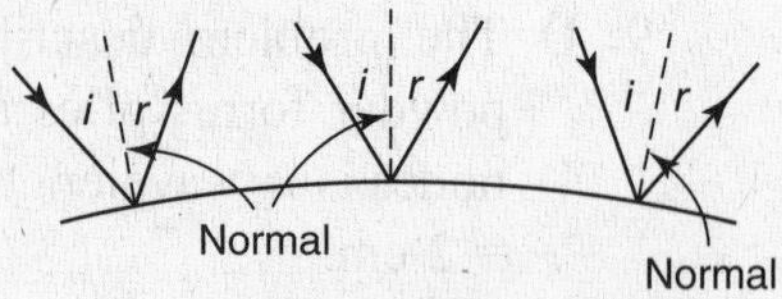

i r i r i r
Normal
Normal

FIGURE 15.1

The *law of reflection* states that, when a wave is reflected, the angle of incidence equals the angle of reflection, and the incident ray, the normal, and the reflected ray lie in one plane. Notice that this law applies not only to light but also to other waves and is true of both smooth and rough surfaces.

If a surface is smooth and flat, we get *regular reflection.* Parallel incident rays are reflected as parallel rays. Most surfaces, however, even this paper, are rough to a degree. From a rough surface we get *diffuse reflection:* Reflected light goes off in all directions even when the incident light is parallel. See Figure 15.2. Nonluminous objects are visible because diffusely reflected light from their surfaces enters our eyes. A perfectly smooth reflecting surface cannot be seen. Luminous objects are seen because light they emit enters our eyes.

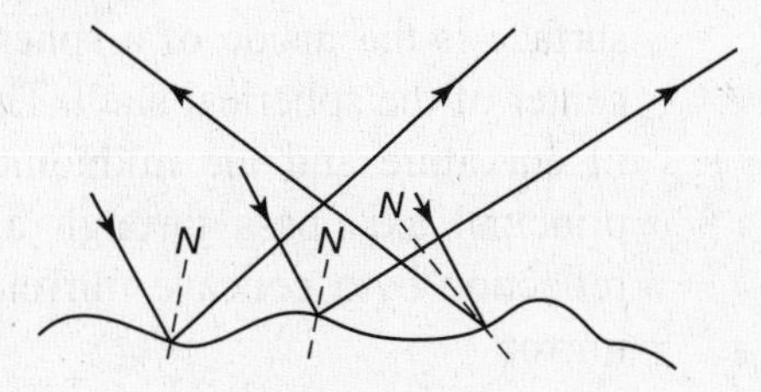

FIGURE 15.2

Image in Plane Mirror A plane mirror is a perfectly flat mirror. The characteristics of an image produced by such a mirror can be determined by means of a *ray diagram* such as Figure 15.3.

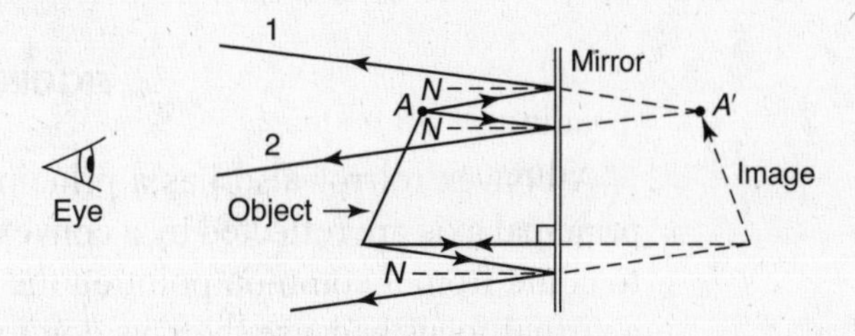

FIGURE 15.3

We can represent the object in some suitable way; often an arrow is used. We take two rays from the top of the object, point *A,* and draw the corresponding reflected rays, making the angle of reflection equal to the angle of incidence. The observer into whose eye these reflected rays go imagines them to come from some point behind the mirror: the point where the extensions of the reflected rays meet, point *A.* This is the image point corresponding to the top point on the object. The extensions of the reflected rays are shown dotted because the rays of light, and the image, are not actually behind the mirror. We do the same with the other end of the object. If we take a ray that is perpendicular to the surface, after reflection it will go right back along the same path. From the diagram we can determine the *characteristics* of an image in a plane mirror:

1. The image is the same size as the object.
2. The image is erect (if the object is erect).
3. The image is as far behind the mirror as the object is in front. (Every point is as far behind the mirror as the corresponding object point is in front.)
4. The image is *virtual;* that is, the image is formed by rays that do not actually pass through it. If a screen were placed at the position of the image, no image would be seen.
5. The image is laterally reversed. This characteristic can be thought of in terms of two people shaking hands: their arms extend diagonally between them. When a person holds his hand out to the mirror, the arm of his image comes straight out at him.

Image in Curved Mirrors

Spherical surfaces are commonly used as the reflecting surfaces of curved mirrors (Figure 15.4). A *convex* mirror is one whose reflecting surface is the outside of a spherical shell. A *concave* mirror is one whose reflecting surface is the inside of a spherical shell. The *center of curvature* (C) is the center of the spherical shell. The *principal axis* is the line through the center of curvature and the midpoint of the mirror. Incident rays parallel to the principal axis pass through a common point on the principal axis after reflection by a concave mirror. This point is the *principal focus* (F) of the mirror.

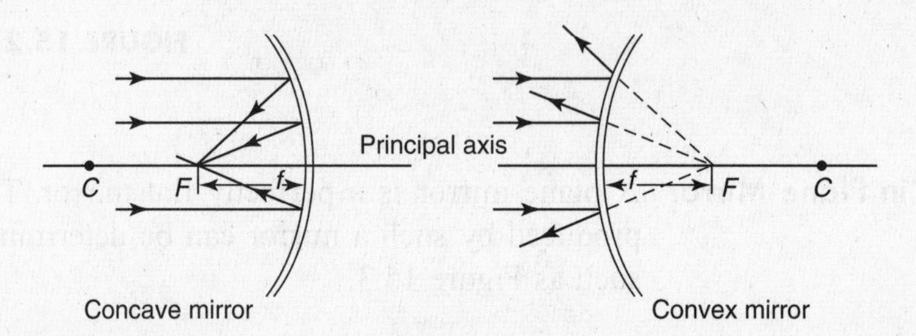

FIGURE 15.4

A convex mirror also has a principal focus. Incident rays parallel to the principal axis are reflected by a convex mirror so that the reflected rays seem to come from a common point on the principal axis. This principal focus is a virtual focus because the rays don't really meet there. The *focal length* (f) is the distance from the principal focus to the mirror. For a spherical mirror the focal length is equal to one-half of the radius of the spherical shell.

$$\boxed{f = \frac{R}{2}}$$

We need one more concept in order to draw ray diagrams of spherical mirrors: A radius is perpendicular to its circle and is the direction of the normal. Therefore, a ray directed along the radius of the spherical shell is perpendicular to the spherical mirror. The angle of incidence is then equal to zero and so is the angle of reflection. Therefore, a ray going through the center of curvature (C) of the mirror is reflected right back upon itself (Figure 15.5).

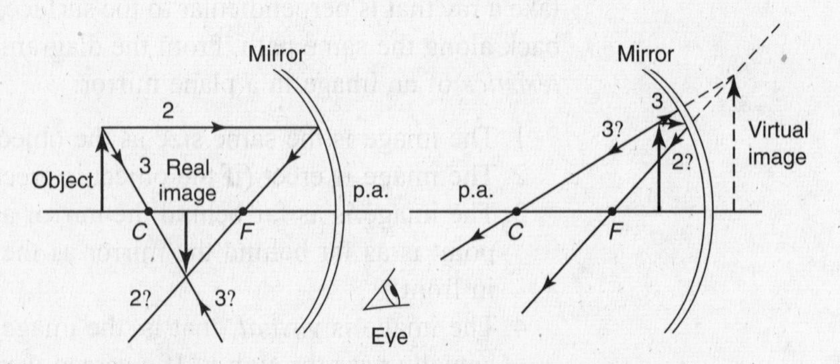

FIGURE 15.5 Images with a Concave Mirror

The technique for drawing the ray diagram follows: We try to find the points of the image that correspond to selected points of the object. We often use an arrow as the object; and if we place the arrow so that its tail touches the principal axis, we need to use rays only from the head of the arrow. The tail of the image will also touch the principal axis. We select two special rays from the head of the arrow; we already know their paths after reflection. Ray 2, parallel to the principal axis, after reflection, passes through the principal focus. Ray 3, whose direction passes through the center of curvature, is normal to the mirror and is reflected right back upon itself. The reflected rays, 2′ and 3′, intersect in the left diagram and produce a *real image* on a screen placed there.

In the diagram on the right, the reflected rays don't meet and therefore do not produce a real image. However, a person into whose eye the reflected light goes will see an image behind the mirror, at the place where the extended reflected rays meet. This is a virtual image, as in the plane mirror. Rays extended behind the mirror are always drawn dotted. The virtual image is often shown dotted, too.

Any other ray starting from the head of the arrow, after reflection, will pass through the head of the image.

Notice that with the concave mirror we can get both real and virtual images. With the convex mirror (Figure 15.6) we can get only virtual images. The procedure for the ray diagram is as described above.

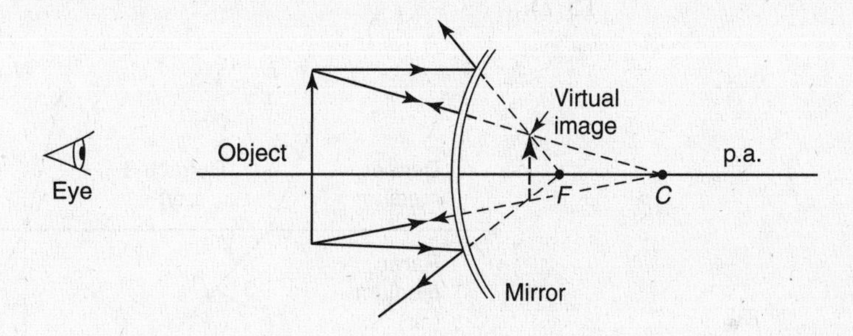

FIGURE 15.6 Image with a Convex Mirror

Notice that the center of curvature and the principal focus are on the inside of the spherical shell. The parallel ray is reflected so that it will seem to come from the principal focus. The ray directed toward the center of curvature is reflected right back upon itself. The rays originally starting from the same point on the object don't meet after reflection and therefore cannot produce a real image. The observer sees a virtual image behind the mirror. The virtual image formed with a convex mirror is smaller than the object; the virtual image formed with a concave mirror is larger than the object.

Careful analysis shows that spherical mirrors don't work precisely as described on the preceding page. The difference is usually negligible. However, such aberrations can cause problems in astronomical telescopes using large mirrors. Instead of spherical mirrors these telescopes usually have parabolic mirrors, which are much more expensive.

Refraction

Refraction is the bending of a wave on going into a second medium. Refraction depends on the difference in speed of the wave in the two media. The medium in which the light travels more slowly is known as the optically denser medium or, for short, the *denser medium.* This term should not be confused with mass density (ratio of mass to volume of a substance). The medium in which the light travels faster is known as the (optically) *rarer* medium.

The *speed of light* is fantastically high. In vacuum it is about 3×10^{10} centimeters per second or about 3×10^8 meters per second. In air the speed of light is only slightly less. Light travels faster in air than in liquids and solids, and faster in water than in glass.

Law of Refraction

A ray of light passing obliquely into a denser medium is bent toward the normal; a ray of light entering a rarer medium obliquely is bent away from the normal; a ray entering a second medium at right angles to the surface goes through without being bent or deviated. The angle formed by the refracted ray and its normal is known as the *angle of refraction* (see Figure 15.7).

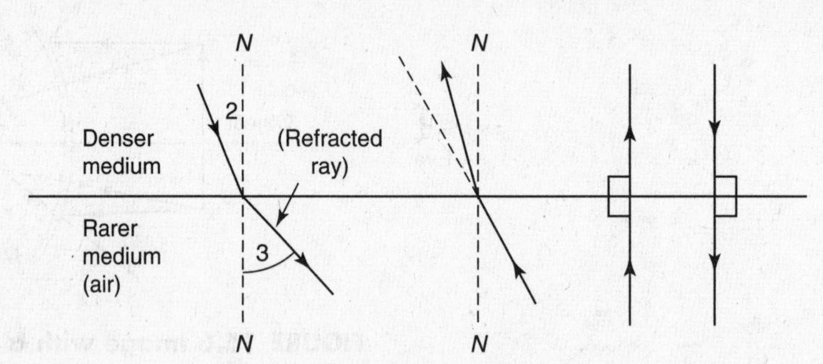

FIGURE 15.7

The *index of refraction* (n) of a substance is defined as the ratio of the sine of the angle of incidence in vacuum to the sine of the angle of refraction in the substance. For many purposes we can use air instead of vacuum (see Figure 15.8).

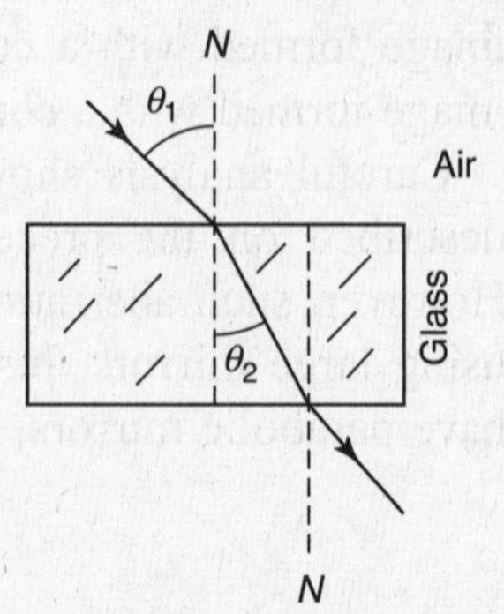

FIGURE 15.8

The index of refraction is also equal to the ratio of the speed of light in vacuum to the speed in the substance:

$$n = \frac{\text{speed of light in vacuum (or air)}}{\text{speed of light in the substance}}$$

The index of refraction is usually given as a number greater than 1. This practice becomes awkward, however, when light goes from the denser substance into air, as in the right of Figure 15.8, where the angle of incidence is in glass and is smaller than the angle of refraction formed as the ray comes out into the air. Therefore, once we understand the principles involved, it is desirable to leave out of the definition of index of refraction the phrases "angle of incidence" and "angle of refraction." To calculate the index of refraction, we divide the sine of the angle in vacuum (or air) by the sine of the angle in the denser medium. In Figure 15.7, for example,

$$n = \frac{\sin \angle 3}{\sin \angle 2}$$

In the preceding formulas for the index of refraction, vacuum is taken as the reference. These formulas give us the index of refraction with respect to vacuum, or the *absolute index of refraction.* If vacuum is not used, the relationship becomes:

$$\frac{n_2}{n_1} = \frac{\sin\ \theta_1}{\sin\ \theta_2}$$

Of course, this can be written as $n_1 \sin \theta_1 = n_2 \sin \theta_2$.

If light goes through parallel surfaces of a transparent substance, such as a flat glass plate, the emerging ray is parallel to the entering ray, provided the two rays are in the same medium. If the two surfaces are not parallel, as in a triangular glass prism, the emerging ray is bent away considerably from its original direction. In drawing the ray diagram you should keep in mind the law of refraction (see Figure 15.9).

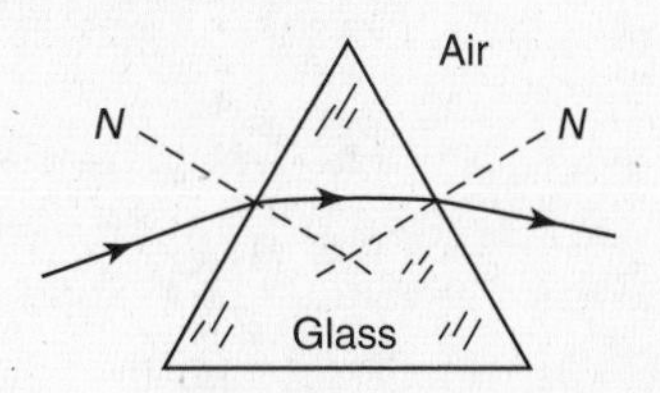

FIGURE 15.9

The diagram assumes that light of a single wavelength (*monochromatic light*) is used.

Critical Angle and Total Reflection

Whenever light enters a second medium whose index of refraction is different from that of the first medium, some of the light is reflected. The greater the angle of incidence, the greater is the amount of reflection. This added reflection is called *partial reflection* and was deliberately omitted in Figure 15.9. In some cases all of the light is reflected. Consider Figure 15.10, in which rays are shown making different angles of incidence; notice that they are all directed from the denser medium toward the rarer one.

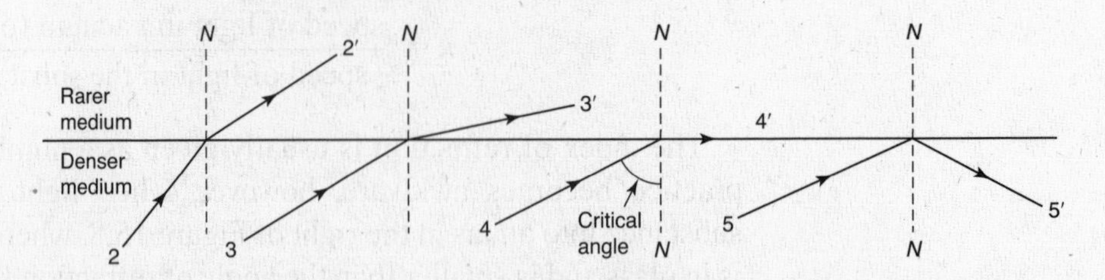

FIGURE 15.10

As we go from ray 2 to ray 3, from ray 3 to ray 4, and so on, the angle (of incidence) in the denser medium keeps increasing, and therefore the angle (of refraction) in the rarer medium increases still more. For each ray the corresponding angle in the rarer medium is greater than the angle in the denser medium. As the angles of incidence increase, the refracted rays 2′, 3′, and so on lie closer and closer to the surface separating the two media. The *critical angle* is the angle of incidence in the optically denser medium for which the refracted ray makes an angle of 90° with the normal. *Total reflection* occurs when the angle of incidence in the optically denser medium is greater than the critical angle; then no light enters the rarer medium. Total reflection is utilized in prism binoculars and in curved transparent fibers and Lucite rods; the light stays inside the curved rod because of successive internal reflections. For all reflections the angle of incidence equals the angle of reflection. For a common variety of glass the critical angle is about 42°. In the triangular glass prism shown in Figure 15.11, total reflection can take place. The acute angles are 45°, and the entering ray shown is perpendicular to the glass surface and thus is not refracted. The angle of incidence on the hypotenuse is therefore 45°, which is greater than the critical angle, and so there is total reflection from the hypotenuse.

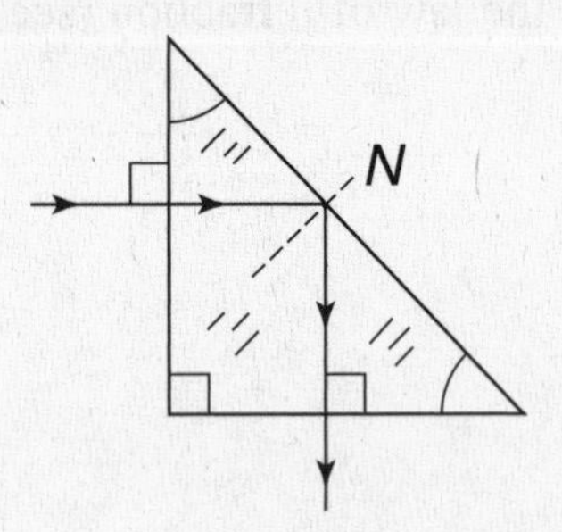

FIGURE 15.11

Lenses

A lens is a device shaped to converge or diverge a beam of light transmitted through it. Lenses described here are thin spherical lenses; one or both of their surfaces are spherical. If there is a nonspherical surface, it is plane. The material is transparent, usually glass. Unless stated otherwise in this description of lenses, the lens is surrounded by air. A *convex* lens is thicker in the middle than at the edges; it is also called a *converging* lens. A *concave* lens is thinner in the middle than at the edges; it is also called a *diverging* lens. The *principal axis* of a lens is the line connecting the centers of its spherical surfaces. A narrow bundle of rays parallel to the principal axis of a convex lens is refracted by the lens so that these rays, after refraction, pass through a common point on the principal axis known as the *principal focus* (F). (See Figure 15.12a).

For the concave lens (Figure 15.12b), rays originally parallel to the principal axis diverge after refraction and seem to come from a common point on the principal axis known as the *principal focus.* It is a virtual focus (F). The extensions of the actual rays are shown dotted and are called *virtual rays.*

For both lenses, the distance from the principal focus to the lens is the *focal length* (f) of the lens. The lenses are drawn quite thick, but since we are discussing thin lenses only, we may show the focal length as a distance to the center line of the lens.

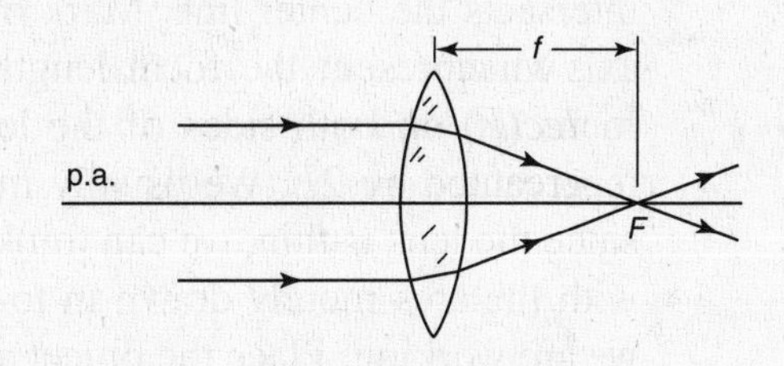

a. Convex Lens

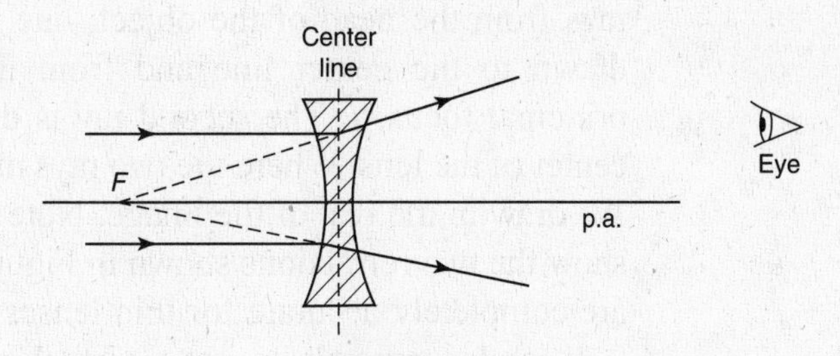

b. Concave Lens

FIGURE 15.12

Notice the similarity between lenses and curved spherical mirrors. The similarity will become even greater as we draw more ray diagrams for the lenses. Don't forget one important difference: lenses let light through and refract it; mirrors reflect light. Also, the focal length of a lens is not so simply related to the radius of curvature of its surfaces.

Images Formed by Lenses

The *convex lens* is similar to the concave mirror in that it can be used to form both real and virtual images, as can be shown by means of ray diagrams.

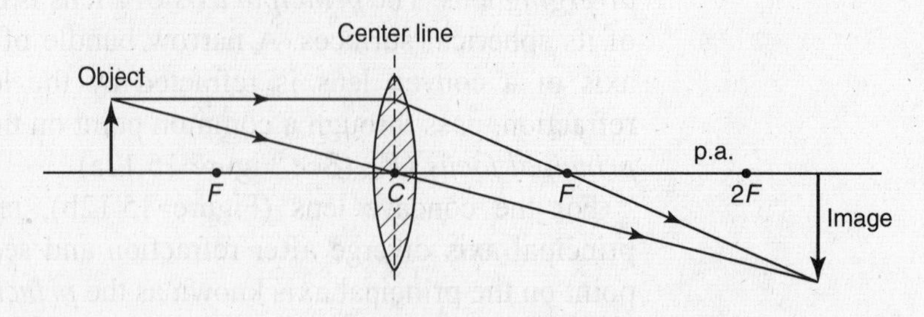

FIGURE 15.13

In Figure 15.13, the formation of a real image is shown. A real image results with the convex lens whenever the object is more than one focal length away from the lens. The procedure for making the ray diagram is as follows. Draw the lens and its principal axis. Then draw the center line of the lens and note the center of the lens (C), where the principal axis intersects the center line. Mark off a convenient length along the principal axis to represent the focal length; this gives the location of the principal focus (F) on both sides of the lens. A point twice as far from the lens is represented by $2F$. We usually measure this distance from the center line; since the lens is thin, we can think of the center line as representing the lens with the arcs merely drawn in to remind us of the type of lens with which we are working. Place the object at some appropriate distance from the lens, depending on the situation. In Figure 15.13 the object is placed at a distance between one and two focal lengths away from the lens. We draw two special rays from the head of the object; one ray, parallel to the principal axis, is drawn to the center line and from there is drawn straight through the principal focus, F. The second ray is drawn as a diagonal line through the center of the lens. Where the two rays meet, the head of the image is formed. We draw in the rest of the image. Note that, although this diagram does not show the two refractions shown in Figures 15.12(a) and 15.12(b), the results are completely accurate for thin lenses (when thickness << diameter.)

It is also possible to get a virtual image with a convex lens. A virtual image results when the object is less than one focal length away from the lens, as shown in Figure 15.14. Notice that the image is larger than the object and erect. A convex lens can be used as a magnifying glass.

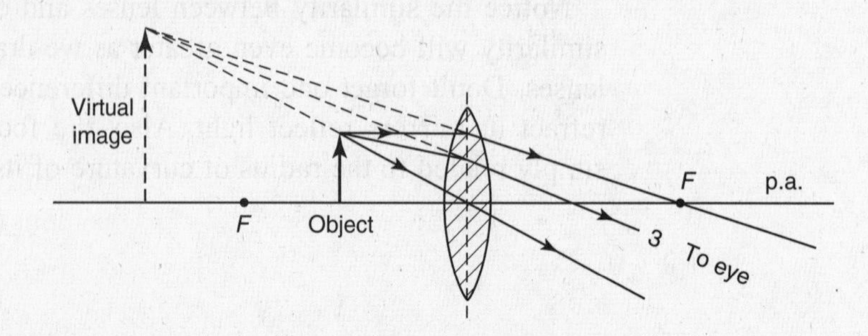

FIGURE 15.14

A *concave lens* always produces a virtual image that is smaller than the object and erect (if the object is erect), as is illustrated in Figure 15.15. Remember: except for aberrations that will be mentioned later, all rays that start from the head of the object will, after refraction by the lens, pass through the same point of the image. This is shown by ray 3, added after the rest of the diagram was drawn.

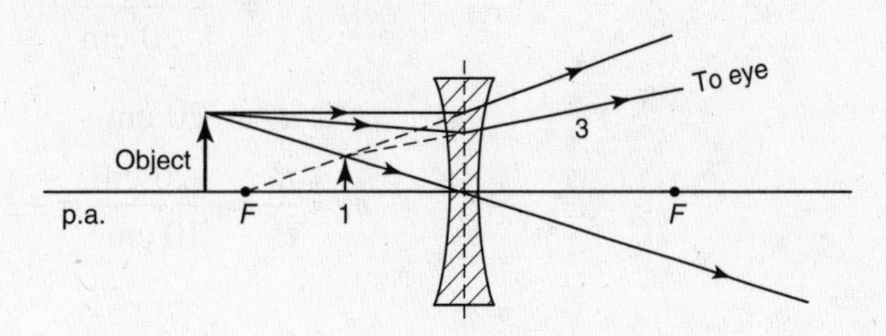

FIGURE 15.15

Numerical computations can be performed with the *lens equation*:

$$\frac{1}{\text{object distance}} + \frac{1}{\text{image distance}} = \frac{1}{\text{focal length}}$$

$$\frac{1}{d_o} + \frac{1}{d_i} = \frac{1}{f}$$

Rules

There are four rules for using lens equation:

1. Distances are measured from the lens; object distance is distance of object from lens, and so on.
2. The object distance (d_o) is always positive.
3. The image distance (d_i) is positive for real images, negative for virtual images.
4. The focal length (f) is positive for convex lenses, negative for concave lenses.

The size (height) of the image can be calculated from this equation:

$$\frac{\text{size of image}}{\text{size of object}} = \frac{\text{image distance}}{\text{object distance}} = \text{magnification}\ (m)$$

The same equations apply to spherical mirrors, but what has been said in the rules about convex lenses applies to concave mirrors, and what has been said about concave lenses applies to convex mirrors.

Example: A convex lens whose focal length is 20 centimeters has an object placed 10 centimeters away from it. Describe the image.

$$\frac{1}{d_i} = \frac{1}{f} - \frac{1}{d_o}$$
$$= \frac{1}{20 \text{ cm}} - \frac{1}{10 \text{ cm}}$$
$$= -\frac{1}{20 \text{ cm}}$$

$$\therefore d_i = -20 \text{ cm}$$

$$m = \frac{d_i}{d_o} = \frac{-20 \text{ cm}}{10 \text{ cm}} = -2$$

The image distance is –20 centimeter. The minus sign means that it is a virtual image, on the same side of the lens as the object. The magnification is minus 2, which means that the virtual image is twice as large as the object.

Some of the characteristics of lenses and spherical mirrors are summarized in the following table.

OBJECT DISTANCE	**IMAGE CHARACTERISTICS**
	Convex Lens (or Concave Mirror)
greater than $2f$	real, smaller, between f and $2f$, inverted
$2f$	real, same size, $2f$, inverted
between f and $2f$	real, larger, greater than $2f$, inverted
less than f	virtual, larger, d_i more than d_o erect
	Convex Lens (or Concave Mirror)
any distance	virtual smaller, erect, d_i less than d_o

For example (referring to the first item in the table), when the object is more than two focal lengths distant from the lens, the image is real, smaller than the object, between one and two focal lengths away from the lens, and inverted.

Note that, as the object gets closer to the lens, the image goes further away. An object "at infinity" will produce a point image a focal length away from the convex lens. A point source of light at the principal focus of a convex lens will produce light parallel to the principal axis after refraction by the lens.

Optical Instruments

The camera and the eye have many points of similarity. The *camera* has a shutter to admit the light from an object at the right time. In the *eye* the eyelid corresponds to the shutter. In the camera the light then goes through an opening in a diaphragm, the aperture, through a convex lens (or combination of lenses) to the film with its surface sensitive to light. In the eye the light goes through the pupil, which is surrounded by the colored iris.

The light then goes through a complicated lens system consisting of the "lens" and some liquids with curved surfaces. The light then falls on the retina, where the image is formed. The image on the retina is real, inverted, reduced in size; appropriate electrical impulses are transmitted along the optic nerve to the brain. The image on the film of the camera is also real and inverted.

The refracting *astronomical telescope* and the *compound microscope* have some points of similarity. Both have convex lenses at opposite ends of a cylinder. The lens closer to the eye is known as the eyepiece; the other lens, closer to the object, is the objective. Keep in mind how these instruments are utilized. The telescope is used to look at very distant objects. Therefore the image produced by its objective lens is small and close to the principal focus of the objective. To make this image as large as possible, we examine it with the eyepiece, which is then used like a magnifying glass. The final image is virtual, if the telescope is used for visual examination.

$$\text{telescopic magnification} = \frac{\text{focal length of the objective}}{\text{focal length of the eyepiece}}$$

With the microscope we look at small things close by. Therefore, the object is placed at just a little more than once focal length away from the objective, to give an enlarged, real image to be examined by the eyepiece. This eyepiece is used as a magnifying glass, as in the telescope.

When we use a *projector,* we want an enlarged image on a screen. A convex lens will produce an enlarged, real image if the object is between one and two focal lengths from the lens. In a projector, the object (e.g., film) is a little more than one focal length away from a convex lens.

When a convex lens is used as a *magnifying glass,* the object is placed just a little less than a focal length away from the lens. The result is a virtual, erect, enlarged image.

In *nearsighted* people the image of an object at an ordinary distance is brought to a focus too soon in the eye. Therefore, by the time the light gets to the retina, the light has diverged again, producing a blurred image on the retina. This blurring is corrected by eyeglasses with concave lenses, which make the light converge more gradually. In *farsighted* people the image of an object at an ordinary distance is in focus beyond the retina. To make the image come to a focus sooner, eyeglasses with convex lenses are used.

A lens is sometimes described by an *f-number,* such as *f*/8, which means that the diameter of the lens (measured along the center line) is one-eighth of the focal length. In a camera this *f*-number would mean that the aperture had been adjusted to one-eighth of the focal length. This measurement is related to the "speed of the lens." If the *f*-number is changed from *f*/8 to *f*/4, the diameter of the aperture is doubled. Therefore, the area of the opening is quadrupled; and, if the illumination of the scene is constant, four times as much light goes through the lens, and only one-fourth of the exposure time is required.

Color and Light

In describing refraction by means of a triangular glass prism we specified the use of monochromatic light. If so-called *white light,* such as light from an incandescent tungsten filament bulb, is used, we get a beautiful display of colors (Figure 15.16). *Dispersion* is the breaking up of light into its component colors. Glass is a dispersive medium for light, meaning that, although the amount of refraction is primarily determined by the type of glass, the index of refraction has a slight—but significant—dependence on the frequency of the incident light.

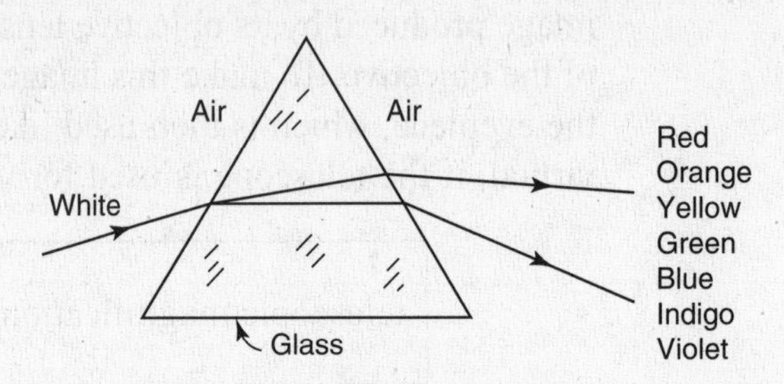

FIGURE 15.16

Seven distinct colors may be detected if white light is used. If a glass prism is used, the order of the colors is, from the one bent the least to the one bent the most, red, orange, yellow, green, blue, indigo, violet. Some students like to remember this sequence by using the first letters to spell a name. The term *spectrum* is used in many different ways. Here we shall define the *spectrum* of visible light as the array of colors produced by dispersion, or the range of waves capable of stimulating the sense of sight. This range of wavelengths is approximately 4×10^{-5} centimeter for violet to 7×10^{-5} centimeter for red. *Infrared* "light" has a greater wavelength than red; *ultraviolet,* a shorter one than violet. Both are invisible to the human eye. Dispersion is produced by the prism because violet light is slowed up more than red light in the glass (or water); the other colors travel at intermediate speeds. In vacuum all electromagnetic waves travel at the same speed.

The *color of an opaque object* is determined by the color of the light it reflects. A red object reflects mostly red; it absorbs the rest. If an object reflects no light, it is said to be black. If a red object is exposed only to blue light, which it absorbs, the red object will appear black. If an object reflects all the light, it is said to be white. A white object exposed to red light, only, will appear red. The color of a transparent object is determined by the light it transmits.

Primary and Complementary Colors

If the above seven colors of light are recombined, white light is produced. However, it is possible to produce white light with fewer colors. *Complementary colors* are two colors of light whose combined effect on the eye is that of white light. They are red and blue-green, yellow and blue-violet, green and magenta (purple). With the proper choice of three colors, known as the *primary colors,* it is possible to reproduce nearly all colors. The three primary colors of light are red, green, and blue-violet. If we mix two primary colors, we get the complementary color of the third. For example, if we mix red and green light we get yellow, which is the complement of blue-violet, the third primary color. Don't confuse these facts with the mixing of *paints.* Paints, like other opaque objects, absorb light and are seen as the color of the light they reflect. If we mix two paints, each paint subtracts the color of light it absorbs, and the color of the mixture is the color of light neither absorbs. Color obtained by the mixing of *lights* is known as an *additive process*.

Chromatic Aberration

Even a perfectly made thin lens doesn't function exactly as described above, if white light is used. The operation of a lens depends on refraction. We saw that refraction of white light by a triangular prism results in dispersion. The different colors in white light are brought to a focus at different points by the lens. The focal length of a lens depends on the color of the light used, just as does the index of refraction of a substance. The result is that images produced by lenses have a slight color fringe, and photographs with such lenses may be slightly fuzzy. This phenomenon is known as *chromatic aberration.* It can be minimized ("corrected") by using a suitable combination of lenses known as an *achromat.*

Types of Spectra

When white light was used with the triangular prism, a *continuous spectrum* was produced: there was no gap as we went from one extreme, the red, to the other extreme, the violet. A continuous spectrum may be obtained from a glowing solid or liquid, or a glowing gas under high pressure. A *spectroscope* is an instrument used to examine the spectrum of light from any source. The light to be examined goes through a narrow, rectangular slit before it reaches the prism, which disperses the light. A telescope is then usually used to look at this dispersed light. A *bright-line spectrum* is obtained from a glowing gas under low or moderate pressure. Narrow rectangles or lines of light appear with gaps in between where there is no light. These lines of light can be thought of as images of the rectangular slit.

Different elements in gaseous form produce different spectra. We say that each element has a *characteristic spectrum* and use it for identification of the element. A *dark-line spectrum* is a continuous spectrum that is interrupted by thin dark lines. The dark lines result from the absorption of the energy of the missing wavelengths by gases that were between the hot source and the spectroscope. When a gas is cool, it can absorb some of the wavelengths that it emits when it is hot. When the Sun's spectrum is examined superficially, it seems continuous. Actually it is a dark-line spectrum. The dark lines in the Sun's spectrum are known as *Fraunholer* lines. They are produced mostly by gases in the Sun's atmosphere that absorb some of the energy emitted by the hotter core. An example of the electromagnetic spectrum is shown in Figure 15.17, and spectral lines are illustrated in Figure 15.18.

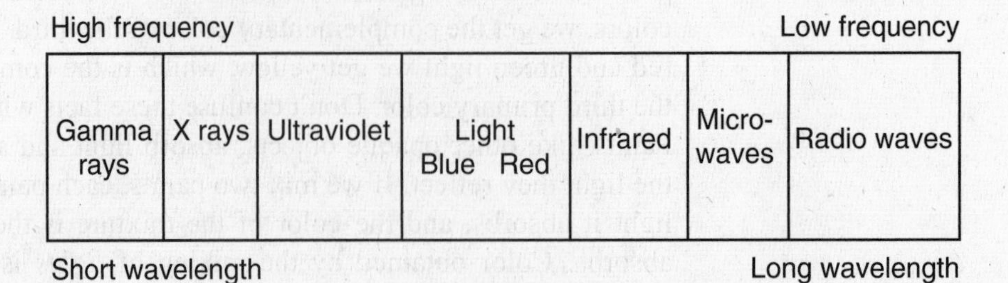

FIGURE 15.17 Electromagnetic Spectrum

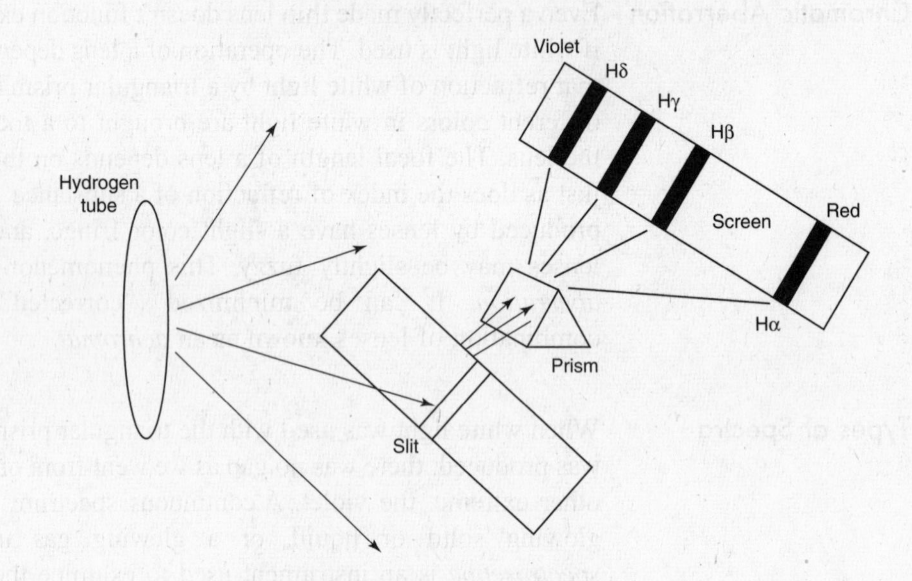

FIGURE 15.18 Spectral Lines

Doppler Effect with Light

When we studied sound, we noted that the pitch we hear depends on the frequency of the vibration of the source. Pitch is affected by the relative motion between the source and the observer. A similar phenomenon is observed in the case of electromagnetic waves.

If the distance between the source and the observer is decreasing, the observed frequency is greater than that of the source. In the case of light, frequency determines the color of light we see. If the frequency is greater than the source frequency (wavelength is smaller), the light is more violet than it would have been if there were no relative motion.

If the distance between the source and the observer is increasing, the observed frequency is lower than that of the source. In the case of light, this difference means that the observed light will be shifted toward the red.

There are some interesting applications of the Doppler effect with electromagnetic waves. Radar may be used to measure the velocity of approach of airplanes, automobiles, and similar objects. At a radar installation a transmitter sends out an electromagnetic wave in a narrow beam. The beam is reflected back by many objects, especially metallic ones. The reflected wave is picked up by a radar receiver. If the object is stationary, the reflected wave has the same frequency as that sent out by the transmitter. If the object's distance from the radar is increasing, the reflected frequency is greater. The change in frequency depends on the velocity of approach. The greater the velocity, the greater is the change in frequency.

The famous *red shift* is explained by the *Doppler principle* as being due to an expanding universe. What does that mean? The light from the Sun, stars, and other heavenly objects has been analyzed with a spectroscope. Present are the characteristic spectra of many elements we know on Earth. By comparing the spectrum of an element on Earth with the corresponding spectra from stars, it has been found that the spectra from some stars are shifted toward the blue end of the spectrum, meaning that these stars must be approaching Earth. The spectra from some other stars have been found to be shifted toward the red end of the spectrum, an indication that these stars are moving away from Earth. There are some very distant objects in the sky known as *nebulae* (singular is *nebula*). Some are galaxies similar to our Milky Way galaxy. The spectra from these nebulae are shifted toward the red. The further away the nebulae are, the greater is the shift. This fact has been interpreted to mean that the distant parts of the universe are moving further and further away from us and from each other: the universe is expanding.

Illumination

A *luminous* body is one that emits light of its own. An *incandescent* object is one that emits light because it has been heated. The filament in electric light bulbs is incandescent; the firefly is luminous, but not incandescent. An *illuminated* object is one that is visible by the light it reflects.

If a point source of light is used, the illumination on a screen at right angles to the light varies inversely as the square of the distance from the source and is proportional to the luminous intensity of the source.

$$\text{illumination} = \frac{\text{intensity of source}}{\text{distance}^2}$$

Answering the Introductory Questions

Let us return to the questions at the beginning of the chapter. The size of an object doesn't change as it moves further away, but it looks smaller to us. The explanation lies in the characteristics of images produced by convex lenses, of which there is one in the eye. The further the object, the smaller is the image on the retina. This is true also of the plane mirror image. Its size is the same as that of the object, but it is further from the eye and so forms a smaller image on the retina.

Incidentally, if you look at the diagram of the plane mirror image (Figure 15.3), you will observe that the virtual image is seen by means of diverging rays that come to the eye. (This is true also of rays coming from any point of a real object.) The convex lens in the eye brings these rays to a focus on the retina, thus producing a real image there. The convex lens in a camera acts in a similar way.

Questions Chapter 15

In each case, select the choice that best answers the questions or completes the statement.

1. In the light wave
   (A) the vibrations of the electric and magnetic fields are transverse
   (B) the vibrations of the electric field are transverse; those of the magnetic field are longitudinal
   (C) the vibrations of the magnetic field are transverse; those of the electric field are longitudinal
   (D) all vibrations are longitudinal
   (E) longitudinal vibrations alternate with transverse vibrations

2. If the distance between a point source of light and a surface is tripled, the intensity of illumination on the surface will be
   (A) tripled
   (B) doubled
   (C) reduced to ½
   (D) reduced to ⅓
   (E) reduced to ⅑

Questions 3–5

X and *Y* are each 2 meters tall. *X* stands 1 meter from a vertical plane mirror; *Y*, 2 meters from the same mirror.

3. The size of *X*'s image as compared with *Y*'s is
   (A) 4 times as great
   (B) 2 times as great
   (C) the same
   (D) ½ as great
   (E) ¼ as great

4. It will seem to *X* that *Y*'s image, as compared with his own, is
   (A) 4 times as great
   (B) twice as great
   (C) the same size
   (D) smaller
   (E) beyond compare

5. The distance between *X*'s image and *Y*'s image
   (A) is 1 m
   (B) is 2 m
   (C) is 3 m
   (D) is 4 m
   (E) can't be calculated because of insufficient data

Questions 6–8

A candle 2 centimeters long is placed upright in front of a concave spherical mirror whose focal length is 10 centimeters. The distance of the candle from the mirror is 30 centimeters.

6. The radius of curvature of the mirror is

(A) 5 cm
(B) 10 cm
(C) 15 cm
(D) 20 cm
(E) 6 cm

7. The image will be

(A) virtual and smaller than the candle
(B) virtual and larger than the candle
(C) virtual and the same size as the candle
(D) real and smaller than the candle
(E) real and larger than the candle

8. If the top half of the mirror is covered with an opaque cloth,

(A) the whole image will disappear
(B) the whole image will be approximately one-half as bright as before
(C) the top half of the image will disappear
(D) the bottom half of the image will disappear
(E) one cannot predict what will happen

---

9. Real images formed by single convex lenses are always

(A) on the same side of the lens as the object
(B) inverted
(C) erect
(D) smaller than the object
(E) larger than the object

10. A virtual image is formed by

(A) a slide projector
(B) a motion-picture projector
(C) a duplicating camera
(D) an ordinary camera
(E) a simple magnifier

11. A person on Earth may see the Sun even when it is somewhat below the horizon primarily because the atmosphere

 (A) annuls light
 (B) reflects light
 (C) absorbs light
 (D) refracts light
 (E) polarizes light

12. Incident rays of light parallel to the principal axis of a convex lens, after refraction by the lens, will

 (A) converge at the principal focus
 (B) converge inside the principal focus
 (C) converge outside the principal focus
 (D) converge at the center of curvature
 (E) diverge as long as they are close to the lens

13. A red cloth will primarily

 (A) reflect red light
 (B) refract red light
 (C) absorb red light
 (D) transmit red light
 (E) annul red light

14. Yellow light of a single wavelength can't be

 (A) reflected
 (B) refracted
 (C) dispersed
 (D) focused
 (E) diffused

15. Some gold is heated to a temperature of 2,000°C. (The melting point of gold is 1,063°C; its boiling point is 2,600°C.) The light emitted by this gold is passed through a triangular glass prism. The result will be

 (A) a continuous spectrum with a golden hue all over
 (B) a line spectrum with a golden hue all over
 (C) a continuous spectrum with a bright gold line
 (D) a line spectrum with a bright gold line
 (E) none of the above

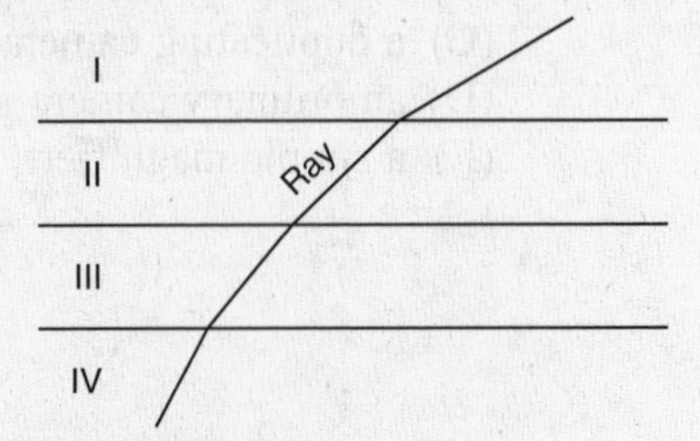

16. The diagram above shows four separate layers of different transparent liquids. The liquids do not mix. The path of an oblique ray of light through the liquids is shown. The medium in which the speed of the wave is lowest is

(A) I
(B) II
(C) III
(D) IV
(E) indeterminate on the basis of the given information

17. The speed of light in a certain transparent substance is two-fifths of its speed in air. The index of refraction of this substance is

(A) 0.4
(B) 1.4
(C) 2.0
(D) 2.5
(E) 5.0

18. A camera of 6-centimeter focal length is used to photograph a distant scene. The distance from the lens to the image is approximately

(A) 6 cm
(B) 1 m
(C) 80 cm
(D) 4 m
(E) 4 cm

19. A film 2.0 centimeters wide is placed 6.0 centimeters from the lens of a projector. As a result a sharp image is produced 300 centimeters from the lens. The width of the image is

(A) 12 cm
(B) 50 cm
(C) 100 cm
(D) 300 cm
(E) 600 cm

20. If an object is placed 30 centimeters from a convex lens whose focal length is 15 centimeters, the size of the image as compared to the size of the object will be approximately

(A) twice as large
(B) more than twice as large
(C) 1.5 times as large
(D) smaller
(E) the same

21. A small beam of monochromatic light shines on a plate of glass with plane parallel surfaces. The index of refraction of the glass is 1.60. The angle of incidence of the light is 30°. The angle of emergence of the light from the glass is

(A) 18°
(B) 30°
(C) 45°
(D) 48°
(E) 60°

22. In an experiment, an object was placed on the principal axis of a convex lens 25 centimeters away from the lens. A real image 4 times the size of the object was obtained. The focal length of the lens was

(A) 20 cm
(B) 25 cm
(C) 33 cm
(D) 50 cm
(E) 100 cm

23. A ray of light goes obliquely from water to glass. The angle of incidence in the water is θ_1, the angle of refraction in the glass in θ_2, the index of refraction of water is n_1, and the index of refraction of the glass is n_2. Which of the following is a correct relationship for this case?

(A) $n_1 \sin \theta_1 = \sin \theta_2$
(B) $n_1 \sin \theta_2 = \sin \theta_1$
(C) $n_2 \sin \theta_1 = \sin \theta_2$
(D) $n_2 \sin \theta_2 = \sin \theta_1$
(E) $n_2 \sin \theta_2 = n_1 \sin \theta_1$

Explanations to Questions Chapter 15

Answers

1. (A)	**7.** (D)	**13.** (A)	**19.** (C)
2. (E)	**8.** (B)	**14.** (C)	**20.** (E)
3. (C)	**9.** (B)	**15.** (E)	**21.** (B)
4. (D)	**10.** (E)	**16.** (D)	**22.** (A)
5. (A)	**11.** (D)	**17.** (D)	**23.** (E)
6. (D)	**12.** (A)	**18.** (A)	

Explanations

1. A Light is an electromagnetic wave. In all electromagnetic waves the electric and magnetic fields vibrate at right angles (transversely) to the direction in which the wave is traveling.

2. E The intensity of illumination from a point source varies inversely as the square of the distance. As the distance from the light source increases, the illumination decreases. When the distance is 3 times as great, the illumination is $(\frac{1}{3})^2$, or $\frac{1}{9}$ as much as before.

3. C The size of the image in a plane mirror is the same as the size of the object. Therefore, *X*'s image is 2 m tall and so is *Y*'s.

4. D Although the actual sizes of the images are the same (see question 4), their apparent sizes will depend on the distances of the images from the observer, just as the apparent sizes of real objects depend on their distances from the observer. The further an object or image is from the observer, the smaller it will *seem.* In this problem, since the image in a plane mirror is as far behind the mirror as the object is in front, *X*'s image is 1 m behind the mirror and 2 m from *X.* For the same reason, *Y*'s image is 2 m behind the mirror and therefore at least 3 m from *X,* who is 1 m in front of the mirror. (Also see question 5.) Therefore, to *X, Y*'s image will seem smaller than his own.

5. A The distance between the two images is the same as the distance between *X* and *Y*. This may be 1 m or anything greater than 1 m, as illustrated in the diagram, depending on how far over to one side *Y* stands.

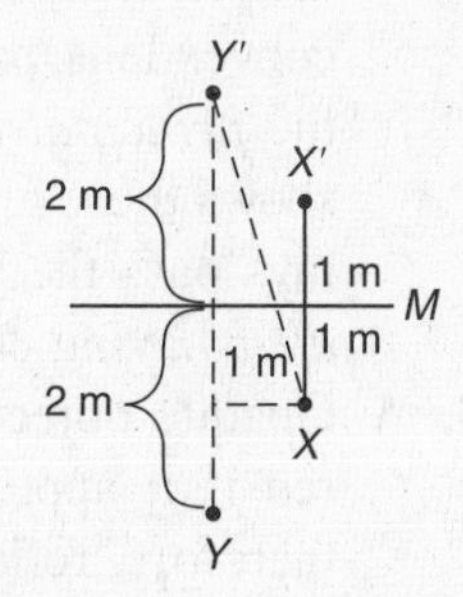

6. D For a spherical mirror, the focal length = ½ of the radius of curvature. $R = 2f = 2 \times 10$ cm = 20 cm.

7. D The situation described in this problem is similar to one for which the ray diagram (Figure 15.13) is drawn and a real image is obtained. Be able to draw such a ray diagram quickly and to deduce pertinent information from it. In this diagram the image is definitely smaller than the object. In Chapter 16 you will see how numerical calculations can be made rather quickly.

8. B All the rays starting from one point on the object will, after reflection from a spherical mirror, converge toward one point (for a real image), or will diverge from one common point (for a virtual image). In this way, points on the object are reproduced as points on the image, producing a sharp image. If part of the mirror is covered, less light will be available for each point reproduction, making the image dimmer.

9. B Note from the ray diagrams in the chapter that real images formed by single convex lenses are always inverted and on the other side of the lens from the object. The real images may be larger or smaller than, or the same size as, the object, depending on the distance of the object from the lens.

10. E Projectors and camera are used to put images on screens or film; images that appear on screens are real images. The simple magnifier (also called magnifying glass or simple microscope) is used to view something without projecting an image on a screen (other than the eye's retina); it produces a virtual image as shown in the ray diagram (Figure 15.14) in the chapter.

11. D Light travels somewhat more slowly in air than in vacuum. Sunlight entering the atmosphere obliquely is refracted toward the normal. Since the atmosphere's density increases as it gets closer to Earth, the path of the light in the atmosphere is slightly curved. The drawing below is not to scale, and the effect is exaggerated. The observer projects a line backward to an imaginary position S', where the Sun will appear to him—somewhat higher in the sky.

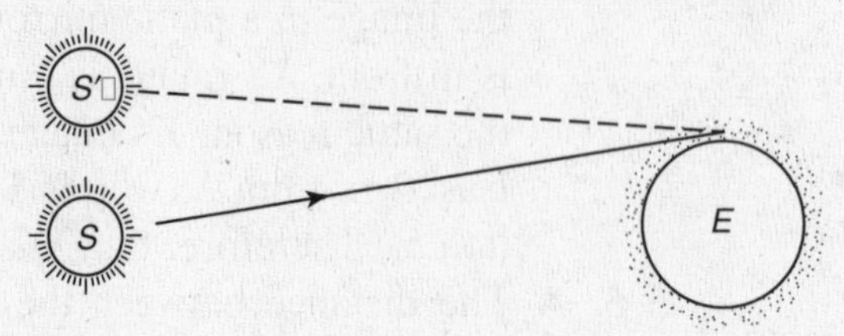

12. A This situation gave us the definition for the principal focus of a convex lens: rays parallel to the principal axis, after refraction by a convex lens, pass through a common point on the principal axis called the *principal focus*. The ray diagram (Figure 15.14) in the chapter shows these refracted rays converging on the principal focus. (If these rays pass the principal focus and are not stopped, they will diverge *after* having converged.) (Figure 15.14)

13. A Opaque objects are seen by the light they reflect. We say a cloth is red if it appears red when viewed in natural white light; it must, therefore, reflect red light, which is a component of the white light.

14. C All light can be reflected, refracted, polarized, and diffused. However, if light contains only one wavelength, it can produce only one color (in this problem, yellow) and can't be dispersed. (Definition of dispersion: breaking up of light into its component colors.)

15. E At the temperature given, gold is an incandescent liquid. The light from an incandescent solid or liquid produces a continuous spectrum without any special bright lines due to the color of the cold solid or liquid. The various colors, such as yellow and orange, gradually change from one to the other.

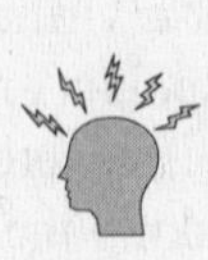

16. D A ray passing obliquely into a medium is bent toward the normal if it travels more slowly in the new medium; it is bent away from the normal if it travels faster in the new medium. In the diagram, if the ray is directed from medium I toward medium IV, it is bent closer to the normal in each successive medium. Therefore the ray travels most slowly in medium IV. If the ray is directed from medium IV toward medium I, we may notice that the ray is bent further away from the normal in each successive medium, again indicating that the wave travels fastest in medium I, slowest in medium IV.

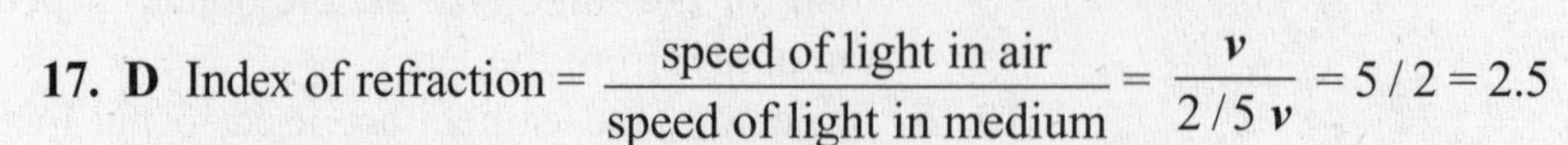

17. D Index of refraction $= \dfrac{\text{speed of light in air}}{\text{speed of light in medium}} = \dfrac{v}{2/5\,v} = 5/2 = 2.5$

18. A You may just recall that, if the object is very far from a convex lens, the image is formed practically 1 focal length away from the lens.

Or you may use the lens equation: $\dfrac{1}{d_o}+\dfrac{1}{d_i}=\dfrac{1}{f}$; the object distance

d_o is very large, so we'll use the infinity symbol (∞) for it.

$$1/\infty + 1\ d_i = 1/f;\ 1/\infty = 0$$

$$1/d_i\ 1/f;\ \text{and } d_i = f$$

In words, the image distance equals the focal length, in this case 6 cm.

19. C $\dfrac{s_1}{s_0} = \dfrac{d_i}{d_o}$

$$s_1 = \frac{d_i}{d_o} \times s_o = \frac{300\text{ cm}}{6\text{ cm}} \times 2\text{ cm} = 100\text{ cm}$$

20. E You should know that, if an object is 2 focal lengths away from a convex lens, a real image is formed 2 focal lengths away from the lens (on the other side from the object) and the image is the same size as the object. Otherwise, you will have to use valuable time on calculations with the lens equation and size relationship.

21. B The angle that the emerging ray makes with the normal is the angle of emergence. We assume here that the entering ray and emerging ray are both in air. Then the two rays are parallel since the glass surfaces are parallel. Therefore, the angle of incidence equals the angle of emergence; both are 30°.

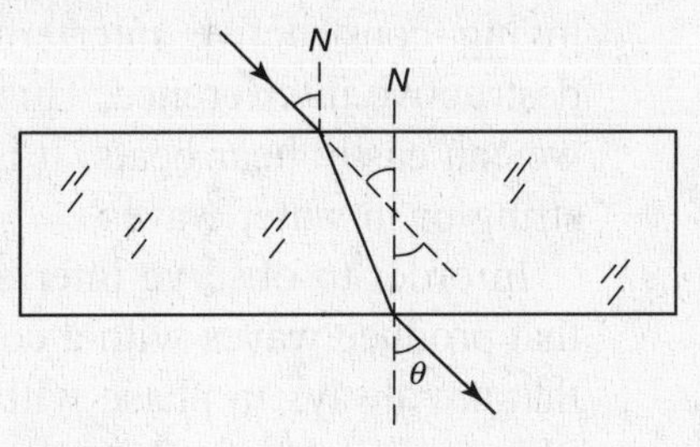

22. A $\dfrac{s_1}{s_0} = 4 = \dfrac{d_i}{d_o}$

$d_i = 4d_o$, and $d_o = 25$ cm

$\therefore d_i = 100$ cm

$1/f = 1/d_o + 1/d_i$

$1/f = 1/25 + 1/100 = 5/100$

$f = 100/5\text{ cm} = 20\text{ cm}$

Also *note* that the object distance is between f and $2f$, since a real, enlarged image was produced. The only choice that fits this is (A).

23. E In refraction, if one of the two media is vacuum, we need only one index of refraction, that of the substance used, which is its absolute index of refraction, or its index of refraction with respect to vacuum. This is also usually adequate if the substance is solid or liquid and we use air instead of vacuum. However, for the general case we have to use the indexes of refraction of both substances, as pointed out in the text.

CHAPTER 16

Physical Optics: Interference and Diffraction

Interference of Light

Coherent Sources

In Chapter 14 on wave motion and sound we noted that, when two waves pass through the same portion of a medium at the same time, interference occurs. The two waves are superposed. They may reinforce each other, giving constructive interference; or they may annul each other, giving destructive interference. This situation is readily observed in sound, where we can easily hear beats. It is also readily observed in standing waves in a string or in water waves.

In order to observe interference we need two *coherent sources,* sources that produce waves with a constant phase relation. For example, two waves that are always in phase will always reinforce each other. If two waves are of opposite phase—that is, one wave is 180° out of step with the other—they will tend to annul each other. We have no trouble maintaining this constant phase relationship with mechanical vibrations. If we strike a tuning fork, both prongs will keep vibrating at the same frequency and in step. If we hold the vibrating fork vertically and rotate it around the stem as an axis, we can observe the interference caused by the waves produced by the two prongs. The two prongs are coherent sources.

Until recently it has been impossible to get two independent light sources that are coherent. The reason is that most sources light emit in short bursts of unpredictable duration. An apt comparison would be a tuning fork that stopped momentarily every few cycles. In the case of light, two sources might be out of phase during one burst but not during the next one. Annulment would be masked by the rapid change in phase relationship. For most purposes, if we want interference of light, we take light from one source and have it travel by two different paths to the same place.

Young's Double Slit—Division of Wavefront

In 1801 Thomas Young performed a famous experiment that suggested the interference of light. Its essential features are similar to the setup illustrated in (Figure 16.1). Light from a small source, *F,* illuminates a barrier that has two narrow slit openings, S_1 and S_2, very close to each other. The light that gets through these two slits falls on a screen. The screen may be thought of as the retina of the human eye, because we usually look at the light source through the two slits. Assume that the light source is monochromatic. What we see on the screen—a whole series of bright lines alternating with dark regions—is surprising. The technique we used in Chapter 15 to draw ray diagrams with lenses and mirrors is inadequate. On that basis we might have expected to see two bright lines, one for each slit that lets light through.

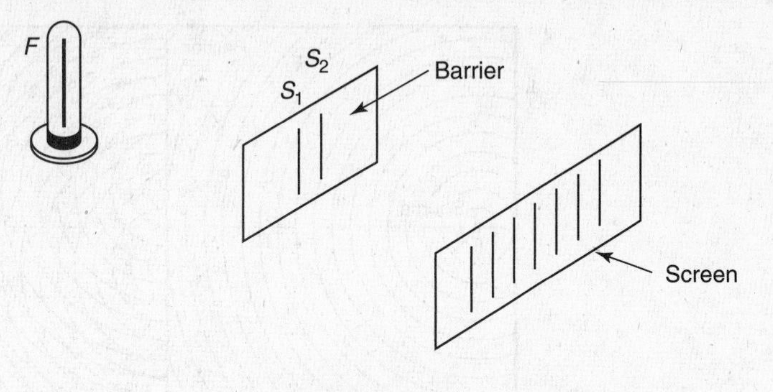

FIGURE 16.1

We apply *Huygens' principle* and the wave theory of light to explain what we actually see (Figure 16.2). Imagine that we are looking down at the setup in Figure 16.1. The light from the source spreads out, and the wave fronts are represented by circles with source *F* at the center. The two slits are equidistant from the source. Each wavefront reaches S_1 and S_2 at the same time. According to Huygens' principle, we can think of each point on a wavefront as a source of a *wavelet.* The *envelope* or curve tangent to all the wavelets is the new wavefront (see Figure 16.2).

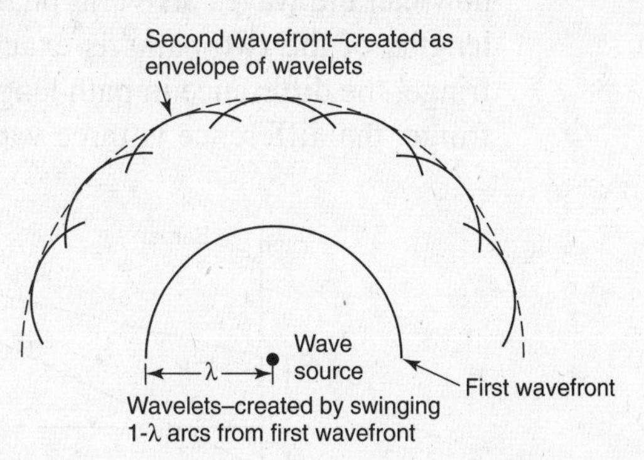

FIGURE 16.2 Huygens' Principle

Since there are two "source points" at the slits, two sets of waves are generated. These are represented as two sets of circles with S_2 being the center of the top set and S_1 the center of the bottom set. Since the light originally came from a single source, *F,* the waves produced by S_1 and S_2 are coherent. (See Figure 16.3.)

The waves overlap on the screen and interfere with each other. In some places the waves arrive in phase, reinforce each other, producing antinodes, and give a bright region or line. One such place is *B,* which is equidistant from S_1 and S_2; at this place we have the central maximum, or the central bright line. There are other places, on either side of the central maximum, where we get bright lines, places where the two waves are in phase and reinforce each other. The waves are in phase, but one is a whole number of wavelengths behind the other. In between the bright lines we have dark regions. The two waves cancel each other, producing nodes, where they are out of step by one-half of a wavelength, one and one-half wavelengths, or any other odd number of half-wavelengths.

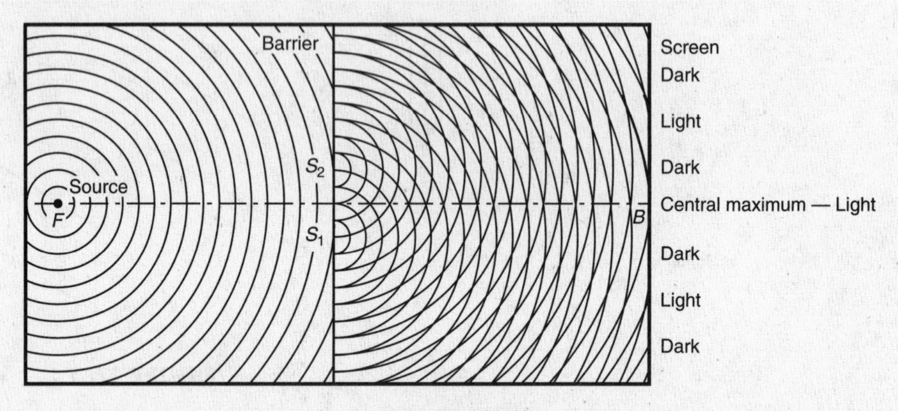

FIGURE 16.3

Let us see if we can calculate the distance between bright lines or fringes. We will look at the right half of Figure 16.3 in a slightly different way. We choose the first bright line on one side of the central maximum, and represent it as point *P.* It is bright because the wave from S_2, traveling in direction S_2P, arrives in phase with the wave from S_1, which travels in direction S_1P. (See Figure 16.4.) The length of path S_1P is greater than S_2P; how can the waves arrive in phase? This is possible if the difference in the lengths of the two paths is exactly one wavelength. For the second bright fringe, the difference in path lengths is two wavelengths; for the third bright fringe, the difference is three wavelengths; and so on.

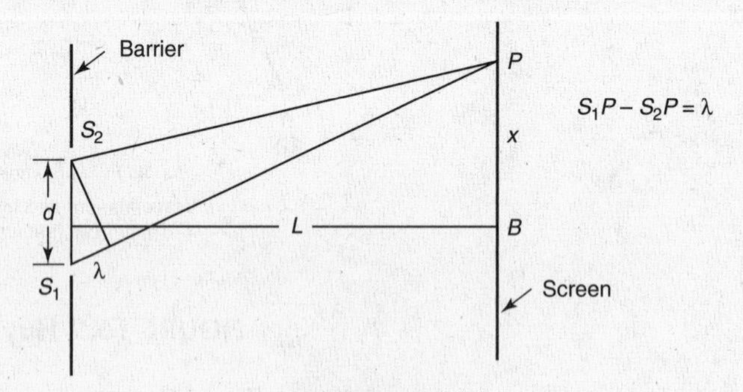

FIGURE 16.4

If *d* is the distance between the two slits, *L* the distance between the barrier and the screen, λ the wavelength, and *x* the distance between the central maximum and the first bright fringe, it can be shown, by selecting two triangles and making some approximations, that

$$\boxed{\frac{\lambda}{d} = \frac{x}{L}}$$

It can also be shown that the distance between any two adjacent bright fringes has the same value of x.

Complete annulment occurs midway between the bright fringes; the distance between adjacent nodes is also x.

Let us look at the important relationship shown above to see what predictions we can make. Since

$$\frac{\lambda}{d} = \frac{x}{L}$$

We can make the following statements:

1. If the wavelength of light (λ) is increased, the distance between fringes increases; if we use red light instead of blue light, the distance between successive red lines on the screen is greater than the distance between successive blue lines.
2. If the distance between the two slits (d) is decreased, the distance between the bright fringes increases.
3. If white light is used, the central maximum (at B) is white, since all the wavelengths are reinforced there. On either side of the central maximum are bright fringes consisting of all the colors of the spectrum, with violet closest to the central maximum.
4. If the distance between the screen and the slits (L) is increased, the distance between the bright fringes increases.

Example: Calculate the distance between adjacent bright fringes if the light used has a wavelength of 6,000 angstroms and the distance between the two slits is 1.5×10^{-3} meter. The distance between the screen and the slits is 1.0 meter.

$$\frac{\lambda}{d} = \frac{x}{L}$$

$$\frac{6{,}000 \times 10^{-10}\text{ m}}{1.5 \times 10^{-3}\text{ m}} = \frac{x}{1.0\text{ m}}$$

$$x = 4{,}000 \times 10^{-7}\text{ m, or } 0.00040\text{ m}$$

In the diagram on wave interference (Figure 16.1), the pattern to the right of the barrier is very similar to the pattern obtained on the surface of water by two small sources vibrating in phase with the same frequency at S_1 and S_2, as in a ripple tank. A succession of nodes (and antinodes) is produced in the whole region as the waves move to the right and interfere with each other. This succession produces *nodal* (and *antinodal*) lines.

Diffraction Grating

If instead of having only two parallel slits, as above, we have many such slits close together, we have a diffraction grating. It is possible now to have cheap gratings with thousands of slits per inch. The full analysis of how a diffraction grating works is complicated but in some respects the process is similar to that with the double slit. However, the more slits, the brighter is the pattern. For approximate calculations, the above formula for the double slit may be used. It provides a convenient method for measuring the wavelength of light. Instead of the triangular glass prism, the grating may be used in the spectroscope.

As with the double slit, the pattern obtained on the screen with a grating has a central maximum that is the same color as the light used. On either side of the central maximum are bright fringes; if white light is used, the bright fringes are continuous spectra. The spectrum closest to the central maximum is known as the *first-order spectrum*. Notice there is a first-order spectrum on either side of the central maximum. If the source of light is a luminous gas, each "bright fringe" is the characteristic spectrum of the element.

Diffraction

Wave Characteristic

When we deal with a wave, we expect to find *diffraction:* the bending of a wave around obstacles. We know that diffraction occurs in the case of sound because, for example, we can hear a speaker even if our direct view of her is blocked by a crowd of people. In the case of a water wave we can see the wave going around objects on the surface of the water, so that after a short distance the wave looks as though there had never been anying in its way.

If light is a wave, why do objects cast sharp shadows? Why doesn't light bend around an object and illuminate the region behind it? Theory shows that the smaller the wavelength, the smaller are the diffraction effects. We already know that the wavelength of light is only a small fraction of a millimeter, while the wavelengths of radio and sound waves are many centimeters (even meters for radio waves). It is due to the very short wavelength of light that shadows are so sharp and that a beam of light tends to travel along straight lines.

However, diffraction of light does occur; because its wavelength is so small, we must look very carefully. For example, if we photograph the shadow of a small ball, using a suitable light source, we find a small illuminated spot in the center of the shadow.

Single-Slit Diffraction

Suppose we shine a beam of monochromatic light on an opaque barrier that has a narrow rectangular opening cut into it, a single narrow slit. Let the light that goes through the slit fall on a screen held parallel to the barrier. On the basis of our experience with ray diagrams, we may expect that the part of the beam that gets through the slit will have the same cross section as the slit and that the illuminated part of the screen will look exactly like the slit in shape and size. (See Figure 16.5.)

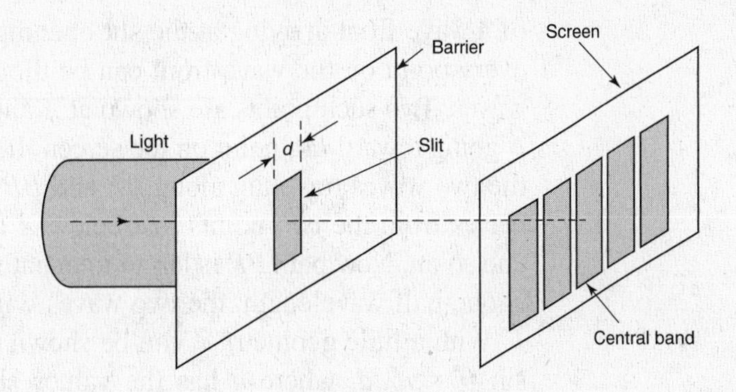

FIGURE 16.5

If the slit is narrow and we look carefully, what we actually see on the screen is many rectangular, parallel bands of light *(images of the slit)*. Each band may be a little wider than the slit. The central band is very bright; to either side are bands of decreasing brightness. (The central band's intensity is about 20 times as great as the first bright band to either side.) Figure 16.6 shows is a rough graph of the intensity of the observed pattern plotted against distance along the screen.

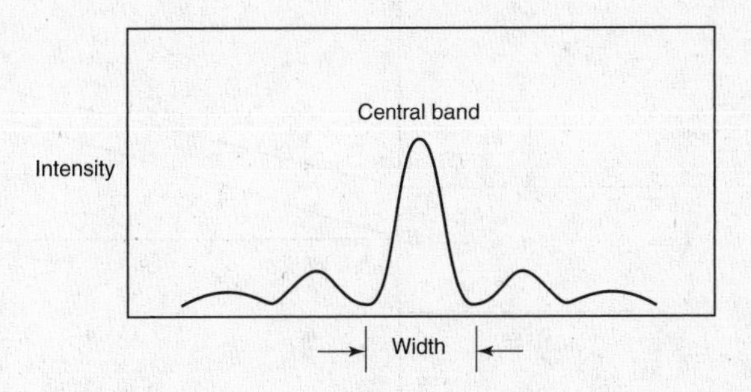

FIGURE 16.6

In the diffraction pattern, the central band is also called the central bright line or the central maximum. Careful analysis shows the following:

1. The width of the central maximum varies inversely as the width of the slit; that is, if we make the slit narrower, the central bright line becomes broader; in other words, there is more and more diffraction, more and more bending of the light around the edges of the slit.
2. The width of the central maximum varies directly as the wavelength; if we use red light, we get a wider or broader central line than if we use blue light.
3. The distances between successive dark bands are approximately the same.

If the opening is large compared with the wavelength, light practically travels along straight lines; diffraction is negligible.

The Wave Theory and Single-Slit Diffraction

As in the case of the double-slit interference pattern, we can apply Huygens' principle and the wave theory of light to explain the single-slit diffraction pattern. Let us view the barrier and screen edge-on (Figure 16.7). Assume that the slit has a width d and that the barrier is a distance L from the screen. Think of a wave front arriving at the slit opening. According to Huygens' principle, every point on the wave front can be thought of as sending out its own little waves. Two such points are shown at *A* and *B*. Think of the waves from *A* and *B* going toward *P*, a point on the screen. If *P* is to be a dark spot on the screen, the two waves traveling along *AP* and *BP* must annul each other, as must the waves from the two points just below *A* and *B* on the wave front at the slit, and so on. Now path *BP* is larger than path *AP*. If this difference in path length is one-half wavelength, the two waves will annual each other.

With a little geometry it can be shown that, for a dark point on the screen, $\sin\theta = n\lambda/d$, where *n* has the values shown in Figure 16.7, and λ is the wavelength of the light used. When $n = 1$, *P* gives us the point of minimum intensity for the central line. In other words, the distance *OP* gives half of the width of the central band. By looking at the relation $\sin\theta = \lambda/d$, we can see that, if the wavelength λ increases, $\sin\theta$, and therefore the width of the central band, must increase. Also, if the width of the slit *d* decreases, $\sin\theta$, and therefore the width of the central band, must increase.

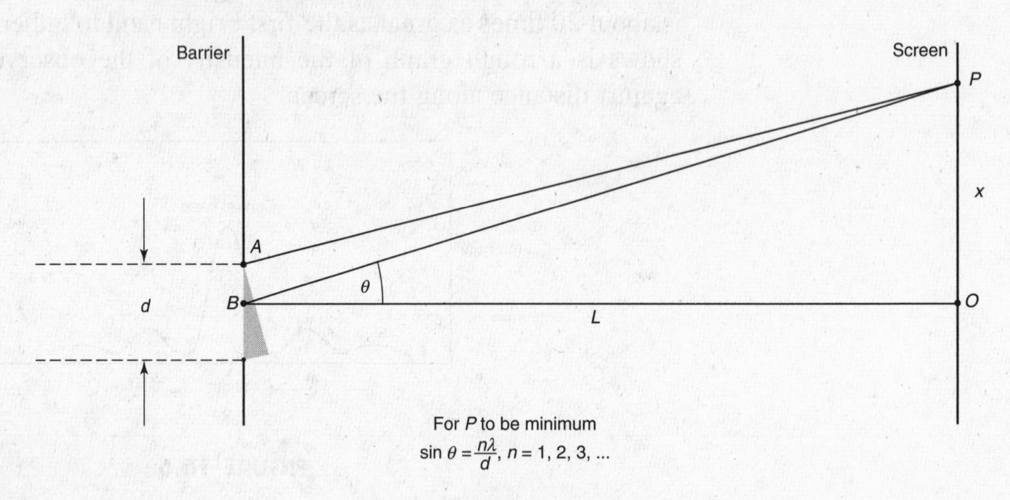

FIGURE 16.7

Diffraction and Lens Defects

We have already indicated that even a perfectly made spherical lens suffers from chromatic aberration. Spherical lenses have other defects even if monochromatic light is used. One of these is due to diffraction. As with a single slit, the light going through a lens is diffracted. The amount of diffraction is usually negligible, but one effect is that parallel light entering the lens is brought, not to a point focus, but to a disk focus. Around this bright diffraction disk we may be able to see bright diffraction rings. The fact that with a lens we get a disk diffraction pattern instead of a point focus sets a practical limit on the magnification obtainable with a lens.

Resolving Power

The ability of a lens or optical instrument to distinguish between two point sources is its *resolving power*. Resolution or resolving power is increased by increasing the diameter of the lens; the process is similar for spherical mirrors. Suppose, for example, we look through a telescope at two stars that are close to each other in the sky. If the objective lens has a small diameter, the diffraction pattern of each star will be large and the two disks will overlap as in Figure 16.8*a*. The resolution is poor: it is hard to tell whether we are looking at one star or two. If we try to use greater magnification, we will not improve resolution, because we will also magnify the size of the disk. If, instead, we use an objective lens with a larger diameter, we will decrease the amount of diffraction, decrease the size of the diffraction disk, and increase the resolving power. The effect of using a lens of larger diameter is shown in Figure 16.8*b*.

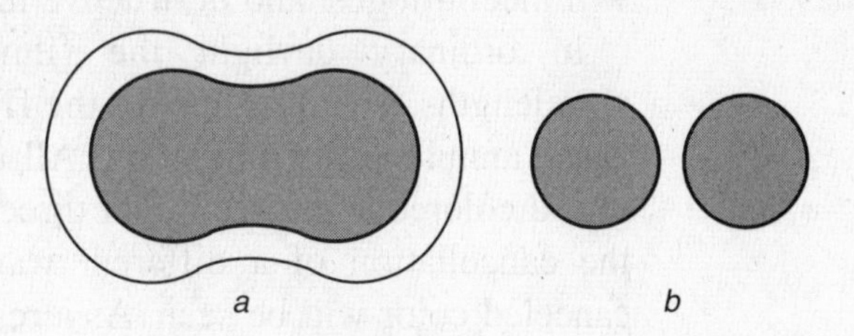

FIGURE 16.8

Interference with Thin Films–Division of Amplitude

The beautiful colors of soap bubbles and of thin films of oil on a city pavement or highway are due to interference of light and give further evidence that light behaves like a wave. Let us assume that we have monochromatic light incident on a thin film (Figure 16.9). The film may be oil, soap solution, or even air trapped between two flat glass plates. Some of the light is reflected at the top surface of the film and is represented by ray 1. Most of the light goes through the film; some of this is reflected at the bottom surface of the film and comes out again. This is represented by ray 2. If we look down on the film, rays 1 and 2 may enter our eye and be focused on our retina. Coherent monochromatic light is reaching our eye by two different paths; ray 2 has had to go back and forth through the film. The two rays (or light waves) arriving simultaneously on our retina will interfere with each other. (Refraction is not shown in the diagram because it is negligible in a thin film.)

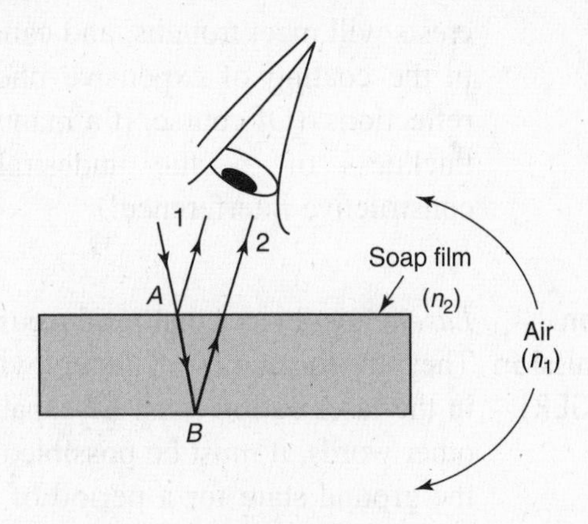

FIGURE 16.9

When a wave encounters a denser medium, its reflection at that point will undergo a phase change of 180°. Thus the ray reflecting at point *A* is exactly out of phase with the ray that continues through the soap film. When the ray gets to point *B* it encounters a less dense medium, so no phase change occurs. The waves continue on out of phase with one another. However, if the thickness of the soap film is ¼ λ and the incident ray is almost perpendicular to the surface, the total path through the soap will be ½. The crest of a wave from *B* will meet the *next* crest of a wave from *A,* and there will be constructive interference. If the film is thicker by any odd number of quarter wavelengths, crests will still meet crests, and constructive interference will occur. If the film is thicker by any even number of quarter wavelengths, crests will meet troughs, and destructive interference will occur.

In ordinary daylight the film is illuminated by many different wavelengths. The thickness of the film may be just right in one direction to cause annulment of one color. All the other wavelengths will combine to give a colored spot. In another direction the thickness may be just right for the cancellation of a different wavelength, and the complement of the canceled color will be seen. As a result, when viewing a soap bubble or an oil film, we see all of the colors of the rainbow.

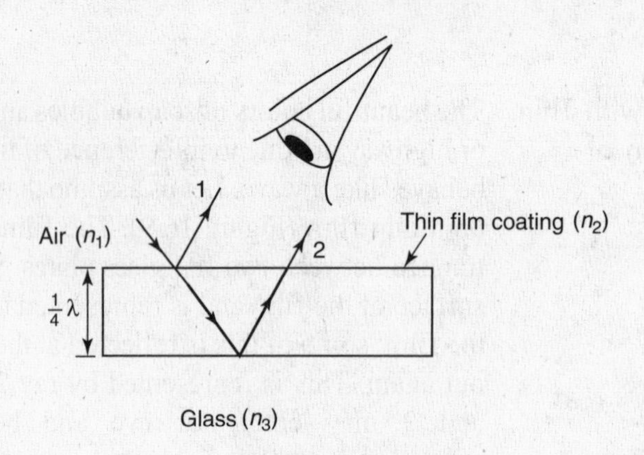

FIGURE 16.10

In Figure 16.10, ray 1 undergoes a phase change as before, but so does ray 2, since $n_3 > n_2$. Thus, if the film thickness is any odd multiple of ¼, crests will meet troughs, and cancellation will occur. This concept is applied in the coating of expensive photographic lenses to minimize undesirable reflections. (Of course, if a manufacturer carelessly applies a coating with a thickness of ½, the undesirable reflections will be exaggerated by constructive interference!)

Light Amplification by Stimulated Emission of Radiation (LASER)

Lasers are devices that produce monochromatic and coherent beams of light. They are formed by cylinders with parallel ends. The atoms that participate in the laser action must be capable of sustaining a *population inversion.* In other words, it must be possible to *pump* more than 50% of the atoms above the ground state for a period of time that is long on an atomic time scale. This is usually done by applying electrical voltage.

When one of the atoms radiates light parallel to the axis of the cylinder, that light will stimulate a nearby atom to emit light, that is coherent with the first wave and is in the same direction. This process continues, with the strength of the beam growing, until an end is reached. At that point, about 99% of the beam will be reflected, but a partially silvered mirror at the end will allow about 1% of the beam to leak through. The reflected 99% continues the reaction, while the output 1% can be used in the laboratory. Because only light parallel to the axis can continue lasing, the output beam is highly *collimated,* that is, it diverges very little from its exit diameter. Because the beam is highly collimated, it has a very high energy density. (See Figure 16.11.)

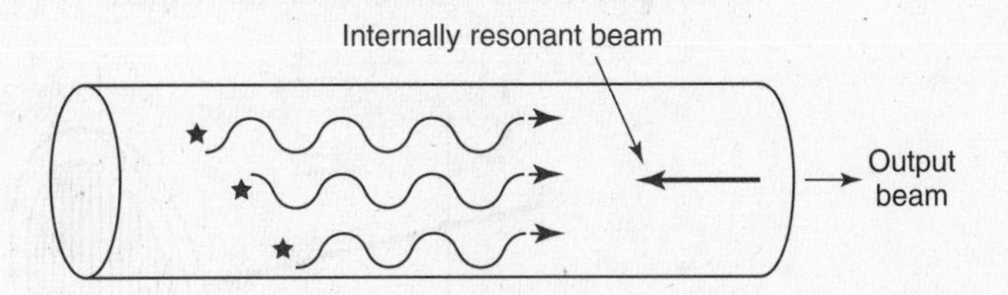

FIGURE 16.11 Light Amplification by Stimulated Emission of Radiation (Laser)

It is now possible to get two lasers to be coherent.

Interference and diffraction experiments are much easier now than they were in 1801!

Polarization

We have been saying that light behaves like a transverse wave. What is the evidence? We have already shown that light behaves like some kind of wave. We explained interference and diffraction of light by making use of wave theory. We can readily see how two waves can get to the same place at the same time and annul each other. We cannot explain this annulment if we think of energy as arriving in the form of streams of particles.

Nothing we have learned so far tells us definitively whether the light wave is longitudinal or transverse. We observed interference and diffraction with sound and decided that the sound wave is longitudinal.

An effect observed with light but not with sound is polarization. *Polarized light* is light whose direction of vibration has been restricted in some way. The direction of a longitudinal wave cannot be restricted; either it vibrates in the direction in which the energy is traveling (the direction of propagation), or it doesn't vibrate at all and there is no wave. On the other hand, in a transverse wave there is an infinite number of ways in which we can have vibrations at right angles to the direction of propagation of the wave. (See Figure 16.12.)

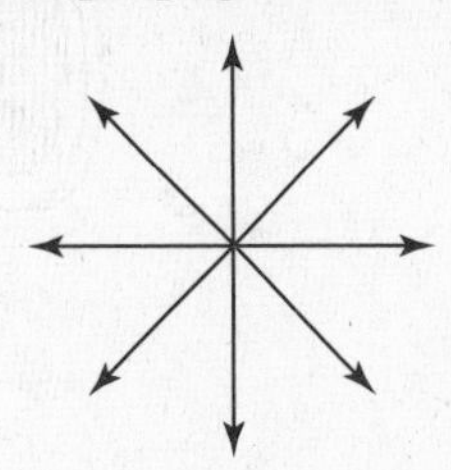

FIGURE 16.12

In Figure 16.13, imagine a ray coming out of the paper. Vibration can be toward the top and bottom edges of the paper, toward the right and left, and in all other directions in between. We think of ordinary natural light, such as we get from tungsten filament bulbs, as a transverse wave with all directions of vibrations present. Some crystals, natural and synthetic, allow only that light to go through whose vibration is in the direction of the axis of the crystal. Such light is then polarized.

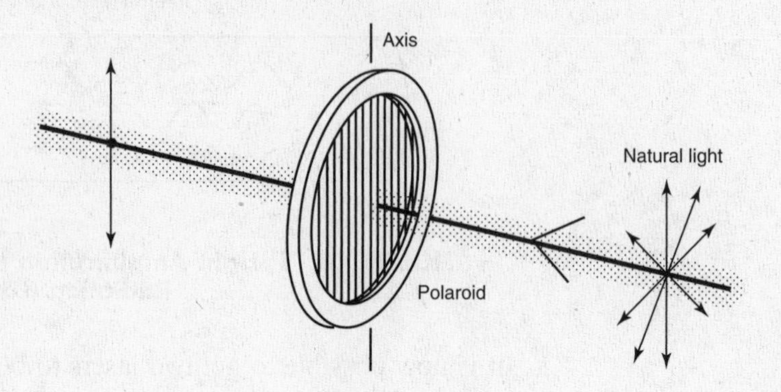

FIGURE 16.13

Polaroid is a synthetic material that polarizes light. It consists of plastic sheets with special crystals embedded in them. If we let ordinary light shine through one Polaroid disk, the light is polarized but the human eye can't tell the difference. If we allow this polarized light to shine through a second Polaroid disk kept parallel to the first one, the amount of light transmitted by the second disk depends on how we rotate the second disk with respect to the first one. If we rotate the second disk so that its axis is at right angles to the axis of the first disk, practically no light is transmitted. The maximum amount of light is transmitted when the two axes are parallel to each other. (See Figure 16.14.)

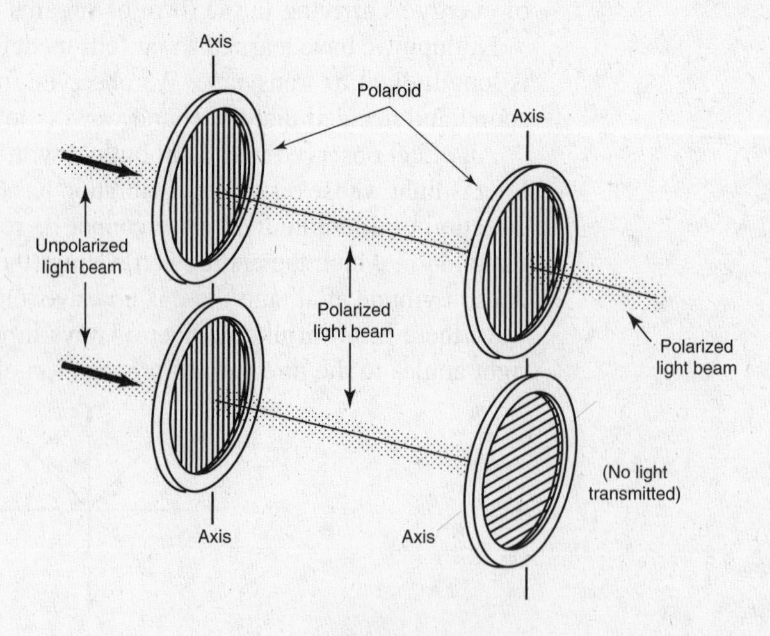

FIGURE 16.14

In many ways light behaves like a wave. We have seen that it is possible to have two light beams interfere with each other so that the combination is dimmer than either beam alone. In fact, polarization experiments suggest that light is a transverse wave. Since light comes to us from the Sun through an excellent vacuum, we would like to know what vibration represents the wave. Until about 1900 it was customary to talk about some ether filling all of space and to describe light as a wave in this ether. Now when we think of light as a wave, we describe it as an *electromagnetic wave;* we think of electric fields and magnetic fields that vibrate (fluctuate in magnitude) at right angles to the direction in which the light travels. These fields can exist in vacuum. Light is one type of electromagnetic wave; its wavelength is rather short: about 5×10^{-5} centimeter. The exact wavelength depends on the color of the light. X rays are also electromagnetic waves—their wavelength is about 10^{-8} centimeter or 10^{-10} meter. (One angstrom unit = 10^{-10} meter.) Radio waves are electromagnetic waves with rather long wavelengths; they may be many meters long.

According to *Maxwell's theory,* light is a transverse wave in which electric and magnetic fields vibrate or fluctuate at right angles to the direction of propagation. Electric and magnetic fields are discussed further in Chapters 17 and 18.

Questions Chapter 16

In each case, select the choice that best answers the question or completes the statement.

1. Which will be produced when blue light with a wavelength of 4.7×10 meters passes through a double slit?

(A) a continuous spectrum
(B) two narrow bands of blue light
(C) alternate blue and black bands
(D) bands of blue light fringed with green
(E) bands of blue light fringed with violet

2. Two pulses approach each other in a spring, as shown.

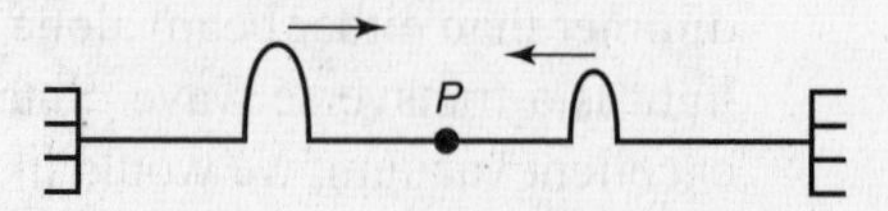

Which of the following diagrams best represents the appearance of the spring shortly after the pulses pass each other at point P?

(A) 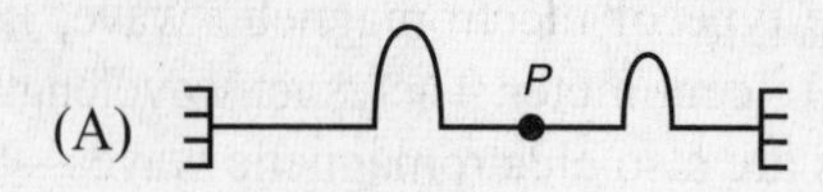

(B) 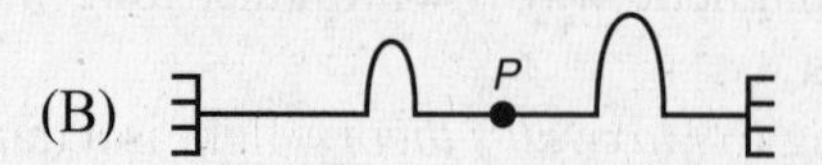

(C) 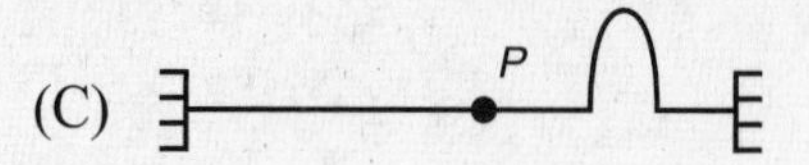

(D) 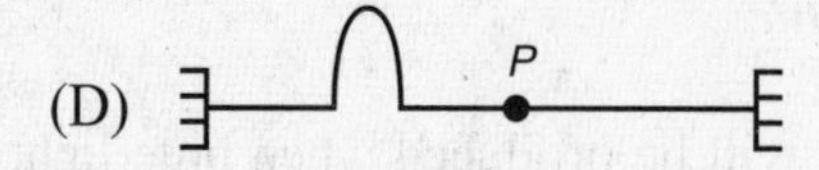

(E) 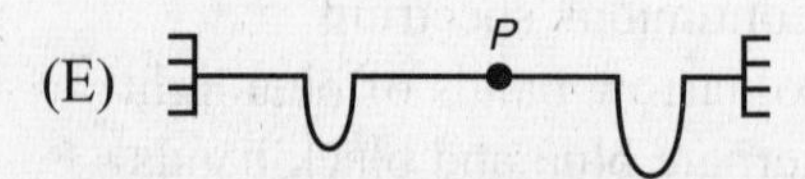

3. The diagram represents straight wave fronts passing through a small opening in a barrier. This is an example of

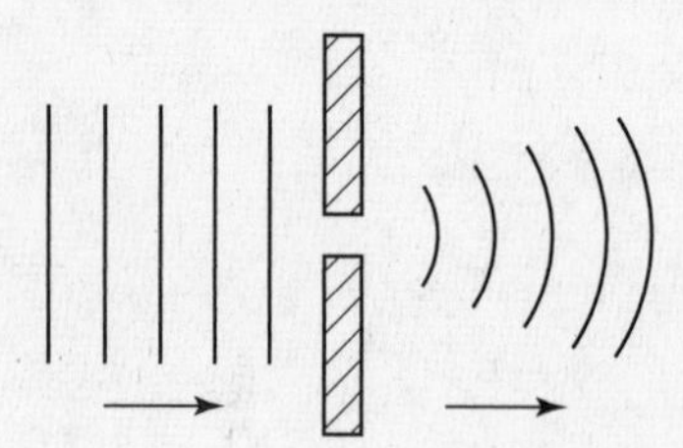

(A) reflection
(B) refraction
(C) polarization
(D) dispersion
(E) diffraction

4. If light of wavelength 5.0×10^{-7} meter shines on a pair of slits 1 millimeter apart, how far from the central maximum will the first bright band appear on a screen 1 meter away?

(A) 5×10^{-4} m
(B) 5×10^{-10} m
(C) 2×10^{3} m
(D) 2×10^{-4} m
(E) 2×10^{-10} m

5. Newton's rings may be observed if you place a curved glass surface on a flat glass surface. A thin film of air is wedged between the glass surfaces. Looking down, you can see several rings around the point where the two glass surfaces touch. Newton's rings are produced by

(A) interference
(B) diffraction
(C) polarization
(D) absorption
(E) shadows

6. Water waves

(A) can be polarized
(B) are polarized
(C) cannot be polarized because they are longitudinal
(D) cannot be polarized because of their frequencies
(E) cannot be polarized because of their amplitudes

Explanations to Questions Chapter 16

Answers

1. (C)	**3.** (E)	**5.** (A)
2. (B)	**4.** (A)	**6.** (B)

Explanations

1. C The light passes through a double slit; the pattern that forms on the screen where the light falls consists of alternating bands of light separated by black regions. If white light is used, the central band is white and on either side each of the bright bands consists of the spectrum of colors of which the white light is composed. In this question blue light of a single wavelength is used. Such light cannot be dispersed. Therefore all bands of light on the screen, including the central band, consist of this blue light. These bands are separated by bands of black regions.

2. B When two pulses or waves pass each other in a medium, each continues on as though there had been no interference from the other. In this question, therefore, the pulse with the larger amplitude will appear on the right side of P and continue traveling toward the right.

The pulse with the smaller amplitude appears on the left and continues traveling toward the left.

3. E *Diffraction* is the bending of a wave around an obstacle. The shaded parts of the diagram represent the obstacle. As the wave goes through the small opening in the obstacle, it spreads behind the obstacle. The small opening acts like a point source of a wave.

4. A

$$\frac{y}{d} = \frac{x}{L}$$

$$= \frac{x}{1\ \text{m}}$$

$$= \frac{5.0\times10^{-7}\ \text{m}}{10^{-3}\ \text{m}}$$

$$x = 5.0\times10^{-4}\ \text{m}$$

Don't forget to change all units to SI!

5. A The production of Newton's rings is an interference phenomenon. The rings are produced by interference between the light reflected from the top surface of the thin air film and the light reflected front the bottom.

6. B While water waves are transverse, and thus can be polarized (A), (B) is the better choice, as water waves are already restricted to one plane of vibration (up and down).

CHAPTER 17

Electricity

The Nature of Electric Charges

The ancient Greeks used to rub pieces of amber on wool or fur. The amber was then able to pick up small objects that were not made of metal. The Greek word for amber was *elektron* (ελεκτρον)—hence the term *electric*. The pieces of amber would retain their attractive property for some time, so the effect appears to have been *static*. The amber acted differently from magnetic ores (lodestones), naturally occurring rocks that attract only metallic objects.

In modern times, hard rubber, such as ebonite, is used with cloth or fur to dramatically demonstrate the properties of electrostatic force. If you rub an ebonite rod with cloth and then bring it near a small cork sphere painted silver (called a "pith ball") and suspended on a thin thread, the pith ball will be attracted to the rod. When the ball and the rod touch each other, the pith ball will be repelled. If a glass rod that has been rubbed with silk is then brought near the pith ball, the ball will be attracted to the rod. If you touch the pith ball with your hand, the pith ball will return to its normal state. This concept is illustrated in Figure 17.1.

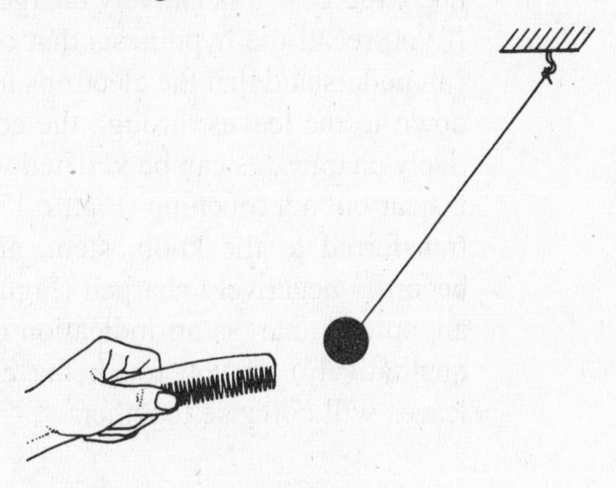

FIGURE 17.1

In the nineteenth century, chemical experiments to explain these effects showed the presence of molecules called *ions* in solution. These ions possessed similar affinities for certain objects, such as carbon or metals, placed in the solution. These objects are called *electrodes*. The experiments confirmed the existence of two types of ions, *positive* and *negative*. The effects they produce are similar to the two types of effects produced when ebonite and glass are rubbed. Even though both substances attract small objects, these objects become *charged* oppositely when rubbed, as indicated by the behavior of the pith ball. Further, chemical experiments coupled with an atomic theory demonstrated that in solids it is the negative charges that are transferred. Additional experiments by Michael Faraday in England during the first half of the nineteenth century suggested the existence of a single, fundamental carrier of *electric charge*, which was later named the *electron*. The corresponding carrier of positive charge was termed the *proton*.

The Detection and Measurement of Electric Charges

When ebonite is rubbed with cloth, only the part of the rod in contact with the cloth becomes charged. The charge remains localized for some time (hence the name *static*). For this reason, among others, rubber, along with plastic and glass, is called an *insulator*. A metal rod held in your hand cannot be charged statically for two reasons. First, metals are *conductors*; that is, they allow electric charges to flow through them. Second, your body is a conductor, and any charges placed in the metal rod are conducted out through you (and into the earth). This effect is called *grounding*. The silver-coated pith balls mentioned in the preceding section can become statically charged because they are suspended by thread, which is an insulator. They can be used to detect the presence and sign of an electric charge, but they are not very helpful in obtaining a qualitative measurement of the magnitude of charge they possess.

An instrument that is often used for qualitative measurement is the *electroscope*. One form of electroscope consists of two "leaves" made of gold foil (Figure 17.2a). The leaves are vertical when the electroscope is uncharged. As a negatively charged rod is brought near, the leaves diverge. If you recall the hypothesis that only negative charges move in solids, you can understand that the electrons in the knob of the electroscope are repelled down to the leaves through the conducting stem. The knob becomes positively charged, as can be verified with a charged pith ball, as long as the rod is near but not touching (Figure 17.2b). Upon contact, electrons are directly transferred to the knob, stem, and leaves. The whole electroscope then becomes negatively charged (Figure 17.2c). The extent to which the leaves are spread apart is an indication of how much charge is present (but only qualitatively). If you touch the electroscope, you will ground it and the leaves will collapse together.

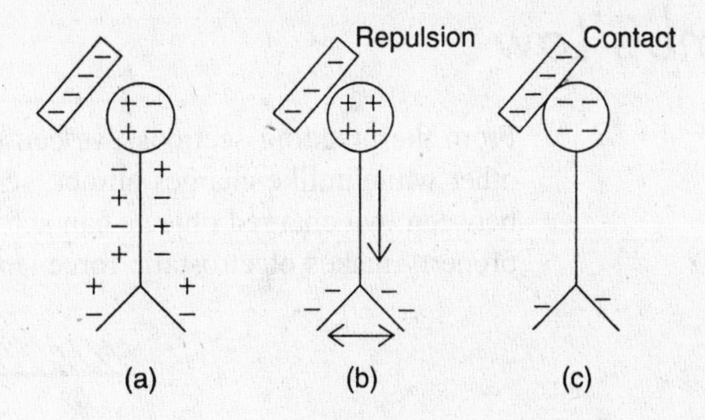

FIGURE 17.2

The electroscope can also be charged by *induction*. If you touch the electroscope shown in Figure 17.2b with your finger when the electroscope is brought near (see Figure 17.3a), the repelled electrons will be forced out into your body. If you remove your finger, keeping the rod near, the electroscope will be left with an overall positive charge by induction (Figure 17.3b).

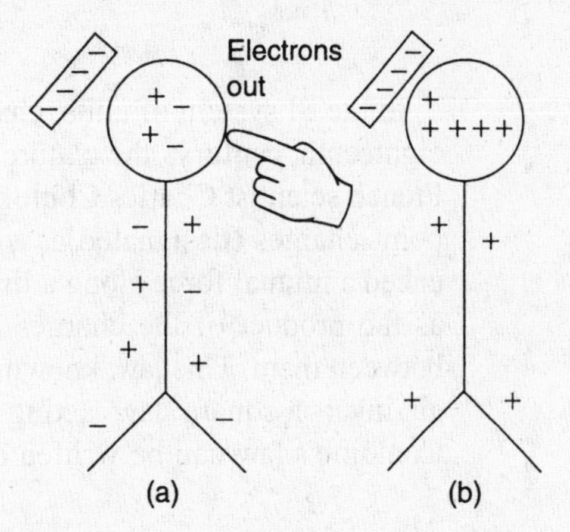

FIGURE 17.3

Finally, we can state that electric charges, in any distribution, obey a conservation law. When we transfer charge, we always maintain a balanced accounting. Suppose we have two charged metal spheres. Sphere *A* has +5 elementary charges, and sphere *B* has a +1 elementary charge (thus both are positively charged). The two spheres are brought into contact. Which way will charges flow? Negative charges always flow from a higher concentration to a lower one, and sphere *B* has an overall greater negative tendency (since it is less positive). Upon contact, it will give up negative charges until equilibrium is achieved. Notice that it is only the excess charges that flow out of the countless numbers of total charges in the sphere! A little arithmetic shows that sphere *B* will give up two negative charges, leaving it with an overall charge of +3. Sphere *A* will gain the two negative charges, reducing its charge to +3. If the two spheres are separated, each will have +3 elementary charges.

Coulomb's Law

From the precding sections, we can conclude that like charges repel each other while unlike charges attract (see Figure 17.4). The electrostatic force between two charged objects can act through space and even vacuum. This property makes electrostatic force similar to the force of gravity.

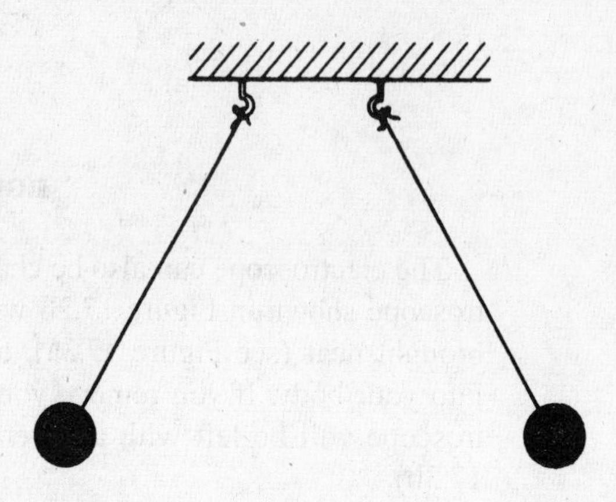

FIGURE 17.4

In the SI system of units, charge is measured in *coulombs* (C). In the late eighteenth century, the nature of the electrostatic force was studied by French scientist Charles Coulomb. He discovered that the force between two point charges (designated as q_1 and q_2), separated by a distance r, experienced a mutual force along a line connecting the charges that varied directly as the product of the charges and inversely as the square of the distance between them. This law, known as *Coulomb's law*, is, like the law of gravity, an inverse square law acting on matter at a distance. Mathematically, Coulomb's law can be written as

$$F = \frac{kq_1q_2}{r^2}$$

The constant k has the value $8.9875 \times 10^9 \approx 9 \times 10^9\ \text{N} \cdot \text{m}^2/\text{C}^2$.

An alternative form of Coulomb's law is

$$F = \left(\frac{1}{4\pi\varepsilon_0}\right)\frac{q_1q_2}{r^2}$$

The new constant, ϵ_0, is called the *permittivity of free space* or *permittivity of the vacuum* and has a value of $8.8542 \times 10^{-12}\ \text{C}^2/\text{N} \cdot \text{m}^2$. In the SI system, one elementary charge is designated as e and has a magnitude of 1.6×10^{-19} coulomb.

Coulomb's law, like Newton's law of universal gravitation, is a vector equation. The direction of the force is along a radial vector connecting the two point sources. It should be noted that Coulomb's law applies only to point sources (or to sources that can be treated as point sources, such as charged spheres). If we have a distribution of point charges, the net force on one charge is the vector sum of all the other electrostatic forces. This aspect of force addition is sometimes termed *superposition*.

Electric Fields

We noticed above that two charged objects exert a force on each other without touching. Physicists have found it desirable to introduce the concept of an *electric field* in order to visualize better and describe more precisely this force acting at a distance. An electric field is said to exist wherever an electric force acts on an electric charge. Note the similarities of the following discussion to the material on gravitational fields.

Electric Field Intensity

The strength of an electric field, or the *electric field intensity,* at a certain point is the force per unit positive charge placed there. It is a vector quantity. Its direction is the direction of the force on a small *positive* test charge; if this is to the right, the direction of the field intensity at that point is to the right. If we use a negative test charge, the direction of the force on it is opposite to the direction of the field intensity.

Here is the equation for electric field intensity:

$$\boxed{\mathbf{E} = \frac{\mathbf{F}}{q}}$$

where $\boldsymbol{E}$ is the electric field intensity (N/C) and $\boldsymbol{F}$ is the force (N) exerted on positive charge q (C).

We can represent the electric field intensity around a charge by drawing vectors of suitable direction and length. Frequently we represent the field by merely drawing lines whose directions are the same as the directions of the field intensity at various points along the lines, but whose lengths are independent of the magnitude. These lines are called *lines of force* or *lines of flux.* The stronger the field, the more lines are drawn in that area.

A few patterns representing electric fields are shown in Figure 17.5.

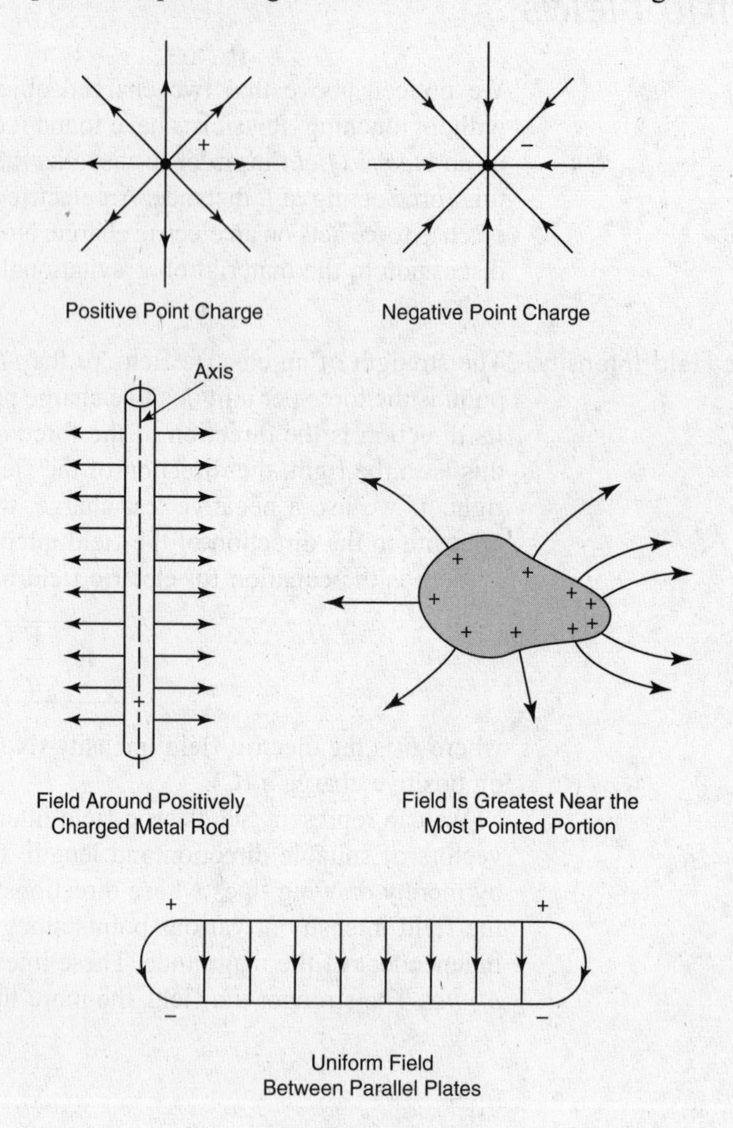

FIGURE 17.5

For point charges and metallic spheres, the electric field intensity varies inversely as the square of the distances from their centers. Note the similarities between these cases and the gravitational field of Earth. Note also that there is no field inside the metallic sphere. In fact, there is no net charge or field inside *any* charged conductor. All excess charge is on the outside surface. This effect, discovered by Michael Faraday, is responsible for the shielding effects of metal buildings and coaxial cables on radio waves.

The field intensity around a long charged rod varies inversely as the distance (not the square of the distance) from the rod. The field between two oppositely charged parallel metallic plates is uniform; the intensity is the same everywhere between the plates except near the edges. Again, note the similarity between this field and the gravitational field at sea level.

Electric Potential and Potential Difference

In Figure 17.6 an electric field is represented by three curved lines. Let A and B be two points in the field. If we put a positive charge q at point A, the field will exert a force F on the charge in the direction of the field. If we want to move the charge from A to B, we must do work. (Recall that work done equals the product of the force and the distance moved in the direction of the force.)

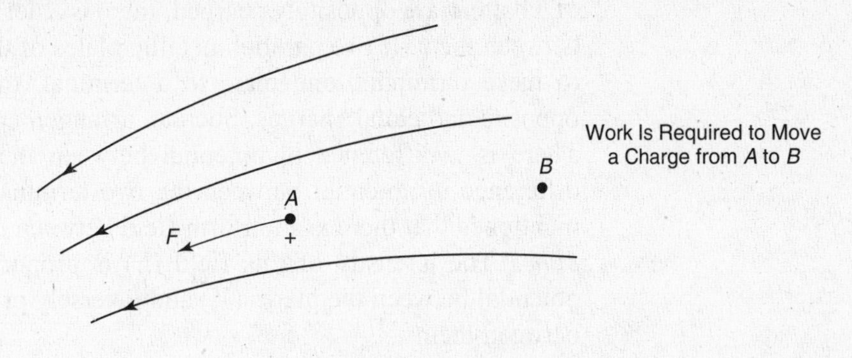

FIGURE 17.6

We define the potential difference, V, between two points in an electric field as the work per unit charge required to move a charge between the points.

$$\text{potential difference} = \frac{\text{work}}{\text{charge}}$$

$$V = \frac{\text{work}}{q}$$

The SI unit of work is the joule, and, as you have learned, the unit of charge is the coulomb. Therefore the unit of potential difference is the joule per coulomb, that is, the volt.

$$1 \text{ volt} = 1 \text{ joule/coulomb}$$

Example: The potential difference between two points is 40 volts. What is the work required to move a charge of 800 microcoulombs between these points?

$$40 \text{ V} = \frac{\text{work}}{(800 \times 10^{-6} \text{ c})}$$

$$\text{work} = 0.032 \text{ J}$$

Electric Potential

Electric potential is really the same as potential difference, except that we arbitrarily select a special point as reference. In some practical work the special reference "point" is the ground. In some of our work the special reference point is any point at infinity. We then define the *electric potential* at any point in the electric field as the work required to bring a unit positive charge from infinity to that point.

Note that, if a positive charge is released in an electric field, by definition of the direction of the electric field, the positive charge will move in the direction of the electric field. On the other hand, a negative charge released in an electric field will move in the direction opposite to that of the field. (If this bothers you, think of the field as being produced by a positive charge.) However, all objects released in Earth's gravitational field move in the direction of the field, namely, toward Earth.

As will be discussed in more detail in the next section, the two terminals of a battery are oppositely charged; there is a definite difference of potential between them. If two parallel metallic plates of the same size are connected to these terminals, one plate to a terminal, the two plates will acquire opposite and equal charges. Such an arrangement is known as a *capacitor*. There is a difference in potential between the terminals, similar to the difference in potential between the two terminals of the battery. We have mentioned that there is a uniform field between *oppositely charged parallel plates.* The intensity of this field (E) is proportional to the difference in potential between the plates (V) and inversely proportional the distance (d) between them:

$$E = \frac{V}{d}$$

Defined in this way, electric field intensity is known as *potential gradient.* As with gravitational field intensity and acceleration, the optimum usage is clear from context. However, both yield correct answers.

$$1 \text{ newton/coulomb} = 1 \text{ volt/meter}$$

Potential Difference; Electric Currents—Direct Current (DC)

Potential Difference

Potential difference is also referred to as difference of potential or voltage. Electromotive force is a special kind of potential difference.

1. The concept: Batteries (also dynamos, electrostatic generators, etc.) have two terminals, one positive, the other negative. The positive terminal has a deficiency of electrons; the negative terminal, an excess of electrons. Energy (chemical energy in the case of the battery) had to be used to attain this charged condition of the terminals; for example, work was required to push electrons onto the negative terminal against the force of repulsion of electrons already there. *Potential difference (V) is the work per unit charge* that was done to get the terminals charged and that is, therefore, potentially available for doing work outside the battery.

Electromotive force (emf) is the potential difference between the terminals of the source when no current flows. When current flows, as when a lamp is connected to a battery, the potential difference between the terminals of the battery is sometimes significantly less than its emf because of the battery's internal resistance. Thus the term *electromotive force* or *emf* is used only when energy is put into a circuit. Potential difference is often referred to as *voltage* when energy is removed.

2. Unit of potential difference: the *volt* (V). For example, the common flashlight cell supplies 1½ volts, the outlet at home, 115 volts.

$$V = \frac{\text{work}}{q}, \qquad \text{volts} = \frac{\text{joules}}{\text{coulomb}}$$

Current

1. The concept: *Current* is the rate of flow of electric charge. In a metallic conductor (e.g., copper wire) the motion of electrons is the electric current; in liquids the electric current consists of moving negative and positive ions (an *ion* is a charged atom or group of atoms); in gases the electric current consists of moving ions and electrons.
2. Direction: If a copper wire is connected to the terminals of a battery or generator, electrons will go through the wire from the negative terminal to the positive terminal. This is the direction of the *electron current* in the wire, and will not be used in this book.

   From the days of Benjamin Franklin, the opposite direction (+ to –) has also been used. Called *conventional current,* the flow of positive charge actually occurs only in specific situations. However, because of its long history, it is often referred to "by convention." Usually on nationwide examinations, questions that depend on a specific definition, however, of current direction have been avoided. Your knowledge of a consistent set of rules based on one definition is expected. One way or the other, you must be able to explain real-world phenomena. The conventional current will be used throughout this book.
3. SI units of current—the *ampere* (A);

$$1 \text{ ampere} = 1 \text{ coulomb/second};$$

$$1 \text{ coulomb} = 6.25 \times 10^{18} \text{ electrons}.$$

Resistance

1. The concept: The amount of current in a device when it is connected to a given source of potential difference depends on the nature of the device. The *resistance* of a device is its opposition to the flow of electric charges, converting electrical energy to heat because of this opposition. (Note that in alternating current (ac) *impedance* is the *total* opposition to the electric current, and the portion of it that converts electrical energy to heat is the resistance.)
2. Unit of resistance: the *ohm* (Ω).

FACTORS AFFECTING RESISTANCE are temperature, length, cross section, and material.

1. Effect of temperature: The resistance of most metallic conductors increases when the temperature rises.
2. Effects of other factors: The resistance of a conductor is directly proportional to its length and inversely proportional to its cross-sectional area, at constant temperature. This is expressed by the formula:

$R = \dfrac{kL}{A}$	L = length in meters R = resistance in ohms A = cross-sectional area in square meters k = a constant for the material, called *resistivity*; unit is ohm-meter

Conductivity. This is the reciprocal of resistivity. The resistivity of good conductors is low; their conductivity is high.

Superconductivity. Since 1911 various conductors have been found whose resistivities become zero or nearly so at extremely low temperatures. Once current is started in a loop made of such a superconductor, it will continue indefinitely without a potential difference being applied to the circuit. The goal is to produce conductors that have this property at room temperature.

Conductance Conductance is the reciprocal of resistance, and its unit is the *mho.* (The higher the resistance, the lower or poorer is the conductance.)

Ohm's Law and Electric Circuits—DC

Ohm's Law The current in a circuit (I_T) is directly proportional to the potential difference applied to the circuit (V_T) and inversely proportional to the resistance of the circuit (R_T).

Formulas

$$I_T = \frac{V_T}{R_T};\ R_T = \frac{V_T}{I_T};\ V_T = I_T R_T$$

Three types of circuits are illustrated in Figure 17.7 and described in the table that follows.

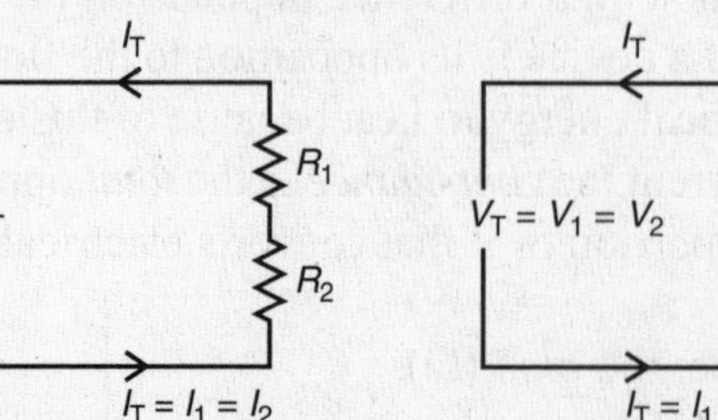

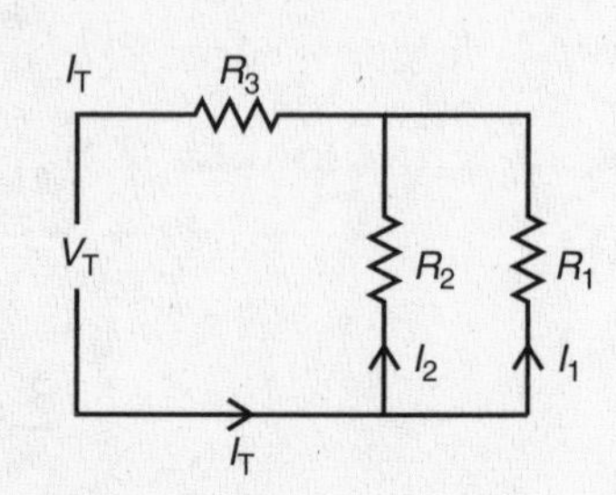

FIGURE 17.7

	Series Circuit	Parallel Circuit	Series-Parallel Circuit
Current	$I_T = I_1 = I_2$	$I_T = I_1 + I_2$	$I_T = I_3 = I_1 + I_2$
Resistance	$R_T = R_1 + R_2$	$\frac{1}{R_T} = \frac{1}{R_1} + \frac{1}{R_2}$ *	$R_T = R_3 + \frac{R_1 R_2}{R_1 + R_2}$ *
Voltage	$V_T = V_1 + V_2$	$V_T = V_1 = V_2$	$V_T = V_1 + V_3 = V_2 + V_3$; $V_1 = V_2$
IR-drop	$V_T = I_T R_T$; $V_1 = I_1 R_1$; $V_2 = I_2 R_2$, etc.		
Symbols	I_1 = current through R_1; V_2 = potential difference across R_2, etc.		

*If only two resistors are connected in parallel, the combined resistance is equal to their product divided by their sum: $R_T = \frac{R_1 R_2}{R_1 + R_2}$.

For a **PORTION OF A CIRCUIT** the same formulas apply, but I, V, and R have to be selected for the relevant part of the circuit. The potential difference across a resistor is sometimes called *potential drop, IR-drop,* or *voltage drop.*

If n equal resistors are connected in parallel, their combined resistance is $R_T = \frac{R}{n}$, where R is the resistance of one of them. The combined resistance of resistors in parallel is always less than the smallest resistance.

Notes: (1) V_T = potential difference at terminals of source = emf of source if internal resistance of source is negligible. (2) The *internal resistance* of the source can be treated as an external resistance in series with the rest of the circuit; to all of this is applied the emf of the source. (3) When the internal resistance is not negligible,

$$\boxed{V_T = \text{emf} - Ir}$$

where I is the total current, r is the internal resistance of the source of electrical energy, V_T is the potential difference across the external circuit = voltage at terminals of source.

LINE DROP is the voltage across wires connecting the source to the devices receiving energy. This voltage may be negligible; if not, it can be represented and treated similarly to R_3 in the series-parallel circuit described above.

Example 1

Series Circuit

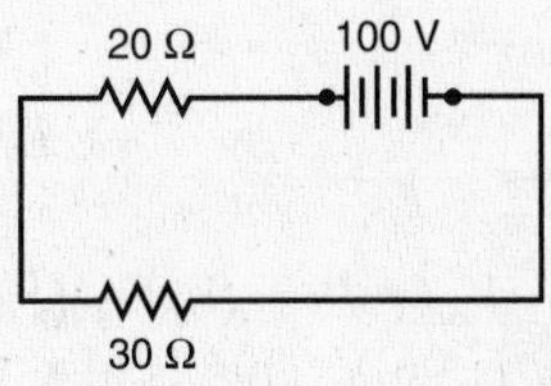

20 Ω
100 V
30 Ω

$V_T = 100\ V$

$R_T = R_1 + R_2 = 20\ \Omega + 30\ \Omega = 50\ \Omega$

$$I_T = \frac{V_T}{R_T} = \frac{100\ V}{50\ \Omega}\ 2.0\ A$$

$I_1 = I_T = 2.0$ A: This is current through each resistor

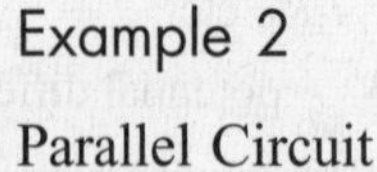

Example 2

Parallel Circuit

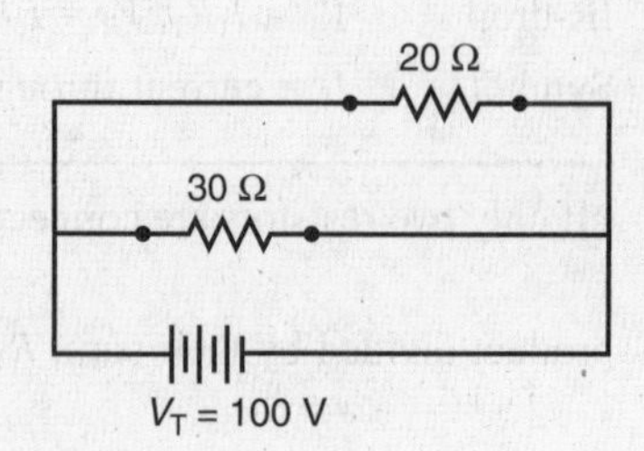

$V_T = 100\ V = V_1 = V_2$

$$R_T = \frac{R_1R_2}{R_1+R_2} = \frac{20\ \Omega \times 30\ \Omega}{20\ \Omega + 30\ \Omega} = 12\ \Omega$$

$$I_T = \frac{V_T}{R_T} = \frac{100\ V}{12\ \Omega} = 8.3\ A \text{ or } 8\tfrac{1}{3}\ A$$

$$I_1 = \frac{V_1}{R_1} = \frac{100\ V}{20\ \Omega} = 5.0\ A \text{ through } 20\ \Omega$$

$$I_2 = \frac{V_2}{R_2} = \frac{100\ V}{30\ \Omega} = 3\tfrac{1}{3}\ A \text{ through } 30\ \Omega$$

Note: $I_1 = I_2 = (5.0 + 3\tfrac{1}{3})\ A = 8\tfrac{1}{3}\ A = I_T$

Example 3

Assume the same circuit as for Example 2, except that the battery has an emf of 120 volts and the internal resistance of battery is 3 ohms.

On the basis of calculations for Example 2, this circuit is equivalent to:

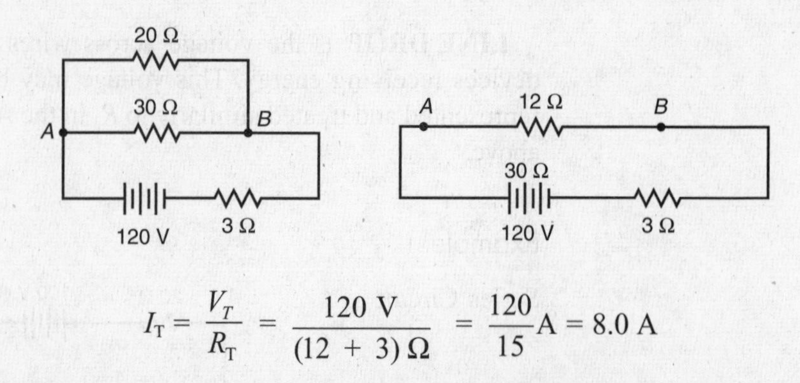

$$I_T = \frac{V_T}{R_T} = \frac{120\ V}{(12 + 3)\ \Omega} = \frac{120}{15} A = 8.0\ A$$

$V_{AB} = IR = (8.0\ A)(12\ \Omega) = 96$ V. The potential difference between A and B is 96 V. This is also the voltage across the 20-Ω resistor, as well as the voltage across the 30-Ω resistor. It is also the terminal voltage of the battery when used in this circuit.

SHORT CIRCUIT: a low-resistance path between two points where there is normally a relatively high resistance. This *may* result in an excessive current, known as an *overload.* If the circuit is overloaded, a fuse properly placed in the circuit will burn out.

In a parallel connection any device may be disconnected without breaking the circuit for the others.

Electrical Energy and Power

Joule's Law

The heat produced by current in a given resistor is proportional to the square of the current through the resistor, the resistance, and the time that the current flows through the resistor:

$$H = I^2Rt$$

where H = heat in joules, I = current in amperes, R = resistance in ohms, t = time in seconds.

Heat produced is proportional to the square of the current if R remains fixed; then, if I is doubled, H is quadrupled, provided that R and t remain the same.

In *parallel circuits* the same formula for heat produced applies, but note that in a parallel circuit the device with the least resistance draws the most current and produces the most heat.

Power

Power is the rate of using or supplying energy. The power consumption P of a resistor is

$$P = VI \qquad P = I^2R \qquad P = \frac{V^2}{R}$$

where P = power in watts, V = potential difference across resistor in volts, I = current through resistor in amperes, R = resistance in ohms.

ADDITIONAL ENERGY AND POWER UNITS Since power = energy/time,

$$\text{energy} = \text{power} \times \text{time}$$

If power is expressed in watts, appropriate energy units are watt-second, watt-minute, and watt-hour.

$$1 \text{ watt-second} = 1 \text{ joule}$$

In the nineteenth century, physicist Gustav Kirchhoff developed some rules for analyzing the flow of electricity in a circuit. His laws are basically a restatement of the laws of conservation of charge and energy as applied to loops in circuits. This is analogous to water flowing through a closed network of pipes with no leaks.

Kirchhoff's First Law: The sum of all currents entering a junction point must be equal to the sum of all currents leaving that junction point.

Kirchhoff's Second Law: The sum of the voltage drops (or rises) across all the elements around any closed loop must equal zero.

Questions Chapter 17

In each case, select the choice that best answers the question or completes the statement.

1. The leaves of a negatively charged electroscope diverged more when a charged object was brought near the knob of the electroscope. The object must have been

(A) a rubber rod
(B) an insulator
(C) a conductor
(D) negatively charged
(E) positively charged

Questions 2 and 3

Two point charges repel each other with a force of 4×10^{-5} newton at a distance of 1 meter.

2. The two charges are

(A) both positive
(B) both negative
(C) alike
(D) unlike
(E) equal

3. If the distance between the charges is increased to 2 meters, the force of repulsion will be, in newtons,

(A) 1×10^{-5}
(B) 2×10^{-5}
(C) 4×10^{-5}
(D) 8×10^{-5}
(E) 16×10^{-5}

4. A proton (electric charge = 1.6×10^{-19} coulomb) is placed midway between two parallel metallic plates that are 0.2 meter apart. The plates are connected to an 80-volt battery. What is the magnitude of the electric force on the proton, in newtons?

(A) 3.2×10^{-20}
(B) 6.4×10^{-17}
(C) 400
(D) 16
(E) 80

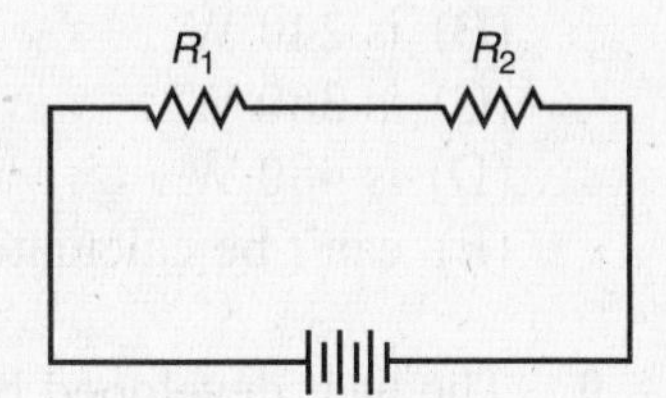

5. In the circuit shown above, R_1 and R_2 are 30 ohms and 60 ohms, respectively, and I_1 = 4 amperes. The potential difference across R_2 is equal to

(A) 30 V
(B) 60 V
(C) 120 V
(D) 240 V
(E) a quantity that can't be calculated with the given information

<u>*Questions 6–9*</u>

In the circuit shown, 4 amperes is the current through R_1.

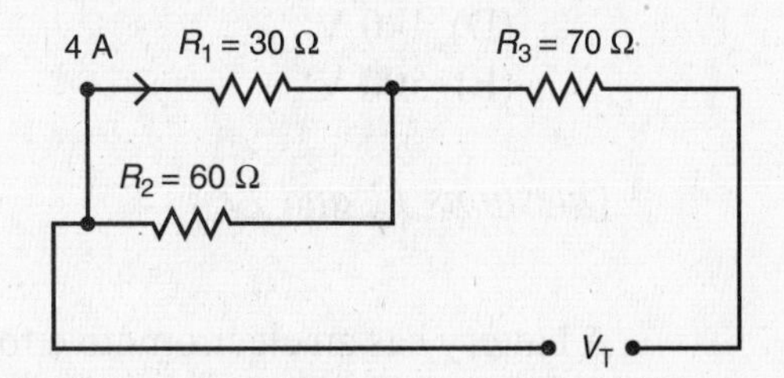

6. The potential difference across R_1 is, in volts,

(A) 7.5
(B) 30
(C) 60
(D) 120
(E) 160

7. V_T is equal to, in volts,

(A) 90
(B) 160
(C) 400
(D) 500
(E) 540

8. The rate at which R_1 uses electrical energy

(A) is 120 W
(B) is 240 W
(C) is 360 W
(D) is 480 W
(E) can't be calculated with the given information

9. The heat developed by R_1 in 5 seconds is

(A) 120 J
(B) 600 J
(C) 1,800 J
(D) 2,400 J
(E) none of the above

---

10. One thousand watts of electric power are transmitted to a device by means of two wires, each of which has a resistance of 2 ohms. If the resulting potential difference across the device is 100 volts, the potential difference across the source supplying the power is

(A) 20 V
(B) 40 V
(C) 100 V
(D) 140 V
(E) 500 V

Questions 11 and 12

A battery has an electromotive force (emf) of 6.0 volts and an internal resistance of 0.4 ohm. It is connected to a 2.6-ohm resistor through a SPST (single pole, single throw) switch.

11. When the switch is open, the potential difference between the terminals of the battery is, in volts,

(A) 0
(B) 0.8
(C) 2.6
(D) 5.2
(E) 6.0

12. When the switch is closed, the potential difference between the terminals of the battery is, in volts,

(A) 0
(B) 0.8
(C) 2.6
(D) 5.2
(E) 6.0

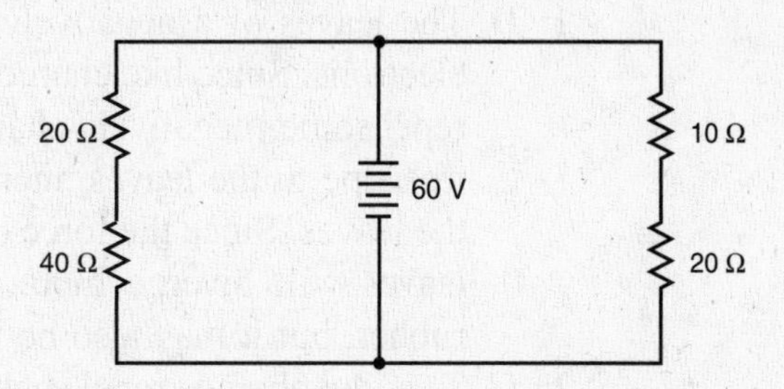

13. The current through the 10-ohm resistor shown above is, in amperes,

(A) 6
(B) 2
(C) 1
(D) ⅔
(E) 0

14. In the circuit shown above, a good lead-acid storage battery (negligible internal resistance) is used. The battery supplies 6.0 volts. F_1, F_2, and F_3 are fuses with ratings of 10 amperes, 10 amperes, and 2 amperes, respectively. When switches S_1, S_2, and S_3 are closed, the fuse(s) that will blow is (are)

(A) F_1 only
(B) F_2 only
(C) F_3 only
(D) F_1 and F_2
(E) F_1 and F_3

Explanations to Questions Chapter 17

Answers

1. (D)	5. (D)	9. (D)	13. (B)
2. (C)	6. (D)	10. (D)	14. (A)
3. (A)	7. (E)	11. (E)	
4. (B)	8. (D)	12. (D)	

Explanations

1. D The leaves of a negatively charged electroscope have an excess of electrons. Since like charges repel, the negatively charged object will repel some (negatively charged) electrons from the knob of the electroscope to the leaves, increasing the amount of negative charge on the leaves. Since the force of repulsion increases with the charge, the leaves will diverge more. The negatively charged object may be rubber, but it may also be some other insulator or a metal.

2. C Two like charges repel each other. Both may be positive or both may be negative. They may be equal or unequal.

3. A The force between two point charges varies inversely as the square of the distance between them. Since the distance is doubled, the force of repulsion becomes one-fourth as great, or 1×10^{-5} N.

4. B The plates are charged oppositely by the battery, producing a uniform electric field between the plates. The intensity of this force is equal to the ratio of the difference of potential between the plates to the distance between the plates:

$$\mathbf{E} = V/d$$

$$= (80 \text{ V}/0.2 \text{ m}) = 400 \text{ N/C}$$

The field intensity is defined as the force on a unit positive charge placed in the field:

$$\mathbf{E} = \mathbf{F}/q$$

$$400 \text{ N/C} = \mathbf{F}/(1.6 \times 10^{-19} \text{ C})$$

$$\mathbf{F} = 6.4 \times 10^{-17} \text{ N}$$

5. D This is a series circuit and the current is the same in every part of the circuit: $I_1 = I_2 = 4$ A. The potential difference across $R_2 = I_2 R_2 = 4 \text{ A} \times 60\ \Omega = 240$ V.

6. D The potential difference across R_1 is merely the IR drop:

$$V_1 = I_1 R_1$$

$$V_1 = 4 \text{ A} \times 30\ \Omega = 120 \text{ V}$$

7. E Since R_1 and R_2 are in parallel, $V_1 = V_2$

$$\therefore I_1R_1 = I_2R_2$$
$$4\text{ A} \times 30\ \Omega = I_2 \times 60\ \Omega$$
$$I_2 = 2\text{ A}$$
$$I_3 = I_1 + 2\text{ A} = 6\text{ A}$$
$$V_3 = I_3R_3 = 6\text{ A} \times 70\ \Omega = 420\text{ V}$$
$$\text{But } V_T = V_3 + V_1 = 420\text{ V} + 120\text{ V} = 540\text{ V}$$

8. D The rate of using energy is power.

$$P_1 = I_1^2R_1 = (4\text{ A})^2 \times 30\ \Omega = 480\text{ W}$$

9. D Energy = power × time = 480 W × 5 s
= 2,400 W-s = 2,400 J

The energy used by a resistor is converted to heat. Therefore the heat developed by the resistor is 2,400 J. Notice that any unit of energy may be used. If desired, we may convert to calories.

10. D The power used by the device can be expressed as $P = VI$.

$$1{,}000\text{ W} = 100\text{ V} \times I$$
$$I = 10\text{ A}$$

Since this is a series circuit, the current in each wire is also 10 A. The voltage across each wire is an IR drop = 10 A × 2 Ω = 20 V. The voltage across the source = voltage across device + voltage across the wires = 100 V + 20 V + 20 V = 140 V.

11. E When the switch is open, the circuit is incomplete and no current flows. When a battery supplies no current, the potential difference between its terminals is the emf; in this problem, 6.0 V.

12. D When the switch is closed, current flows. The terminal voltage = emf – Ir, where I is the current flowing and r is the internal resistance of the battery.

$$I = \frac{\text{total voltage}}{\text{total resistance}} = \frac{6.0\text{ V}}{(2.6 + 0.4)\ \Omega}$$
$$I = 2\text{ A}$$
$$\text{terminal voltage} = 6.0\text{ V} - (2\text{ A} \times 0.4\ \Omega) = 5.2\text{ V}$$

13. B This is really a parallel circuit with the two branches drawn on opposite sides of the battery. You may be able to see it better if you draw both branches on the same side of the battery. In any case, the full 60 V is applied across the branch containing the 10-Ω and the 20-Ω resistors in series. For that branch $R_T = 30\ \Omega$, $V_T = 60$ V, and

$$I = V/R = 60\text{ V}/30\ \Omega = 2\text{ A}.$$

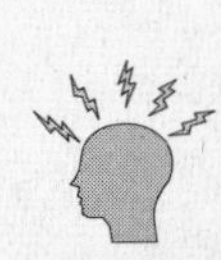

14. A The fuse rating gives the maximum current that the fuse can carry continuously. The resistance of fuses is usually quite small—a fraction of an ohm at the current ratings in this problem. When switches S_1 and S_2 are closed, F_1 is directly across the battery; with practically no resistance in the circuit, a current much larger than 10 A will flow momentarily through F_1, burning out the fuse. The additional closing of S_3 closes the circuit for F_2 and the 3-Ω resistor. The current in this circuit is $V/R = 6 \text{ V}/3\ \Omega = 2$ A. This current is handled safely by fuse F_2.

CHAPTER 18

Magnetism

Some Basic Terms

A magnet attracts iron and steel; when freely suspended it assumes a definite position (because of Earth's magnetism). *Ferromagnetic* substances (iron, nickel, and cobalt) and alloys of iron (e.g., Alnico) are strongly attracted by a magnet. All other materials are often referred to as nonmagnetic, yet every atom exhibits some form of magnetism. *Paramagnetic* substances are attracted feebly; *diamagnetic* ones are repelled feebly. A *magnetized* substance is a ferromagnetic substance that has been made into a magnet. A magnetic *pole* is the region of a magnet where its strength is concentrated; every magnet has at least two poles. The *north* pole (*N-pole,* or *north-seeking pole*) of a suspended magnet points toward Earth's magnetic pole in the Northern Hemisphere. A *lodestone* is a natural magnet. Usually we use artificial (laboratory-made) magnets.

Facts and Theory of Magnets

The *law of magnets* states that like poles repel; unlike poles attract. If the poles are concentrated at points, this force between two poles is proportional to the product of their strengths and varies inversely as the square of the distance between them. (Note similarity to Coulomb's law of electric charges.) The old *molecular theory of magnetism* states that every molecule of a *magnetic* substance is a magnet with a north and a south pole. When the substance is fully magnetized, all the like poles face in the same direction. When it is not magnetized, the molecules are arranged in helter-skelter fashion. A more modern theory speaks about *domains* rather than molecules. A domain is about 0.01 millimeter long and consists of many, many atoms; some of its properties are like those of the above-described molecular magnets. In general, a moving electric charge produces a magnetic field. In the atoms the spinning of the electrons produces the magnetism. (The explanation of para- and diamagnetism involves the orbital motion of the electrons around the nucleus.)

Magnetic Field and Magnetic Flux

The *magnetic field* is the region around the magnet where its influence can be detected as a force on another magnet. The direction of the field at any point is the direction in which the north pole of a compass would point. The magnetic field (see Figure 18.1) can be represented by fictitious *lines of flux* with the following properties: they are closed curves; they leave the magnet at a north pole and enter at a south pole; they never cross each other; and they are more crowded at the poles: in general, the greater the crowding, the greater is the force on the pole located at that point. A *permeable* substance is one through which magnetic lines of force go readily; an example is soft iron. A representation of the magnetic field can be obtained by sprinkling iron filings on a horizontal surface in the field. A compass needle takes a position in the direction of the magnetic field in which it is located and can be used to obtain a point-by-point representation of the magnetic field. (By now, this discussion should be sounding very familiar. Indeed, you should be finding many similarities among gravitational, electric, and magnetic fields.)

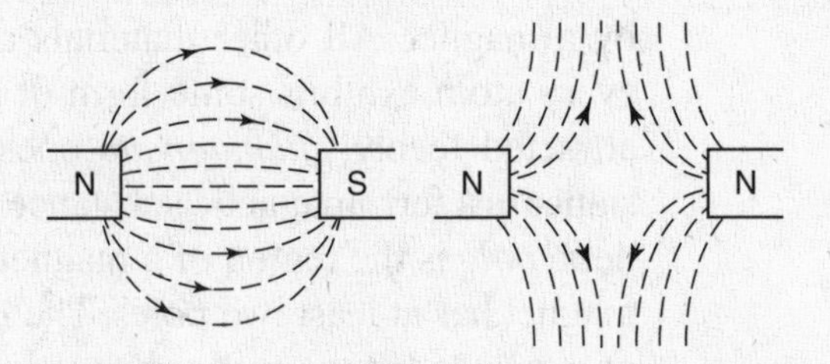

FIGURE 18.1. Typical Magnetic Field Diagrams

Terrestrial Magnetism

In a rough way, Earth acts as if it had a huge bar magnet inside with its south-seeking pole in the Northern Hemisphere. (However, this pole is called the North Magnetic Pole.) Remember that it attracts the north pole of the compass needle and that the magnetic poles do not coincide with the geographic poles, which are the points where Earth's axis of rotation intersects the surface of Earth.

The angle between the direction of the compass needle at a particular location and the direction to geographic north (or true north) is the *angle of declination* or compass *variation* of that location. *Isogonic lines* are the irregular lines drawn through locations having the same angle of declination. The *agonic line* is the isogonic line drawn through points having zero declination. Places east of the agonic line have west declination because the compass needle points west of the direction to true north.

The *dipping needle* is a compass needle mounted on a horizontal axis; it measures the *angle of dip* or angle of *inclination* at a given place. At the magnetic poles the angle of dip is 90°. The magnetic equator is the line drawn through points of zero dip. *Isoclinic lines* are lines drawn through points having the same dip.

Electromagnetism

Oersted discovered that a wire carrying current has a magnetic field around it. The magnetic field around a long, straight wire carrying current is represented by lines of magnetic flux, which are concentric circles (in the plane perpendicular to the wire) with the wire at the center. (The magnetic lines of flux are somewhat analogous to the electric lines of force.) To determine the direction of these magnetic lines of flux, use the *right-hand rule* (Figure 18.2): Grasp the wire with the right hand so that the thumb will point in the direction of *electron flow;* the fingers will then circle the wire in the direction of the lines of flux. Remember: this discussion deals with conventional current.

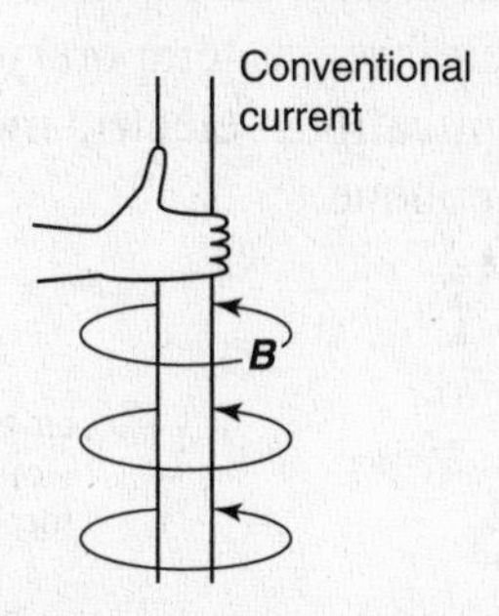

FIGURE 18.2. Right-Hand Rule for Straight Conductor

The strength of a magnetic field can be given in a number of ways. *Magnetic flux density,* or *magnetic induction,* can be specified with its SI unit (the *tesla*), or as the number of flux lines perpendicular to a unit area (*webers/meters squared*), or with reference to its causative factor: a magnetic force created by a current flowing through some distance. In thia case, the SI unit is the newton/ampere • meter:

$$1\text{T} = 1\frac{\text{Wb}}{\text{m}^2} = 1\frac{\text{N}}{\text{A}\cdot\text{m}}$$

The symbol **B** is used to denote magnetic induction. Note that the term *magnetic field intensity* cannot be used here. For historic reasons, it pertains to a slightly different quantity.

In a metallic solid the electric current does consist of moving electrons, and the right-hand rule is natural for determining the direction of the magnetic field. In a liquid or gas the electric current usually consists of moving negative and positive ions.

A *solenoid* is a long, spiral coil carrying electric current. It produces a magnetic field similar to that of a bar magnet, and has a north pole at one end and a south pole at the other (Figure 18.3). Its strength is increased by the use of a permeable core such as soft iron. The location of the poles can be found by the use of another *right-hand rule* for solenoids and electromagnets: Grasp the coil with the right hand so that the *fingers* will circle the coil in the direction of *electron flow;* the extended thumb will then point to the north pole and in the direction of the lines of flux.

An *electromagnet* is a solenoid with a permeable core such as soft iron. The *strength* of an electromagnet depends on the number of turns in the coil, the current through the coil, and the nature of the core. The first two factors are combined in the concept of *ampere-turn,* which is the product of the number of turns and the number of amperes of current through the wire. A solenoid with 50 turns and 2 amperes going through it has 100 ampere-turns. Over a wide range, the strength of an electromagnet is proportional to the number of its ampere-turns, which is affected by the nature of the core. Iron is said to have a high permeability; it helps to produce a strong electromagnet. Air, wood, paper, and so on have low permeabilities; their presence does not affect the strength of the solenoid. (*Strength* here refers to the magnet's ability to attract iron.) An electromagnet can be made very strong. Its strength can be altered by changing the current and becomes practically zero when the current is turned off. The electromagnet is, therefore, a temporary magnet. The electromagnet has many practical applications: electric motor, meter, generator, relay, telegraph sounder, bell, telephone.

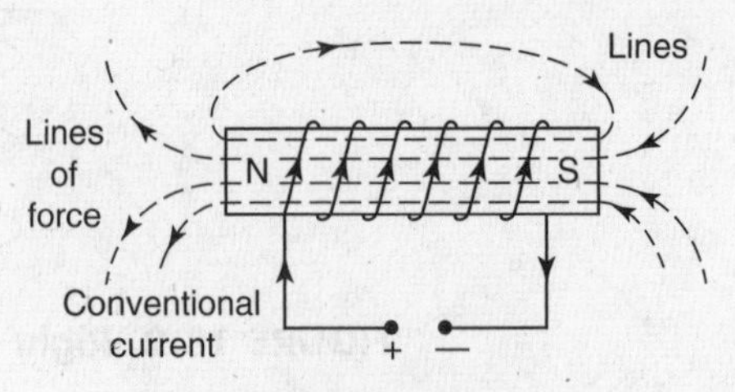

FIGURE 18.3

It is now appropriate to reconsider bar magnets and terrestrial magnetism. Note that we said that Earth behaves *as if* it had a huge bar magnet inside. Note also that bar magnets have fields shaped like objects with circulating currents. This is appropriate, because bar magnets *have* circulating currents—in their constituent atoms. In ferromagnetic substances, it is possible for the atoms to line up in domains. Of course there is no huge bar magnet in Earth. Terrestrial magnetism arises from circulating currents of ions in magma, deep in Earth.

Circular Loop

The magnetic field around a long, straight wire carrying current does not have any poles. As we saw above, when the wire is wound in the form of a coil, magnetic poles are produced. The magnetic field around a single circular loop carrying current is such that the faces show polarity. We may apply the same right-hand rule as above to see this. (In Figure 18.4), notice that both the left half and the right half of the loop produce lines of flux that have the same direction near the center of the loop. In this region the two fields combine to give a relatively strong field. The face of the loop from which the lines are directed is the north pole: in this diagram, it is the face away from us.

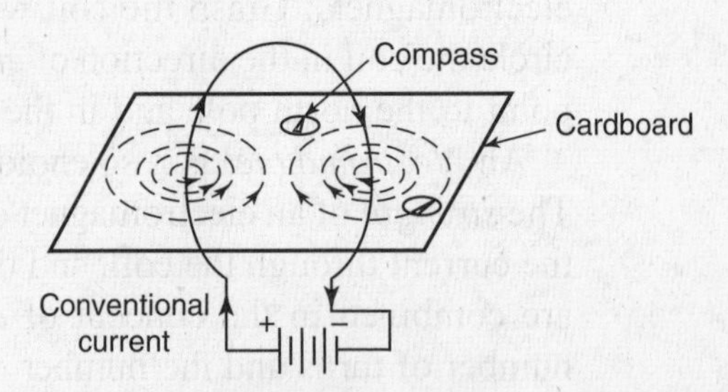

FIGURE 18.4

Force on a Current-Carrying Conductor

When electric charges move across a magnetic field, a force acts on the moving charges that is not present when the charges are stationary. In Figure 18.5, imagine the magnetic lines of flux between the north and south poles of some magnet. In this magnetic field is located a wire perpendicular to the paper and to the lines of flux. The current through the wire is shown going into the paper. We can determine the direction of the force on the wire by using the first right-hand rule mentioned in this chapter. Grasp the wire with the right hand with the thumb pointing (in this case) out of the paper. The fingers then indicate that the lines of flux due to the current are counterclockwise around the wire; they oppose the magnet's lines of flux about the wire, but reinforce them below the wire.

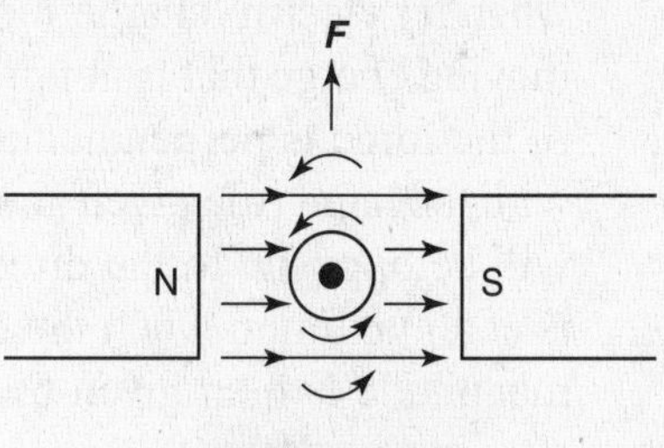

FIGURE 18.5

Since opposite fields cause an attraction and like fields produce a repulsive force, the wire is pushed up. There is a hand rule that gives this result directly. If the fingers of the flat right hand point in the direction of the field (from north to south), and the thumb is in the direction of electron flow, the hand will push in the direction of the magnetic force. Note in Figure 18.6 that the direction of force on the wire is at right angles to both the direction of the electron current and the lines of flux of the magnet.

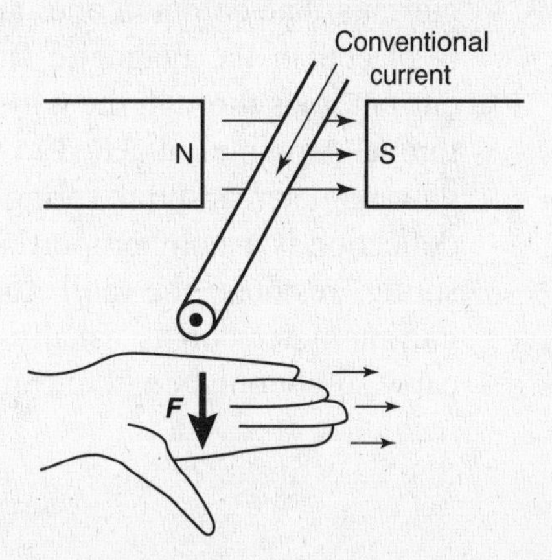

FIGURE 18.6. Right-Hand Rule for Force on a Current-Carrying Conductor

The *magnitude of the force* depends on the direction of the wire in the magnetic field. It is maximum when the wire is at right angles to the field and zero when the wire is parallel to it. It is proportional to the current and the strength of the magnetic field. When the wire is at right angles to the field, the force on the wire is given by the following formula:

$$\mathbf{F} = \mathbf{I}L\mathbf{B}$$

where **I** is the current in the wire, L is the length of wire in the magnetic field, and **B** is the flux density.

When two parallel wires carry current, a force acts on both wires. The reason for this is that each wire is in the magnetic field produced by the other. When the two currents are in the same direction, the two wires attract each other. When the currents are in opposite directions, the wires repel each other. The force is proportional to the product of the two currents and inversely proportional to the distance between them:

$$\mathbf{F} \propto \frac{\mathbf{I}_1\mathbf{I}_2}{d}$$

Force on a Moving Charge

The electric current consists of moving charges. We mentioned before that, when an electric charge moves across a magnetic field, a force acts on the moving charge that is not present when the charge is stationary. The direction of the force is perpendicular to the field and the velocity, and can be found with the hand rule given above. If the moving charge is positive, the direction will be opposite to the direction predicted by the right-hand rule. The force is at a maximum when the charge's velocity is at right angles to the flux. Its magnitude is then given by:

$$\mathbf{F} = qv\mathbf{B}$$

where q is the magnitude of the charge (in coulombs), v the velocity (in meters per second), and **B** the flux density (in teslas). The force will then be given in newtons. The path of this charge will be circular. Since the force will be at right angles to the velocity, the magnitude of the velocity will not change.

Direct-Current Meters

A *galvanometer* (Figure 18.7) measures the relative strengths of small currents. The common ones are of the moving-coil type; a coil free to rotate is placed in the magnetic field provided by a permanent magnet. When current goes through the coil, the force on two sides of the coil produces a torque. A pointer attached to the coil rotates with it. In the Weston type of galvanometer a spiral spring keeps the coil from rotating too much; the deflection is usually proportional to the current. When the current stops, the spring restores the coil to its zero position. The resistance of the galvanometer coil is usually small (about 50 ohms); usually the coil can carry safely only small currents (a few milliamperes).

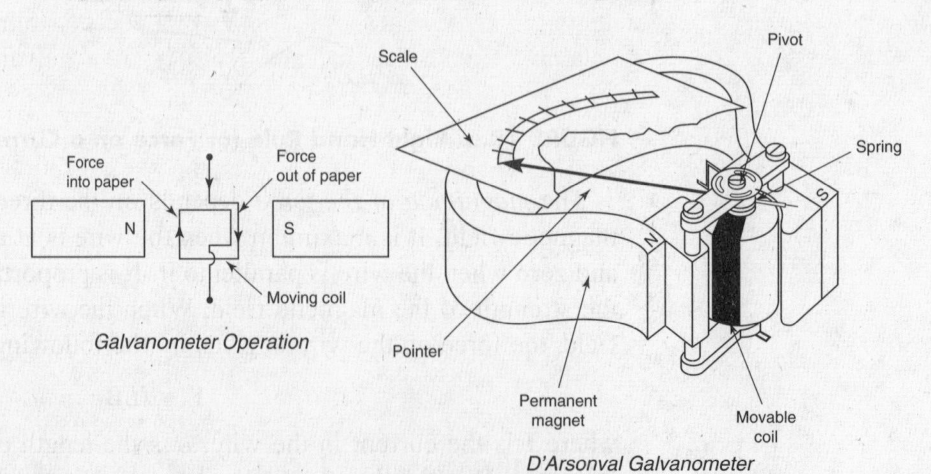

FIGURE 18.7

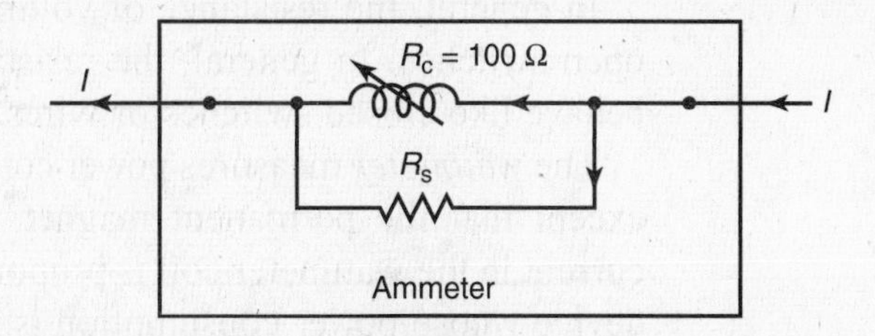

FIGURE 18.8

An *ammeter* (Figure 18.8) is an instrument for measuring current; it is calibrated to give the actual magnitude of the current. An ammeter usually consists of a galvanometer with a low resistance in parallel with the moving coil; this resistance is called a *shunt.* The ammeter is connected in series with the device whose current is to be measured. All the current to be measured enters the ammeter. Internally the current divides: part of it goes through the coil; the rest, through the shunt. We can calculate the required resistance of the shunt (R_S) if we apply the rules for a parallel circuit. The shunt is in parallel with the coil (R_C), and therefore the voltage drops are equal: $I_S R_S = I_C R_C$. If the coil has a resistance of 100 ohms and gives full-scale deflection for 10 milliamperes, we can convert the scale to a 0- to 1-ampere ammeter: remember that the coil gives full-scale deflection when 0.01 ampere passes through it. The other 0.99 ampere must then go through the shunt:

$$0.99\ R_S = 0.01 \times 100$$

and R_S is approximately 1 ohm.

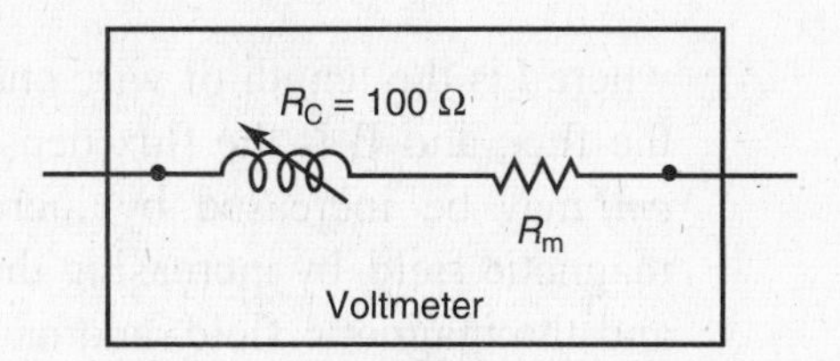

FIGURE 18.9

A *voltmeter* (Figure 18.9) is an instrument calibrated to measure the potential difference connected to its terminals. If we want to measure the potential difference across a lamp, we connect a voltmeter in parallel with the lamp. A voltmeter usually consists of a galvanometer with a high resistance connected in series with it. This resistance is often called a *multiplier.* The range of a voltmeter can be changed by changing the multiplier. Suppose we want to convert the galvanometer described above to a voltmeter with full-scale deflection at 100 volts. (Range: 0–100 volts.) At full-scale deflection 0.01 ampere goes through both the multiplier (R_m) and the coil. The IR drop across both together is 100 volts:

$$I_T R_T = 100$$

$$0.01\ R_T = 100$$

and R_T = 10,000 ohms. The coil supplies 100 ohms of this total resistance; therefore, the required multiplier resistance is 9,900 ohms.

In general, the resistance of voltmeters is high; ideally they behave like open switches. In general, the resistance of ammeters is low; ideally they behave like closed switches or wires.

The *wattmeter* measures power consumption. It is similar to the voltmeter except that the permanent magnet is replaced by an electromagnet; the current in the wattmeter coil is proportional to the current going through the device whose power consumption is being measured.

The *kilowatt-hour meter* measures energy consumption.

Electromagnetic Induction

Faraday in England and Henry in the United States independently discovered that a magnetic field can be used to produce an electric current. If a wire moves so that it cuts magnetic lines of flux, an electromotive force is induced in the wire. An *induced emf* is produced in a wire whenever there is relative motion between a wire and a magnetic field. If the wire is part of a complete conducting path, current flows in this circuit in accordance with Ohm's law; this current is referred to as *induced current.*

No emf is induced when the wire moves parallel to the magnetic flux. The emf is maximum when the wire moves at right angles to the magnetic flux, as in Figure 18.11. In SI units, the magnitude of the maximum induced emf is given in volts by the following formula:

$$\text{emf} = l\mathbf{v}\mathbf{B}$$

where l is the length of wire cutting the flux, $\mathbf{v}$ is the *relative* speed cutting the flux, and $\mathbf{B}$ is the flux density in teslas. The *magnitude* of the *induced emf* may be increased by increasing the length of the wire cutting the magnetic field, by increasing the relative speed of motion between the wire and the magnetic field, and/or by increasing the strength of the magnetic field. Note that the same result is obtained if the magnet is stationary and the wire moves or if the magnet moves and the wire is stationary.

An emf is induced in a *single loop* of wire when the magnetic flux going through the loop changes. The magnitude of the induced emf is proportional to the speed with which the flux changes. The emf is 1 volt when the change in flux is 1 weber per second. If a coil has several turns of wire, the emf is proportional to the number of turns in the coil.

Lenz's Law

The *direction* of the induced current can be determined by the use of *Lenz's law:* The direction of the induced current is such as to produce a magnetic field that will hinder the motion that produced the current. For example, if a permanent magnet is moved away from a stationary coil, an emf is induced in the coil. In Figure 18.10, the motion of the magnet is opposed if the induced current produces a south pole to attract the retracting north pole, as will happen if the induced current has the direction shown. Of course, a north pole will be produced simultaneously at the left end of the coil. If a single wire is moving across a magnetic field, the direction of the current can be determined

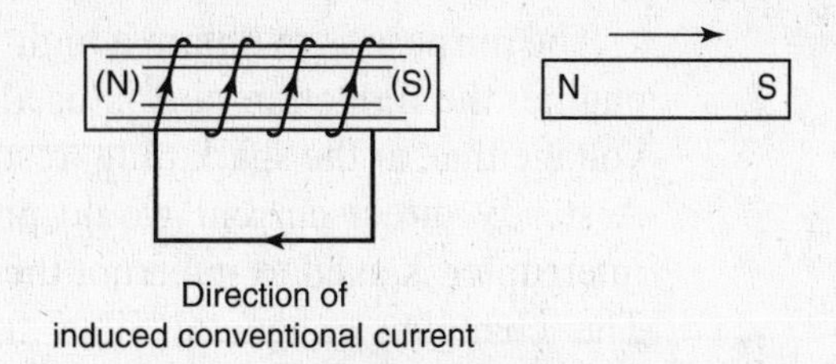

FIGURE 18.10

in this way: Assume that the wire in the diagram is perpendicular to the page and pulled toward the bottom edge of the paper (Figure 18.11). If the induced current is directed into the paper, the right-hand rule tells you that the magnetic field is weakened above the wire, strengthened below it (closer to you). This tends to push the wire toward the top edge of the paper, opposing the motion that produced the induced current. Therefore, you guessed the direction of the induced current correctly: into the paper. Note that Lenz's law is a special case of the law of conservation of energy.

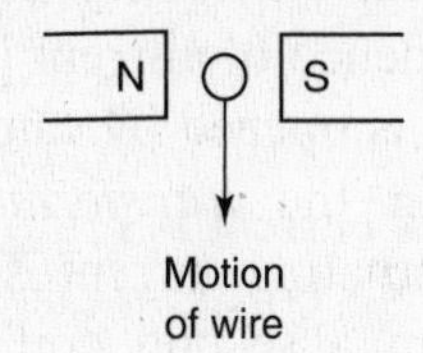

FIGURE 18.11

Mutual Induction and Transformers

We saw before than an emf is induced in a coil if the current in that coil is changing. This is self-induction. We have *mutual induction* if an emf is induced in a coil because of current changes in a second coil. The current may be alternating or direct. As long as there is a change in the current, there will be a change in the magnetic field surrounding the coil. If this changing magnetic field cuts the second coil, an emf is induced in that coil. The *primary* coil (P) is the coil that is directly connected to the source of electrical energy, such as a dynamo. The *secondary* coil (S) is the coil from which electrical energy may be taken as a result of the emf induced in it. In the *induction coil* the source of energy is direct current, usually a battery. (See Figure 18.12.)

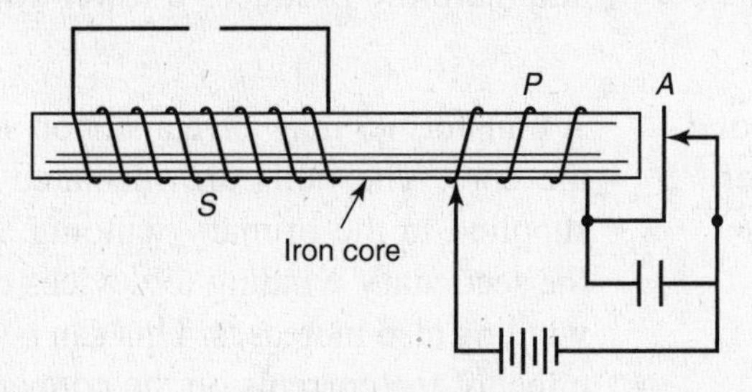

FIGURE 18.12

The purpose is to obtain a high voltage. In an automobile using a gasoline engine, the induction coil is used as the *ignition coil:* it provides the high voltage that, at the spark plug, results in a spark that ignites the fuel mixture. A steady direct current would produce no emf in the secondary coil. An interrupter is used to interrupt the current in the primary coil automatically. The changing magnetic field then cuts across the many turns of the secondary coil. The greater the number of turns, the greater is the induced emf. In Figure 18.12, *A* is springy, magnetic material. When the iron core is sufficiently magnetized, it will pull the springy armature (*A*) to the left, thus breaking the primary circuit. When, as a result, the current decreases, the magnetism becomes less and the spring (*A*) returns to its original position. This completes the primary circuit again; the primary current and the magnetic field increase, inducing an emf in the secondary coil. Notice that the increase and decrease of the current result in an alternating emf in the secondary coil. Usually this emf is greater in one direction than in the other, and the gap in the secondary circuit can be arranged so that there will be current in only one direction in the secondary coil. A spark at the gap is evidence of current in the secondary coil. (In the automobile the armature (*A*) is opened by cams or projections on a rotating shaft.)

In the *transformer,* we also depend on mutual induction; usually alternating current is applied to one coil and alternating current is obtained from a second coil. In transformers used on low-frequency alternating current (up to few thousand hertz), the following formulas apply:

$$\frac{\text{secondary emf}}{\text{primary emf}} = \frac{\text{number of turns on secondary}}{\text{number of turns on primary}}$$

power supplied by secondary coil = efficiency × power supplied to primary coil

$$V_s I_s = V_p / I_p \times \text{efficiency}$$

When the efficiency is 100%, $V_s I_s = V_p / I_p$.

The efficiency of practical transformers is high and constant over a wide range of power. When more power is used in the secondary circuit, more power is supplied to the primary circuit automatically. If we want a higher voltage than the generator supplies, we use a *step-up transformer;* which has more turns on the secondary coil than on the primary coil. A *step-down* transformer has fewer turns on the secondary coil than on the primary coil and therefore produces a lower voltage than the generator supplies.

Eddy Currents and Hysteresis Losses

A transformer may have a silicon steel core with two coils of wire wound on the core. The coils are insulated from each other. Alternating-current is supplied to the primary winding. As more and more power is supplied by the secondary winding to devices connected to it, the current in the primary winding also increases. This can be explained by considering the magnetized effect of the currents on the core and the emf induced in the coils. However, the changing magnetic field in the core also has two undesirable effects:

eddy currents and hysteresis losses. *Eddy currents* result from the emf induced in the core. The current is reduced by making the core out of thin laminations insulated from each other. *Hysteresis losses* result from the fact that the magnetic domains tend to be twisted back and forth as the direction of the applied emf changes.

Electric Generators

The term *dynamo* is sometimes used to refer to an electric generator that converts mechanical energy into electrical energy. Usually a coil is made to rotate in a magnetic field, and an emf is induced in the coil. The source of the energy for turning the coil may be a waterfall, steam under pressure, and so on. Image the coil in Figure 18.13 turning counterclockwise. The axis of the coil's rotation is perpendicular to the paper and is represented by the dot. The coil is rotated counterclockwise. The stationary magnet provides a magnetic field that is cut by the rotating coil; this magnet is called a *field magnet.* Because the rotating coil cuts across magnetic lines of flux, an emf is induced in it and electric energy can be taken from it; this coil is called an *armature* coil.

We can use Lenz's law to figure out the direction of the current in the armature. At the instant shown, the direction of this current must be such as to produce the poles marked in Figure 18.13; then the motion will be opposed by the repulsion of like poles. If we use our right-hand rule for a solenoid, we will find that the direction of current in the armature is as shown in the diagram. *Slip rings A* and *B,* insulated from each other, are on the same shaft as the armature and rotate with it. *Brushes C* and *D* are stationary and make wiping electrical contact with the slip rings. Current can therefore flow through an external circuit connected to the brushes.

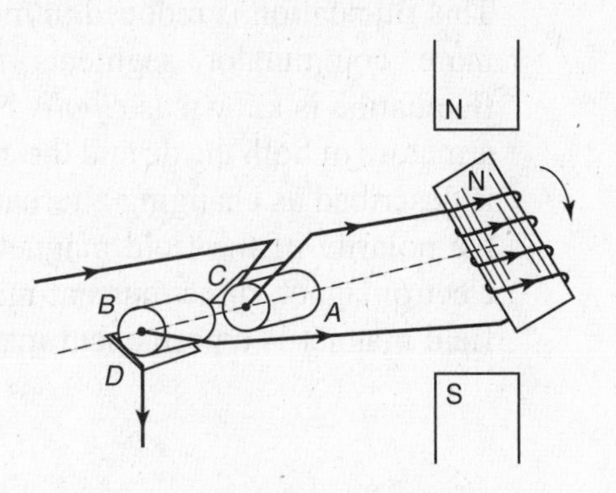

FIGURE 18.13

However, a moment later the armature will be in a new position. In order for the motion to be opposed now, the polarity of the armature, and therefore the direction of the current in the armature, must change (Figure 18.14). This is repeated as the armature keeps rotating. Current whose direction is constantly reversing is known as *alternating current* (ac). In the generator described, using slip rings, the current in the external circuit is also alternating. One complete back-and-forth variation is known as a *cycle.* In the simple ac generator shown, there will be one cycle for each revolution of the armature. The alternating current ordinarily used in the home is 60 hertz. The conservation of energy principle, of which Lenz's law is a special case, indicates that as more current is drawn from the generator, more energy has to be used to turn the armature.

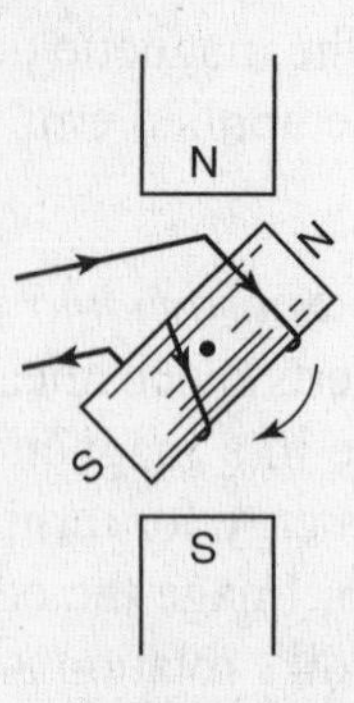

FIGURE 18.14

A *direct-current (dc) generator* can be constructed with a slight modification of the above ac generator (Figure 18.15). We use the same field magnet and armature coil, but instead of slip rings we use a *commutator,* a metallic ring split into two insulated parts of equal size. (In more complicated dc generators the ring is split into several commutator segments.) Each end of the armature coil is connected to a commutator segment. As the armature rotates, the stationary brushes alternately make contact with one segment and then with the other. The brushes are set so that this change in contact occurs just when the direction of current in the armature coil changes. As a result, the current in the external circuit is always in one direction (dc), but its magnitude is constantly changing; such current is known as *pulsating* or *fluctuating direct current* (Figure 18.16). This fluctuation is reduced in more complicated dc generators by the use of more commutator segments and more armature coils. The residual fluctuation is known as *ripple.* Notice that there is alternating current in the armature of both the dc and the ac generator. The function of the commutator is described as changing alternating current to direct current. Also note that the polarity of the field magnet does not change, meaning that, if it is an electromagnet, direct current must be used. A *magneto* is a dynamo whose field magnet is a permanent magnet.

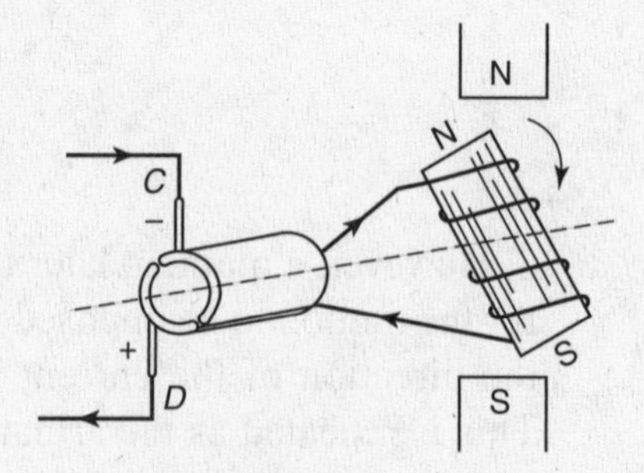

FIGURE 18.15

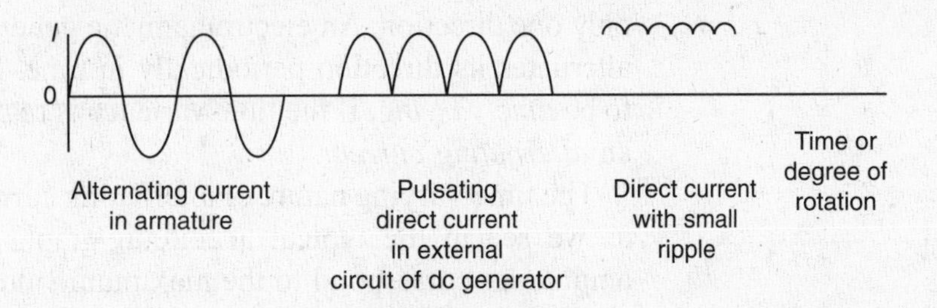

FIGURE 18.16

Electric Motors—DC

The dc motor is similar to the dc generator except that electrical energy is supplied to the motor to be converted to mechanical energy. Figure 18.17 for the dc motor is similar to Figure 18.15 for the dc generator. A *field winding* or field coil has been shown for the magnet. The diagram is for a *shunt-wound motor;* again the commutator rotates with the armature.

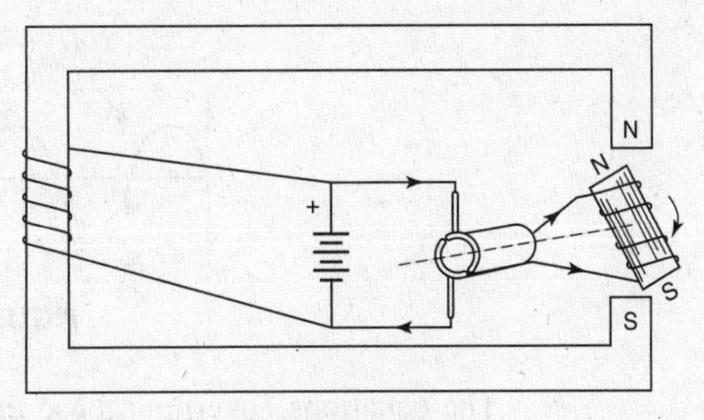

FIGURE 18.17

The field winding is in parallel with the armature coil. (In a series-wound motor the field coil and armature coil are in series.) Notice that the connection has been made that will cause the armature to rotate counterclockwise. To keep the coil rotating in that way, the current in the armature must reverse at the right instant to change the polarity of the armature. The interaction between the poles of the field magnet and the poles of the armature provides the force to keep the armature rotating. However, because the armature coil rotates in a magnetic field, an emf is induced in the armature. According to Lenz's law, this emf will be in such a direction as to oppose the rotation; therefore this induced emf is called a *counter-emf* or *back emf.* This counter-emf tends to reduce the current in the armature. Because the counter-emf is greatest when the armature is rotating fastest, the current in the armature is less when the motor is running at full speed than when it is starting up.

Alternating Current (AC) Circuits

It is worthwhile to discuss briefly the nature of ac circuits at this time. In a dc electric circuit, the emf of the battery produces a current that flows in only one direction. An electromagnetic generator can produce a current that alternates its direction periodically in time. In this case the current is said to be *time-varying*. If the time variance is regular, we simply call the current an *alternating current.*

The time-varying nature of alternating currents and voltages is sinusoidal, as we see in the typical ac circuit graphs shown in Figure 14.18. The amplitudes correspond to the maximum values I_{max} and V_{max}. The period of alternation, T, is related to the angular frequency, ω:

$$\omega = 2\pi / T = 2\pi f$$

where f is the frequency in hertz.

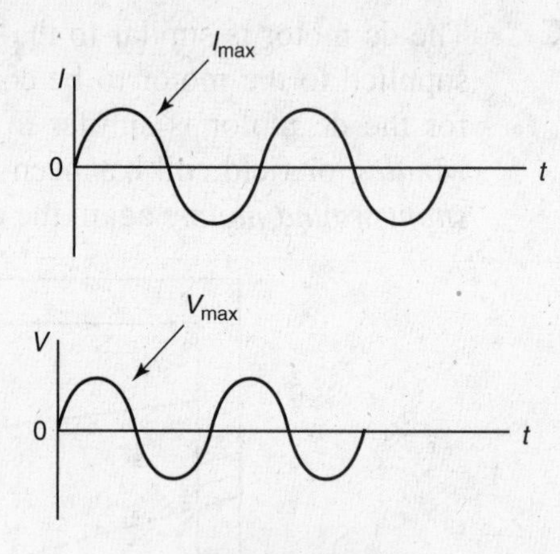

FIGURE 18.18

The equations governing these time-varying quantities are as follows:

$$I = I_{max} \sin \omega t$$

$$V = V_{max} \sin \omega t$$

If a single resistor is added to the circuit (Figure 18.19), we have the following conditions. The power consumed in the resistor is dissipated as heat (as in the dc circuit), but at a varying rate. Comparisons with dc equations are risky since the maximum current lasts for only a brief interval of time. The instantaneous voltage is given by a version of Ohm's law:

$$V = I_{max} R \sin \omega t$$

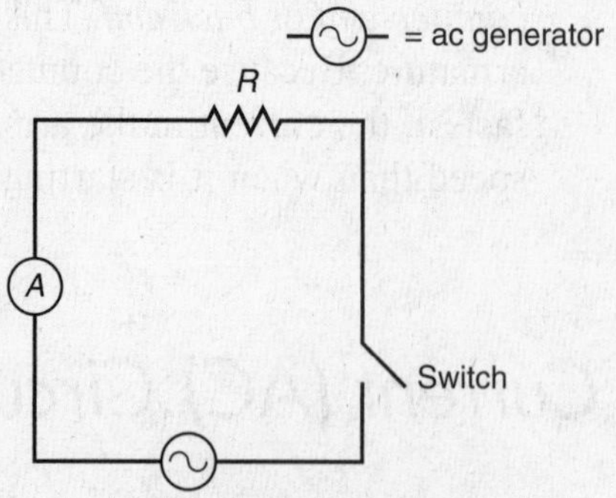

= ac generator

R

A

Switch

FIGURE 18.19

The instantaneous power in the resistor is given by this equation:

$$P = I^2 R = I^2_{max} R \sin^2 \omega t$$

It is often practicable to work with average values. To obtain the time average value of the power generated in a resistor, we need to obtain the average value of the square of the sine function. Recall that a sine function, when graphed, is symmetrical about the *zero* equilibrium position. The average value of $I = I_{max} \sin \omega t$ is equal to zero. However, if we were to square this current equation, the magnitude of the function would always remain positive. This is illustrated in Figure 18.20.

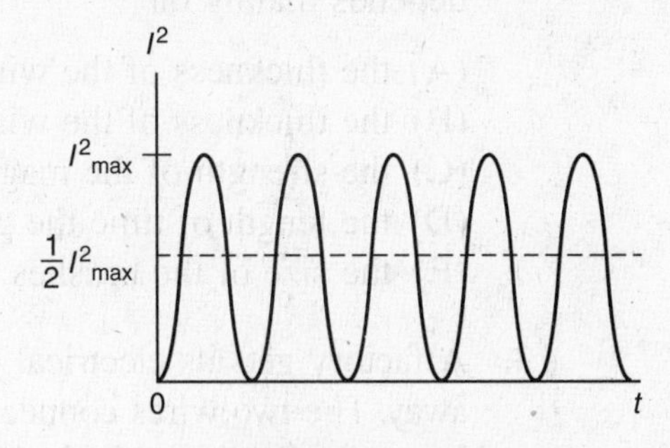

FIGURE 18.20

You can see that the average of the square of the alternating current is given by one-half the value of the square of the maximum current. What we need is the *average* or *root-mean-square* value of this current; that is, we need the square root of the average of the square of the function and not the average of the function itself! We must take the square root to get the root-mean-square value:

$$\overline{I^2} = \frac{1}{2} I^2_{max}$$

$$I_{rms} = \sqrt{\tfrac{1}{2} I^2_{max}} = 0.707\, I_{max}$$

The average power is therefore given by

$$P_{avg} = I^2_{rms} R = \frac{1}{2} I^2_{max} R$$

In a similar way, the root-mean-square value of the alternating emf can be shown to be equal to

$$V_{rms} = 0.707\, V_{max}$$

Some textbooks refer to these quantities as *effective* values, and they are equivalent to what we have called root-mean-square values.

Questions Chapter 18

In each case, select the choice that best answers the question or completes the statement.

1. Heating a magnet will

   (A) weaken it
   (B) strengthen it
   (C) reverse its polarity
   (D) produce new poles
   (E) have no effect

2. The emf produced by a generator operating at constant speed depends mainly on

   (A) the thickness of the wire on the armature
   (B) the thickness of the wire on the field magnet
   (C) the strength of the magnetic field
   (D) the length of time the generator operates
   (E) the size of the brushes

3. A factory gets its electrical power from a generator 2 kilometers away. The two wires connecting the generator to the factory terminals have a resistance of 0.03 ohm per kilometer. When the generator supplies 50 amperes to the factory, the terminal voltage at the generator is 120 volts. A voltmeter connected to the factory terminals should then read

   (A) 100 V
   (B) 110 V
   (C) 114 V
   (D) 118 V
   (E) 120 V

<u>*Questions 4–6*</u>

In the circuit below, each of the three resistors has a resistance of 30 ohms.

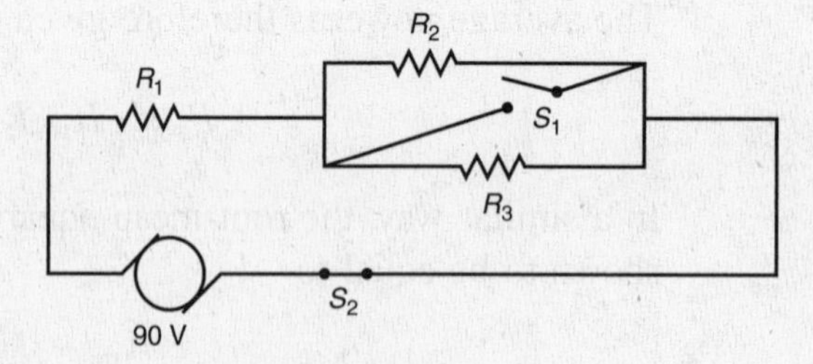

4. When switch S_1 is open and S_2 is closed as shown, the potential difference across R_1 is, in volts,

   (A) 0
   (B) 30
   (C) 45
   (D) 60
   (E) 90

5. When switches S_1 and S_2 are closed, the potential difference across R_1 is, in volts,

(A) 0
(B) 30
(C) 45
(D) 60
(E) 90

6. When switches S_1 and S_2 are open, a voltmeter connected across S_2 reads

(A) 0
(B) 30 V
(C) 45 V
(D) 60 V
(E) 90 V

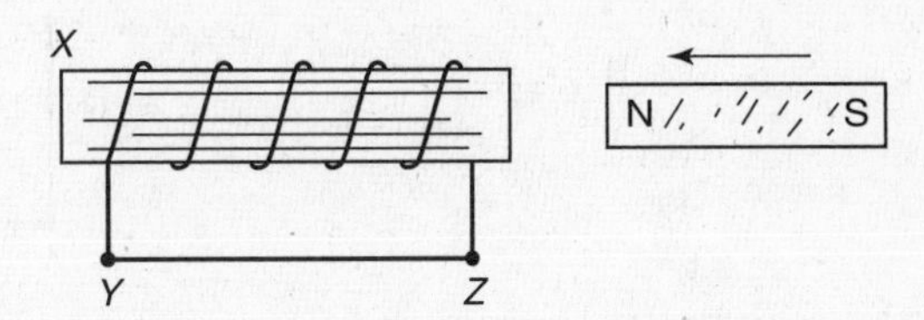

7. In the diagram above, X is a coil wire with a hollow core. The permanent magnet is pushed at constant speed from the right into the core and out again at the left. During the motion

(A) there will be no current in wire YZ
(B) current in wire YZ will be from Y to Z
(C) current in wire YZ will be from Z to Y
(D) current in wire YZ will be from Z to Y and then from Y to Z
(E) current in wire YZ will be from Y to Z and then from Z to Y

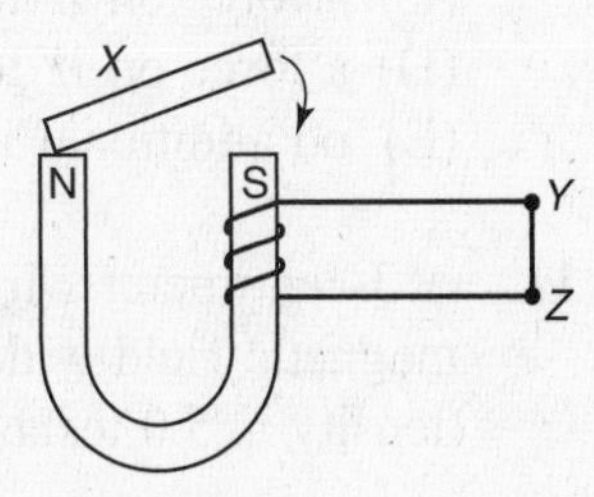

8. The soft-iron armature X in the above diagram is allowed to fall into position on top of the poles of the horseshoe magnet. While it falls,

(A) there will be no current in wire YZ
(B) current in wire YZ will be from Y to Z
(C) current in wire YZ will be from Z to Y
(D) current in wire YZ will be from Z to Y and then from Y to Z
(E) current in wire YZ will be from Y to Z and then from Z to Y

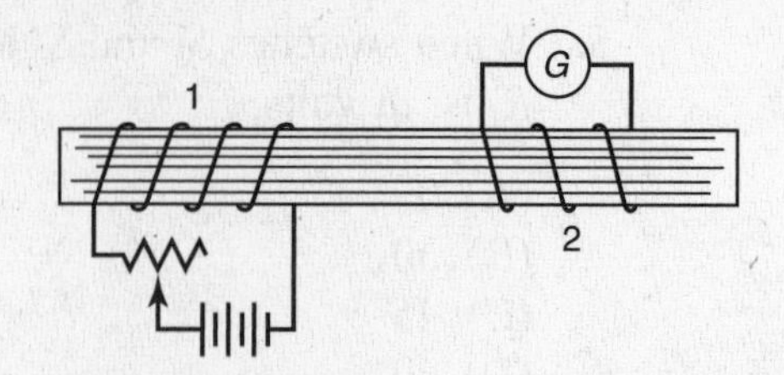

9. In the above diagram, G is a sensitive galvanometer connected to coil 2. Coil 1 is insulated from coil 2. Both coils are wound on the same iron core and are insulated from it. When the variable contact on the rheostat is moved halfway to the left, the needle on the galvanometer

(A) moves to the left
(B) moves to the right
(C) moves momentarily and then returns to its starting position
(D) doesn't move because coils 1 and 2 are insulated from each other
(E) doesn't move because coils 1 and 2 are insulated from the iron

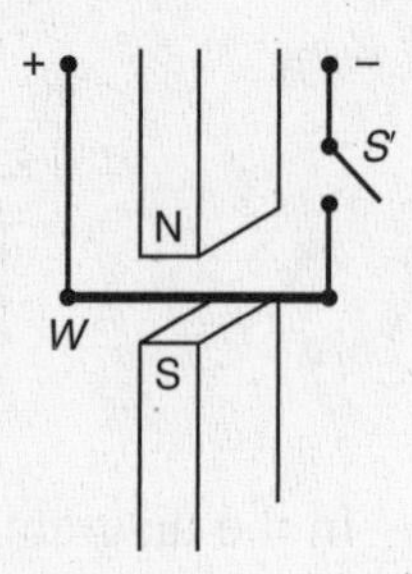

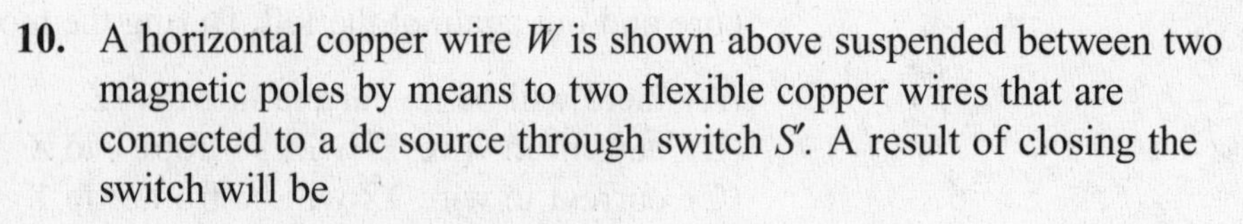

10. A horizontal copper wire W is shown above suspended between two magnetic poles by means to two flexible copper wires that are connected to a dc source through switch S'. A result of closing the switch will be

(A) a force on W toward the front
(B) a force on W toward the back
(C) a force on W toward the left
(D) a force on W toward the right
(E) no additional force on W

11. A 3-centimeter wire is moved at right angles across a uniform magnetic field with a speed of 2.0 meters per second. If the flux density is 5.0 teslas, what is the magnitude of the induced emf?

(A) 0.03 V
(B) 0.3 V
(C) 0.6V
(D) 10 V
(E) 20 V

12. If a step-up transformer were 100% efficient, the primary and secondary windings would have the same

(A) current
(B) power
(C) number of turns
(D) voltage
(E) direction of winding

Questions 13 and 14

When a transformer is connected to 120-volt alternating current, it supplies 3,000 volts to a device. The current through the secondary winding then is 0.06 ampere, and the current through the primary is 2 amperes. The number of turns in the primary winding is 400.

13. The number of turns in the secondary winding is

(A) 16
(B) 30
(C) 1,000
(D) 2,000
(E) 10,000

14. The efficiency of the transformer is

(A) 75%
(B) 80%
(C) 85%
(D) 90%
(E) 95%

Explanations to Questions Chapter 18

Answers

1. (A)	**5.** (E)	**9.** (C)	**13.** (E)
2. (C)	**6.** (E)	**10.** (B)	**14.** (A)
3. (C)	**7.** (E)	**11.** (B)	
4. (D)	**8.** (B)	**12.** (B)	

Explanations

1. A Heating increases the random motion of the particles in the magnet. This leads to a less orderly arrangement of molecular magnets (or domains) in the magnet, making the magnet weaker.

2. C The emf induced in a wire depends on the length of the wire cutting the magnetic field, on the strength of the magnetic field being cut, and on the speed with which the field is being cut at right angles. Of these, only the second is offered as a choice (C). In the practical design of the generator, the thickness of the wire used is important but does not affect directly the emf produced because it does not affect the three factors mentioned. Once the generator reaches constant speed, the length of time the generator operates does not affect these three factors.

3. C Four kilometers of wire are used to connect the generator to the factory terminals. Since the wire has a resistance of 0.03 Ω/km, 4 km will have a resistance of 0.12 Ω. This resistance is in series with the factory load, which gets 50 A. The *IR* drop in the connecting wires = (50 A) × (0.12 Ω) = 6 V. Since the generator supplies 120 V to the circuit and the drop in the connecting wires is 6 V, 114 V will be left for the factory load. (In a series circuit, $V_T = V_1 + V_2$.) A voltmeter should measure this voltage at the factory terminals.

4. D When S_1 is open and S_2 is closed, S_1 has no effect. We then have a series-parallel circuit: R_2 and R_3 are in parallel; their combination is in series with R_1 and the generator supplying 90 V. Since R_2 and R_3 have the same resistance, 30 ohms, their combined value is 30/2 ohms, or 15 ohms. The equivalent circuit is then as shown. The total resistance of the circuit then is 45 ohms. (In a series circuit, $R_T = R_1 + R_2$.)

$$I_T = \frac{V_T}{R_T} = \frac{90\text{ V}}{45\ \Omega} = 2\text{ A}$$

$$V_1 = \frac{I_1}{R_1} = 2\text{ A} \times 30\ \Omega = 60\text{ V};\quad I_T = I_1$$

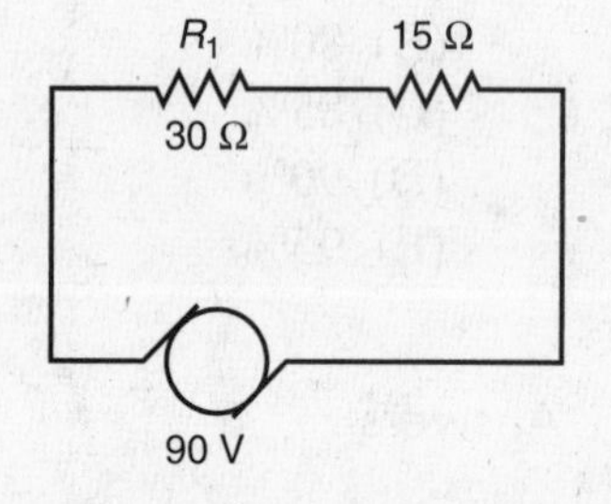

5. E When switch S_1 is closed, we have a short-circuit across R_2 and R_3. Their combined resistance is 0 Ω, because the switch, in parallel with them, is assumed to have zero resistance. The circuit is then the equivalent of having the generator connected directly to R_1. The potential difference across R_1 is then the same (90 V) as the voltage supplied by the generator.

6. E When switch S_2 is open, there is a break in the whole circuit and no current flows in any part. We can find the answer to the problem in different ways. One way is to realize that, when no current flows through a resistor, there is no voltage drop across it. Therefore the voltage supplied by the generator will be available at the break, and this is what the voltmeter should measure. Another way is to think of the voltmeter as completing the circuit; the voltmeter than acts as a resistor as well as a meter. The resistance of the voltmeter is high: several thousand ohms. The current through it and the other resistors in the circuit will be small. The IR drop across the other resistors in the circuit (having small resistance) will be negligibly small. Practically the full generator voltage will be across the voltmeter, and the voltmeter will record that voltage (90 V).

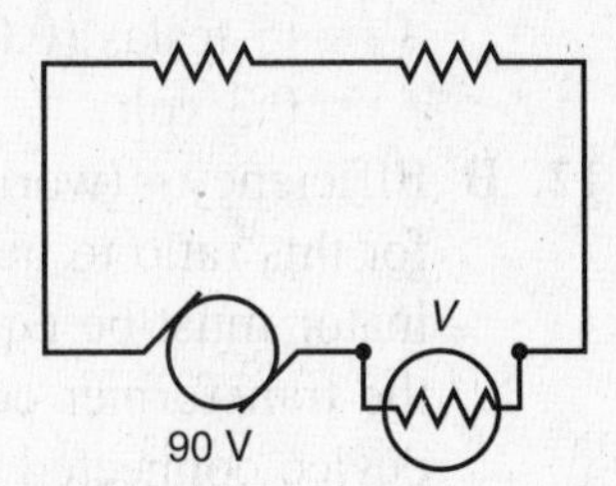

7. E When the magnet approaches the coil, conventional current will flow through the coil and from Y to Z. This follows from Lenz's law. An emf is induced in the coil (because the magnetic field moves with respect to the coil); the direction of the induced current must be such as to oppose the approach of the magnet by producing poles in the coil, as shown. The right-hand rule applied to the coil (thumb pointing in direction of N-pole) then gives the direction of electron flow indicated. When the magnet has moved through the coil and is moving away from it, the direction of the induced current must be reversed; in that case the poles of the coil will be reversed, as needed to oppose the motion. The conventional current will then be from Z to Y.

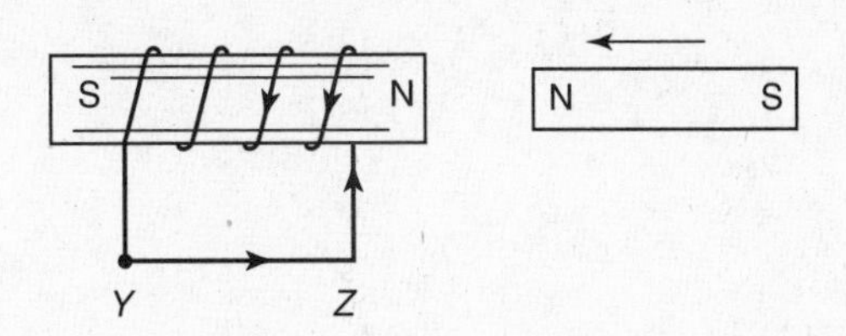

8. B The soft-iron armature X is magnetized by induction: The part on the left becomes an S-pole; the part on the right, an N-pole. The falling of X is then equivalent to a magnet moving near a coil and a current will be induced in the coil. According to Lenz's law, the direction of the current should be such as to oppose the falling of X. This will be the case if the top of the coil becomes an N-pole, opposing the falling N-pole. The right-hand rule for the coil indicates that electron flow in wire Yz is from Y to Z.

9. C There will be a change of current in coil 1; the resulting change in the magnetic field in the iron core affects coil 2 in spite of any insulation. The current is momentary, because an emf is induced only while there is a change in the magnetic field. We can't tell whether the needle moves momentarily to the left or the right because we don't have enough information about the connection to the galvanometer.

10. B The closing of the switch results in conventional flow through *W* from left to right. The right-hand rule for a wire tells us the direction of the resulting magnetic field around the wire: in front of the wire downward, behind the wire upward. The magnetic field due to the two poles is thus reinforced in front, weakened in back of the wire. The resulting force on *W* is toward the back.

11. B $V = \mathbf{B}l\mathbf{v}$

$= (5 \text{ teslas})(0.03 \text{ m})(2 \text{ m/s})$

$= 0.3 \text{ volt}$

12. B Efficiency = (work or power output)/(work or power output). In order for this ratio to be equal to 1 (or 100%), the numerator and denominator must be equal; that is, the power supplied to the primary of the transformer equals the power supplied by the secondary to the device connected to it.

13. E In such a transformer,

$$\frac{\text{number of turns on the secondary}}{\text{number of turns on the primary}} = \frac{\text{secondary emf}}{\text{primary emf}}$$

$$\frac{N_s}{400} = \frac{3{,}000\text{V}}{120\text{V}}; N_s = 10{,}000$$

14. A $\text{Efficiency} = \dfrac{V_s I_s}{V_p I_p} = \dfrac{3{,}000\text{V} \times 0.06 \text{ A}}{120\text{V} \times 2 \text{ A}}$

$= 0.75 = 75\%$

CHAPTER 19

Elements of Electronics

Capacitors and Capacitance

A *capacitor* consists of two conductors separated by an insulator or *dielectric.* The capacitor was formerly called a condenser. Common dielectrics are paper, mica, air ceramic. Capacitors are often named by the dielectric used, such as paper capacitors. In some capacitors the relative positions of the two conductors can be changed. These capacitors are known as *variable capacitors;* the others are called *fixed capacitors.* The *Leyden jar* is an old type of capacitor in which the insulator is a glass jar. The *function* of a capacitor is to store an electric charge. Capacitors are used in tuning a radio, in reducing hum in a radio by smoothing the rectified output of the power supply, and in reducing sparks at electrical contacts. Recall that we use conventional current directions.

How does a capacitor store a charge? Let us first imagine a capacitor in a dc circuit—connect a capacitor to a battery (Figure 19.1). On the negative terminal of the battery is an excess of electrons. These electrons repel each other and we have provided a path for them. They start moving along the connecting wire *A* to *B,* the right plate of the capacitor. At the same time, electrons are attracted from wire *D* and plate *C* of the capacitor to the positive terminal of the battery, where there is a deficiency of electrons. As a result there is a difference of potential between the plates of the capacitor. For every electron that reaches plate *B*, another electron leaves plate *C.* Notice that no electrons move through the dielectric. A motion of charge occurs through the wires until the potential difference between the plates of the capacitor is equal to the emf of the battery. Usually this equality is reached in only a fraction of a second; the capacitor is charged almost immediately.* For a given capacitor the amount of charge that is stored depends on the voltage; if we double the emf of the battery, we double the charge on the capacitor and we double the voltage across the capacitor. If we disconnect the capacitor from the battery, the capacitor can stay charged indefinitely and have a difference of potential between its plates.

*If a resistor is connected in series with the capacitor, the charging current is reduced, so the capacitor gets charged more slowly. The greater the resistance, the slower is the rate of charging.

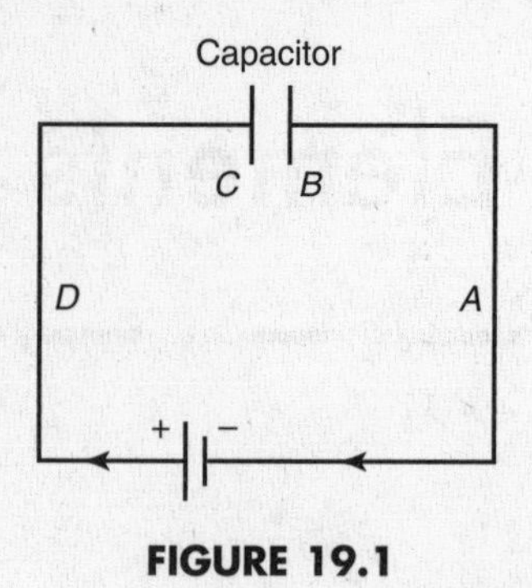

FIGURE 19.1

$$Q = CV$$

where Q is the charge on either plate of the capacitor, C is the capacitance of the capacitor, and V is the voltage or potential difference of the capacitor. If the voltage is expressed in volts and the charge in coulombs, then the capacitance is given if *farads* (F). The farad is a rather large unit. Capacitance is usually expressed in microfarads (μF).

$$1 \text{ farad} = 10^6 \text{ microfarads}$$

The capacitance of a capacitor is proportional to the area of its plates. If we double the area of the plates, we provide twice as much space for storing electrons. The capacitance of a capacitor is inversely proportional to the distance between the plates; if we double the distance between the plates, we reduce the capacitance to one-half of its original value. The capacitance also depends on the nature of the dielectric. Since capacitors store charge at some potential difference (energy per charge), it is appropriate to think of capacitors as storing energy. Indeed, that is one of their functions in photographic strobe lights and other electronic equipment. The actual value of the electric potential energy stored in a capacitor is given by the equation

$$\text{potential energy}_{(\text{capacitor})} = \frac{1}{2}CV^2$$

From the above discussion and formulas, it is evident that

$$\text{potential energy}_{(\text{capacitor})} = \frac{1}{2}CV^2 = \frac{1}{2}QV = \frac{1}{2}\frac{Q^2}{C}$$

Additionally with two capacitors in *series,*

$$\frac{1}{C_{eq}} = \frac{1}{C_1} + \frac{1}{C_2}$$

and with two capacitors in *parallel,*

$$C_{eq} = C_1 + C_2$$

Cathode Ray Tubes

Edison Effect Thomas A. Edison discovered that electric current will flow in a vacuum between a heated filament and a cold metal plate if the positive terminal of a battery is connected to the plate and the negative terminal to the filament.

Thermionic Emission Thermionic emission is the giving off of electrons by a heated metal. The higher the temperature of the metal, the greater is the electron emission. The television picture tube is an example of a vacuum tube that depends on this principle. In such tubes, which are sometimes called *thermionic* tubes, the part that is heated so thatF it will give off electrons is called the *cathode.* The *directly heated* cathode is connected directly to a source of emf. The *indirectly heated* cathode is a sleeve slipped over a filament from which it is insulated. The filament is connected to a source of emf; the hot filament heats the cathode by radiation.

Electron Beams Once a cathode has released electrons by thermionic emission, the electrons can be accelerated to high energies. This is done by an electric field, established between the negative cathode and positive anode (plate). In a cathode ray tube (CRT), when the electrons reach the anode, those moving along the axis find a hole. With their now considerable momentum, they continue on to the face of the tube. No particles inhibit their motion, as the air has been evacuated. Such a tube is called a *vacuum tube.* When the beam's energy is released, the phosphor glows, producing the dot seen when many television sets are turned off.

Television The control circuitry provides for horizontal and vertical movement of the beam. Figure 19.2 shows a tube utilizing electrostatic deflection. If the upper vertical deflection plate is charged positively and the lower one negatively, the beam will be deflected upward. Similar control can be exerted horizontally, giving the beam access to the entire portion of the tube on which a picture must be displayed. Instead of deflecting plates, many television sets use perpendicular coils, which exert a magnetic force according to $\mathbf{F} = q\mathbf{v}\mathbf{B}$.

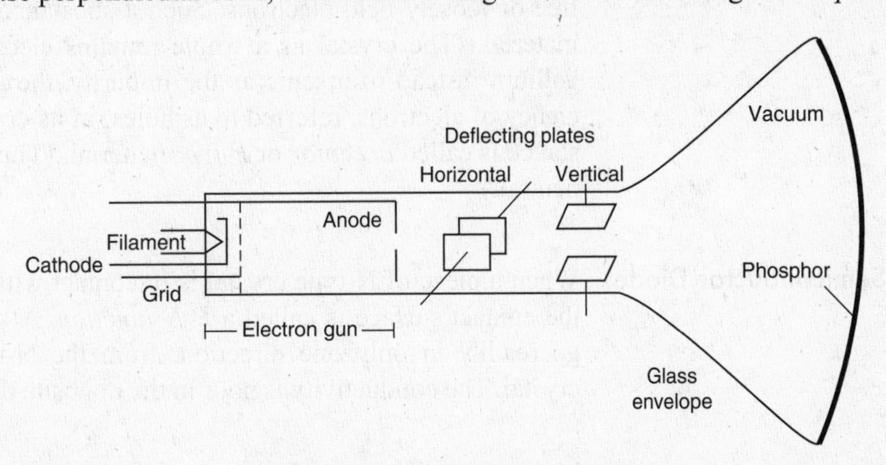

FIGURE 19.2 Cathode Ray Tube

But so far we have only a beam drawing horizontal lines from top to bottom; the result would be a white screen. The part of a television set controlled by the television signal is the grid. When the grid is charged highly negative, most electrons are repelled back to the cathode, and the screen is black. By varying the negativity of the grid, white, black, and shades of gray may be produced. Color television has a similar basis in theory, but the tubes are much more difficult (and expensive!) to manufacture.

In the United States, television pictures consist of 525 horizontal lines drawn every 1/30 second; that is, 30 complete pictures are produced every second. Note that the diagram below (Figure 19.3) is recognizable even though it is drawn with only 11 lines. Television pictures are much sharper! In high-definition television (HDTV), with more than twice as many lines as the current standard, the images are of nearly photographic quality.

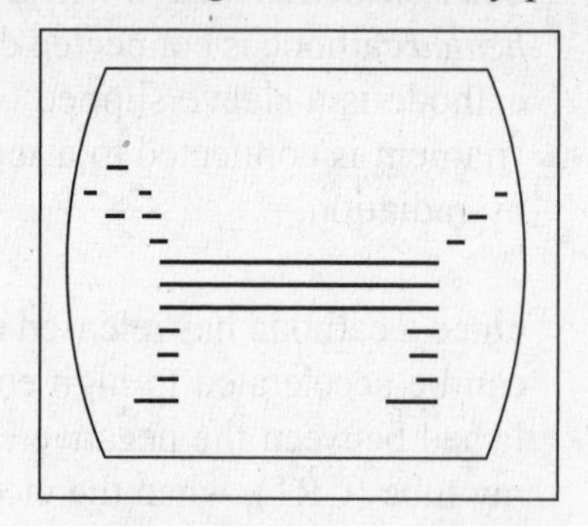

FIGURE 19.3 Scanned Image

Semiconductor Devices

A *semiconductor* is a material whose conductivity is very low in comparison with a conductors such as copper, but greater than that of an insulator such as glass. Common semiconductors are germanium and silicon. In practice a small, precise amount of a selected impurity is added to the pure semiconductor to give it the desired characteristics. For example, 1 part per million of arsenic may be added to a germanium crystal to provide it with free or loosely held electrons. Such a substance is called *donor* or *N-type* material. (The crystal as a whole remains electrically neutral.) If we use gallium instead of arsenic as the impurity, the germanium has an insufficiency of electrons, referred to as holes, in its crystal structure. Such a substance is called *acceptor* or *P-type* material. (This crystal, too, as a whole is neutral.)

Semiconductor Diodes

When a piece of N-type crystal is in contact with a piece of P-type crystal, the contact surface is called a *P-N junction.* At this junction electrons can go readily in only one direction, from the N-type crystal to the P-type crystal. The conductivity is poor in the opposite direction. (See Figure 19.4.)

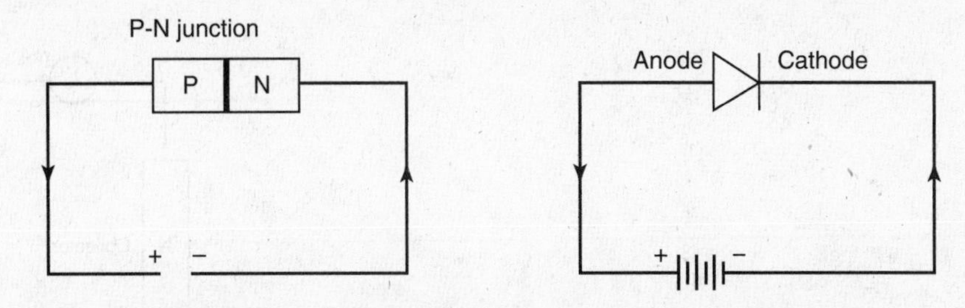

FIGURE 19.4 Conducting Arrangement of Crystal Diode

One of the chief uses of a diode is to *rectify*—that is, to change alternating current to direct current. Figure 19.5 shows the circuit for a *half-wave rectifier,* with the output as seen on a cathode ray oscilloscope.

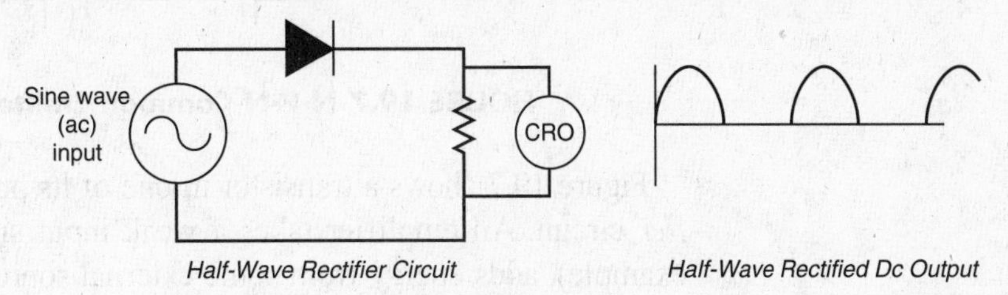

FIGURE 19.5

Transistors

A *transistor* has two junctions, which may be formed by a P-type wafer between two sections of N-type, giving an N-P-N transistor. There can also be a P-N-P type transistor, in which an N-type wafer is between two P-types. (See Figure 19.6.)

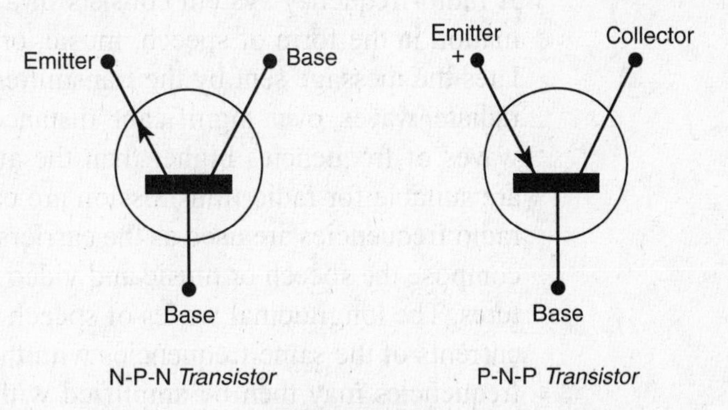

FIGURE 19.6

The miniaturization of radios, calculators, and computers has become widespread because of many factors. The semiconductor diode and transistor require no heat, they can be made very small on a mass-production basis, they require little power, and they can be readily combined in useful combinations in *integrated circuits.*

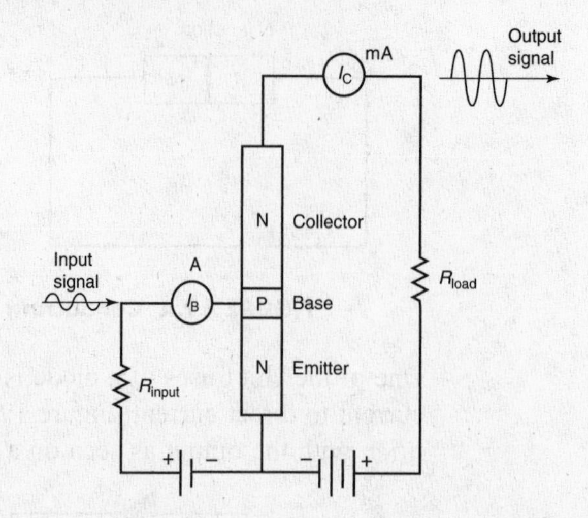

FIGURE 19.7 N-P-N-Common Emitter Amplifier Circuit

Figure 19.7 shows a transistor in one of its primary applications: an *amplifier* circuit. An amplifier takes a weak input signal (from a microphone for example), adds energy from some external source (battery or power line), and delivers an output signal that is a faithful reproduction of the input signal in frequency, but at a much greater amplitude (sufficient to drive a loudspeaker, for example).

Transmitter-Receiver

A radio-frequency system consists of a *transmitter,* which sends out information in the form of speech, music, or video, and a *receiver,* which translates the message sent by the transmitter. It has been found that, in order to radiate waves over significant distances through space, electromagnetic waves of frequencies higher than the audio range are required. Those that are suitable for radio transmission are called *radio frequencies* (RF). These radio frequencies are used as the carriers for the audio frequencies (AF) that compose the speech or music and video frequencies (VF) that compose pictures. The longitudinal waves of speech or music are converted to electrical currents of the same frequencies with the aid of a microphone. These audio frequencies may then be amplified with the aid of transistors. These voltages are then impressed on the RF carrier; the AF is said to *modulate* the carrier.

Amplitude Modulation

There are different ways in which the audio-frequency voltages can modulate the carrier. One of these is *amplitude modulation* (AM): The amplitude of the carrier is changed. The greater the amplitude change of the AF, the greater is the amplitude change of the RF. Notice, in Figure 19.8, that both the top and the bottom of the RF are changed in amplitude to the same degree, and that the outline of both top and bottom of the modulated RF is the shape of the AF. The carrier frequencies used for commercial AM broadcasting are 550 to 1,650 kilohertz (kHz).

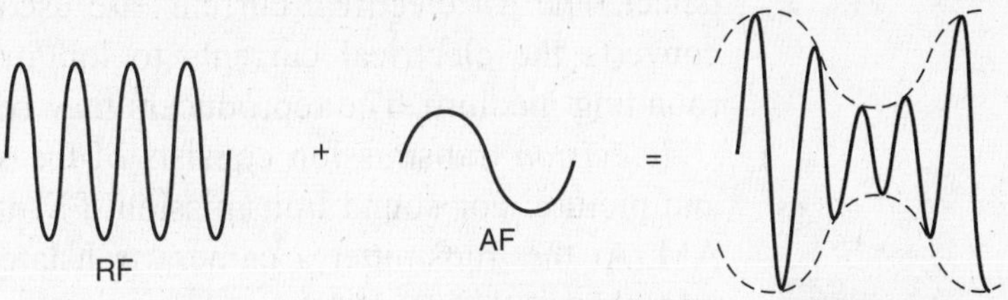

FIGURE 19.8

Frequency Modulation In *frequency modulation* (FM) the frequency of the carrier is changed (Figure 19.9). The greater the amplitude of the AF, the greater is the frequency change of the carrier; that is, the amplitude of the AF determines how much the frequency changes, while the frequency of the AF determines the rate at which this change takes place. The FM band of frequencies is about 88 to 108 megahertz. (One mega- is equal to one million.) An FM system can be designed readily to be relatively static-free as compared with AM.

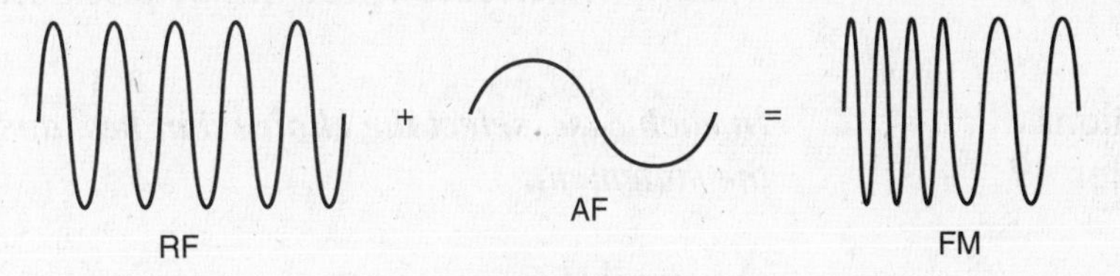

FIGURE 19.9

The *antenna* at the transmitter is used to radiate electromagnetic waves whose frequency and variation are the same as those of the modulated RF described before. These electromagnetic radio waves travel through space with the same speed as other electromagnetic waves such as light: about 3×10^{10} centimeters per second or 3×10^8 meters per second. The length of the antenna at the transmitter is often made about one-half that of a wavelength. This is easily calculated. For example, if we want to transmit a frequency of 100 megahertz:

$$\text{wavelength } (\lambda) = \frac{\text{speed of wave}}{\text{frequency}} = \frac{3 \times 10^8 \text{ m/s}}{100 \times 10^6 \text{ Hz}}$$

$$\lambda = 3 \text{ m}$$

$$\frac{1}{2}\lambda = 1.5 \text{ m}$$

the antenna will be about 1.5 meters long.

At the *receiver,* electromagnetic waves of many frequencies pass the antenna and induce electric currents in it of the same frequency as those of the carrier. When we tune the receiver, we usually vary a capacitor to establish resonance for the desired carrier frequency. In the receiver we need a *detector* circuit in which a semiconductor diode may be used in a circuit similar to the half-wave rectifier described before. In the detector, however, after rectifying the modulated RF, we use a filter circuit to recover the AF that was used to modulate the RF. The detector is said to *demodulate* or detect. The AF electrical currents are used to operate a *reproducer,* which converts the electrical currents to longitudinal sound waves in the surrounding medium. The reproducers may be loudspeakers or earphones.

Television transmission consists of the sending of two messages: sound and picture. For sound transmission, FM is used; for picture transmission, AM. At the transmitter a camera translates information about the picture into electrical signals whose frequency range reaches about 4 megahertz (video frequencies). This is done by a scanning process that is repeated in a somewhat different way in the picture tube of the television receiver.

In television cameras, and elsewhere, an important principle is the photoelectric effect; when light shines on some metals at room temperature electrons are emitted. A *photoelectric cell* is a vacuum tube in which the cathode is made of such metal; another metal is used for the plate or collector to which the electrons go when an external voltage serves to keep the collector positive with respect to the cathode. In a photovoltaic cell a somewhat similar process takes place when light falls on two specially selected materials that are in contact with each other; no external voltage is required.

The photoelectric effect will be discussed further in Chapter 20.

Questions Chapter 19

In each case, select the choice that best answers the question or completes the statement.

Questions 1–4

A certain capacitor consists of two metal plates in air placed parallel and close to each other but not touching. The capacitor is connected briefly to a dc generator so that the difference in potential between the two plates becomes 12,000 volts. The capacitance of the capacitor is 1×10^{-6} microfarad.

1. The charge on either plate is, in coulombs,

(A) 1.2×10^{-8}
(B) 1.2×10^{-6}
(C) 1.2×10^{-4}
(D) 1.2×10^{-2}
(E) 1.2

2. If a thin, uncharged glass plate is then slipped between the two metal plates without touching either, the charge on each plate

(A) will increase by an amount depending on the thickness of the glass
(B) will decrease by an amount depending on the thickness of the glass
(C) will increase by an amount depending on the distance the glass plate was inserted
(D) will decrease by an amount depending on the distance the glass plate was inserted
(E) will not change

3. As a result of the insertion of the glass plate, the potential difference between the plates

(A) increases by an amount depending on the thickness of the glass
(B) increases by an amount depending on the distance the glass was inserted
(C) decreases somewhat
(D) decreases to zero
(E) does not change

4. How much energy, in joules, is stored on the plates?

(A) 6.0×10^{-21}
(B) 6.0×10^{-9}
(C) 1.2×10^{-8}
(D) 7.2×10^{-5}
(E) 1.44×10^{-4}

---

5. When a metal is heated sufficiently, electrons are given off by the metal. This phenomenon is known as

(A) thermionic emission
(B) photoelectric effect
(C) piezoelectric effect
(D) secondary emission
(E) canal ray emission

6. When a radio station is broadcasting a musical program, the antenna of its transmitter

(A) radiates RF electromagnetic waves
(B) radiates AF electromagnetic waves
(C) radiates RF longitudinal waves
(D) radiates AF longitudinal waves
(E) none of the above

7. Cathode rays are

(A) sharp projections on the cathodes of diodes
(B) sharp projections on the cathodes of triodes
(C) striations in a gas discharge tube
(D) proton beams
(E) electron beams

8. A cathode ray oscilloscope has two pairs of deflection plates. If no voltage is applied to one pair, while the voltage applied to the other pair is alternating current of the type used for power in most homes, the pattern on the face of the oscilloscope will be most like

(A) a square wave
(B) a sine wave
(C) a saw-tooth wave
(D) an ellipse
(E) a straight line

9. Two good students, X and Y, connect a diode in series with a battery and a milliammeter. When X performs the experiment, he gets a zero reading on his milliammeter. Y gets the desired half-scale deflection on her milliammeter. If both use identical equipment, the most probable reason for the difference is

(A) X has reversed connections to the milliammeter
(B) Y has reversed connections to the milliammeter
(C) Y has reversed connections to the battery
(D) X has reversed connections to the battery
(E) X has used wire of the wrong color

10. In a cathode ray tube,

(A) electrons go from cathode to plate and then from plate to cathode
(B) electrons go from cathode to plate only
(C) electrons go from plate to cathode only
(D) electrons go from cathode to plate and protons go from plate to cathode
(E) protons go from cathode to plate and electrons go from plate to cathode

11. In a cathode ray tube, electrons for the electron current come from

(A) the grid
(B) the plate
(C) the cathode
(D) the tube
(E) ionization of residual gases in the tube

12. An N-P-N transistor is connected in a common emitter amplifier circuit. A milliammeter is inserted between the collector and emitter. The base current (current between base and emitter) is 30 micro-amperes and the milliammeter reads 10 microamperes. When the base current is changed to 60 microamperes, the milliammeter reading will

(A) not change
(B) become 20 mA
(C) become 5 mA
(D) decrease to some value between 5 and 10 mA
(E) decrease, possibly even to zero

13. The operation of a transistor requires

(A) that the emitter be heated
(B) that the base be heated
(C) that the collector be heated
(D) that the collector be enclosed in a vacuum
(E) none of the above

Explanations to Questions Chapter 19

Answers

1. (A)	**5.** (A)	**9.** (D)	**13.** (E)
2. (E)	**6.** (A)	**10.** (B)	
3. (C)	**7.** (E)	**11.** (C)	
4. (D)	**8.** (E)	**12.** (B)	

Explanations

1. A The capacitance equals the charge on either plate divided by the resulting difference of potential:

$$C = \frac{Q}{V};\ Q$$
$$= CV = (1 \times 10^{-6} \times 10^{-6}\ \text{F}) \times (1.2 \times 10^{4}\ \text{V})$$
$$= 1.2 \times 10^{-8}\ \text{C}$$

2. E Since the glass is an insulator, and the air is an insulator, no path is provided for electrons to go to or from the metal plates. Even accidental touching of the metal by the glass will not result in a significant transfer of electrons, so the charge on each plate will not change.

3. C Since $C = Q/V$, $V = Q/C$. In question 3 we saw that the charge Q did not change. What happened to the capacitance C? By inserting the glass we replaced some of the air by a solid dielectric. This increased the capacitance. Since the denominator (C) increases, while the numerator (Q) remains the same, the value of the fraction (V) decreases.

4. D $\text{P.E.}_{(\text{capacitor})} = \frac{1}{2}CV^2$

$$= \frac{1}{2} \times 1 \times 10^{-6} \times 10^{-6}\ \text{F} \times (1.2 \times 10^{4}\ \text{V})^2$$
$$= 7.2 \times 10^{-5}\ \text{J}$$

5. A Thermionic emission is the giving off of electrons by a metal when it is heated. Secondary emission may occur in vacuum tubes; this is the emission of electrons from a cold electrode, such as the plate, when it is bombarded by fast electrons.

6. A All the waves radiated by an antenna are electromagnetic waves; these are transverse. It is not practical to radiate and transmit AF electromagnetic waves over long distances. When a radio station is on the air, it radiates an RF electromagnetic wave. This wave may be modulated or not, but it is RF.

7. E The term *cathode ray* came into use before the actual nature of the "ray" was known. Further study showed that the rays are fast-moving electron beams.

8. E In the cathode ray oscilloscope a fine beam of fast-moving electrons is accelerated toward a fluorescent screen, as in a TV picture tube. Where the electrons hit the screen there is a bright spot of light. One pair of deflection plates deflects the electrons horizontally, the other pair vertically, when a difference of potential is applied to the plates. The amount of deflection depends on the magnitude of the applied voltage. When an ac voltage is applied to one pair of plates, the fast-moving electrons will be deflected varying amounts, depending on the instantaneous voltage. Therefore, on the screen there will be a succession of bright spots of light, which will appear as a line. If no voltage is applied to the other pair of plates, the deflection is along one straight line, and the pattern on the fluorescent screen (the face of the oscilloscope) will be a straight line.

9. D Study the diagram of the half-wave rectifier (Figure 19.5). You will see that reversing the connections to the meter would result in a negative reading. However, no current can go from donor to acceptor unless the donor is connected to the negative terminal of the battery (through the milliammeter). If the battery connections were reversed, the acceptor would be negative with respect to the donor and repel electrons. Therefore, X has probably reversed connections to the battery.

10. B A cathode ray tube is like a check valve; electrons can go only one way through it, from cathode to plate. Ordinarily there are no protons in motion in a vacuum tube.

11. C In the cathode ray tube we depend on thermionic emission to yield the electrons that move through the space in the tube. The cathode is the only element that can act as the emitter of electrons in the vacuum tube.

12. B The common emitter amplifier is a *proportional* amplifier; that is, small changes in the base-emitter circuit are reproduced proportionally in the collector-base circuit. Thus, with an increase from 30 to 60 μA in the base current, the milliameter reading will change from 10 to 20 mA.

13. E One of the advantages of a transistor is that no heated cathode is used to supply electrons. Electrons are made available by using semiconductors, which have free electrons. The transistor is a solid-state device; the motion of electrons in it takes place within a solid and no vacuum is needed.

CHAPTER 20

Quantum Theory and Nuclear Physics

In discussing geometrical optics, we were concerned only with effects that could be explained by light traveling in straight lines. In physical optics, we discovered that light could diffract around obstacles, implying that light has wave characteristics. In 1905, Albert Einstein gave a surprising explanation of an effect that implies that light can exhibit particle characteristics.

Photoelectric Effect

When visible light hits most metals, electrons remain bound to the atoms. However, when high-frequency electromagnetic radiation hits metals, electrons are emitted. This is called the *photoelectric effect* (see Figure 20.1), and the electrons emitted are termed *photoelectrons*. Einstein's theory invokes the *quantum theory* of light, which states that light is emitted and absorbed in little bundles of energy. Historically, the notion of the quantum of energy was introduced by Max Planck in 1900 to explain the continuous spectrum produced by a hot blackbody. This ideal object is also a perfect absorber of light.

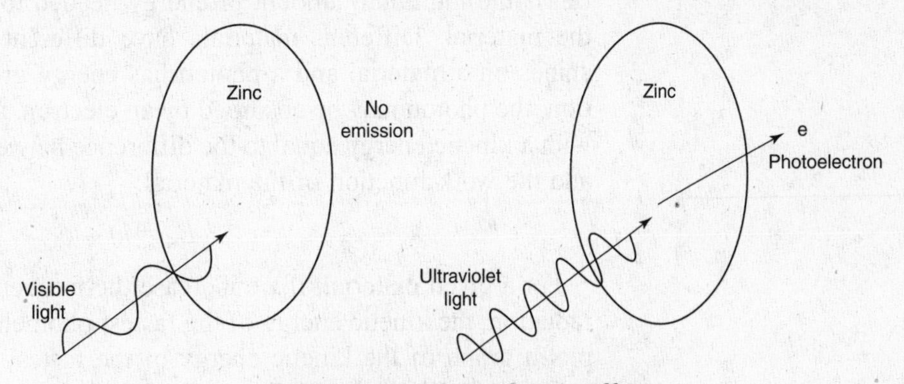

FIGURE 20.1 Photoelectric Effect

The photoelectric effect has some surprising features. If careful measurements are made with monochromatic light, the following observations can be made:

1. If the intensity of the light is increased, the *number* of emitted electrons increases.
2. If the intensity of the light is increased, the *speed* of the emitted electrons does not increase.

3. If we decrease the frequency of the light gradually, we reach a frequency below which no electrons are omitted. This is known as the *threshold frequency*. Not matter how much we increase the intensity of the radiation, if it is below the threshold frequency, no electrons will be emitted by the substance.
4. For each substance, the speed of the emitted electrons varies. The maximum speed and kinetic energy of the electrons increase with the frequency of the incident radiation.
5. Different substances have different threshold frequencies.
6. If the frequency that is used is above the threshold frequency, even the weakest radiation will result in the emission of electrons, and their maximum speed will be the same as that for very intense radiation of the same frequency.

Only the first of these observations can be easily explained by the wave theory. Albert Einstein explained them all simply by using an expanded concept of the quantum. He suggested that we think of electromagnetic energy as being granular, traveling through space as little "grains," and being emitted and absorbed in this way. The amount of energy, or quantum, in each grain is given by Planck's formula, $E = hf$, where f is the frequency of the radiation. (The term *photon* is often used to refer to the grain, and *quantum* to the amount of energy of the grain; sometimes the terms are used interchangeably). All photons in a monochromatic radiation of a different frequency are the same, but they are different from the photons in radiation of a different frequency. The higher the frequency, the greater is the energy of the photon. All photons travel with the same speed, the speed of light.

Energy has to be used to remove an electron from a material. Some electrons are removed more readily than others. The *work function* (ϕ) of a material is the minimum amount of energy needed to remove an electron from the material. Different materials have different work functions. If light shines on a material and a photon has energy greater than the work function, the photon may be absorbed by an electron, and the electron is emitted with a kinetic energy equal to the difference between the quantum of energy and the work function of the material:

$$E_K = hf - \phi$$

For a given material, if we increase the frequency of the monochromatic radiation, the kinetic energy of the fastest photoelectron will increase. If we plot a graph of the kinetic energy of the fastest electrons against the frequency of radiation used, the result is a straight line (Figure 20.2).

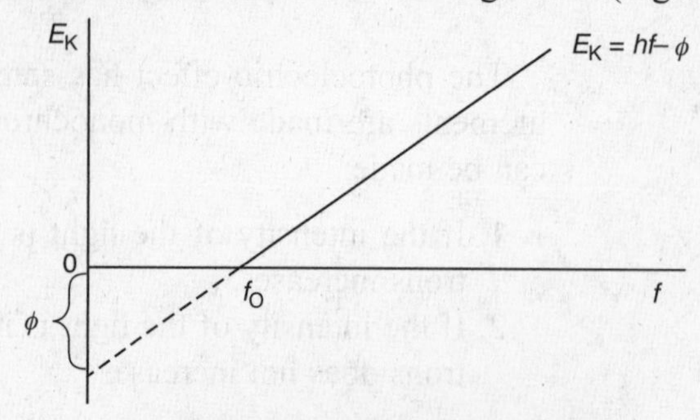

FIGURE 20.2 Photoelectric Effect

The plots for other substances will be straight lines parallel to this one. The slope of the straight line is Planck's constant, $h = 6.63 \times 10^{-34}$ joule-second. The magnitude of the y-intercept is the work function of the material. The photon with that amount of energy is barely able to remove an electron from this material and has a frequency equal to the threshold frequency:

$$\phi = hf_o$$

Example: Suppose that we have a material whose work function is 3.0×10^{-19} joule, and light with a wavelength of 4,000 angstrom units is shining on it. Let's calculate a few relevant quantities.

What is the energy of each photon?

$$\begin{aligned} E &= hf = hc/\lambda \\ &= (6.63 \times 10^{-34}\ \text{J} \cdot \text{s})(3 \times 10^8\ \text{m/s})/(4{,}000 \times 10^{-10}\ \text{m}) \\ &= 5.0 \times 10^{-19}\ \text{J} \end{aligned}$$

What is the speed of the fastest photoelectron?

$$\begin{aligned} E_K &= hf - \phi \\ &= 5.0 \times 10^{-19}\ \text{J} - 3.0 \times 10^{-19}\ \text{J} \\ &= 2.0 \times 10^{-19}\ \text{J} \end{aligned}$$

$$\begin{aligned} E_K &= \tfrac{1}{2}mv^2 \\ 2.0 \times 10^{-19}\ \text{J} &= \tfrac{1}{2}(9.1 \times 10^{-31}\ \text{kg})\ v^2 \\ v &= 6.6 \times 10^5\ \text{m/s} \end{aligned}$$

When dealing with small energies such as those possessed by electrons and photons, we often find it convenient to introduce another unit of energy. The *electron volt* (eV) is the energy that one electron acquires when it is accelerated through a difference of potential of 1 volt.

$$1\ \text{eV} = 1.6 \times 10^{-19}\ \text{J}$$

With the quantum theory verified and expanded by Einstein's explanation of the photoelectric effect, more discoveries came rapidly.

Atomic Models

The existence of atoms was first theorized by the ancient Greeks. They proposed a unit of matter so small as to be indivisible. When J. J. Thompson discovered the electron, he proved that the "atom" discussed in laboratories is, in fact, divisible. He proposed an atomic model wherein electrons were scattered through a medium of positive charge like raisins in a pudding.

Ernest Rutherford performed a number of experiments in which alpha particles given off by radioactive materials were shot at thin gold foil. On the basis of observations and theoretical calculations he drew (1910) the following conclusions:

1. Atoms are mostly empty space.
2. Practically all of an atom's mass is concentrated in a very small nucleus.
3. The nucleus is positively charged because of its protons.

4. The neutral atom has as many electrons outside the nucleus as there are excess protons in the nucleus. (Neutrons were not discovered until 1932. Rutherford assumed that there were enough other protons and an equal number of electrons in the nucleus to provide the rest of the mass that an atom has.)

The electrons are relatively far from the nucleus. Alpha particles shot into the atom are scattered at various angles because of the coulomb force of repulsion between its positive charge and the positive charge of the nucleus. The closer its path is to the nucleus, the greater is the angle of scattering. Unfortunately, the Rutherford model of the atom was unstable. If atoms were so constructed, the universe would collapse in a small fraction of a second as electrons spiralled into positive nuclei, radiating energy as they went.

Niels Bohr (1913) used the Rutherford model of the atom and quantum theory to develop a stable atomic model. He specifically considered hydrogen because it consists of only two objects (an electron and a proton) and because the known spectrum of hydrogen at that time consisted of a few lines in the visible and infrared whose wavelengths were well known. Bohr postulated:

1. The electron of the hydrogen atom revolves around its nucleus (a proton) in only definite, allowed orbits. No energy is radiated by the atom while the electron is in these orbits.
2. The most stable orbit is the one closest to the nucleus. In this condition the atom (or the electron) is said to be in the *ground state.* The electron then has the least amount of energy with respect to the nucleus.
3. The atom can be raised to higher or excited energy states that are fairly stable. Only definite, quantized amounts of energy can be absorbed for changes to higher energy states or levels. (Bohr thought of this energy in terms of the amount of work required to pull the negative electron away from the positive nucleus.) For example, 10.2 electron volts are needed to raise an atom to the second energy level, and 12.08 electron volts to reach the third level from the ground state.
4. When the atom returns to a lower energy state, it can jump back only to one of the allowed energy states, and the difference in energy is given off as a photon. The jump may be to the ground state or an intermediate one. The hydrogen lines in the visible spectrum had been observed by Balmer; they involve electrons falling to the $n = 2$ state.
5. Only specified orbits are permitted. For all orbits of the electron; $mvr = \frac{nh}{2\pi}$, where m is the mass of the electron, v its orbital speed, r the radius of the orbit, h Planck's constant. The produce mvr is *known as the angular momentum* of the electron due to its orbit around the nucleus.

Figure 20.3 shows some of the energy levels of hydrogen. On the left, under *n,* the allowed levels are numbered, starting with the ground state as number 1. These numbers are known as the *principal quantum numbers.* On the right, under "eV," are two columns of numbers. Both give the energy of each level in electron volts. The left column uses the ground state as the reference level and labels it the *zero level.* Since it is the lowest energy level, the other levels are positive with respect to it. Frequently the ionization state is taken as the reference and is assigned the zero value. The other levels are lower in energy than the ionization state, and therefore the numbers are negative. The difference between any two energy levels is, of course, the same (allowing for significant figures) in both columns. Notice, for example, that both columns tell us that 13.6 electron volts are needed to ionize the hydrogen atom from the ground state. If the atom is in the $n = 3$ state, only 1.5 electron volts are required to ionize the atom; but if a photon whose energy is 1.5 electron volts hits a hydrogen atom in the ground state, it will not be absorbed. The collision is elastic.

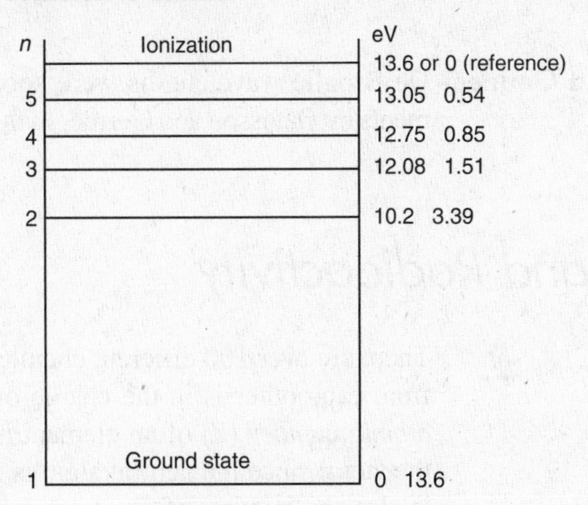

FIGURE 20.3 Hydrogen

Slow Collisions

When a stream of electrons passes through a gas, the electrons may lose energy as they collide with the atoms of the gas. In some collisions practically no energy is lost; in other collisions practically all the energy may be lost by the electron. For example, an electron having an energy of 7 electron volts may lose nearly all of it's energy when hitting the same atom. Careful experiments show that during such collisions atoms can absorb only definite bundles—quanta—of energy.

Compton Effect

In the description of the photoelectric effect we stressed a very important characteristic. In order to raise the hydrogen atom to a higher energy state, such as from the ground state to $n = 2$, the photon that arrives must have the exact quantum of energy, 10.2 electron volts. If it has 11.5 electron volts it will not raise the electron to a higher level. In 1923, A. H. Compton discovered that, if very high energy photons hit a material like carbon, a different effect is observed. When he used high-energy X rays, he found that some of the scattered X rays had a lower frequency than the original X rays. He was able to explain this effect quantitatively by reasoning that each photon has a momentum of hf/c ($= h/\lambda$); the photon collides with an electron in the carbon atom and loses some of its energy and momentum to the electron; energy and momentum are conserved. The photon behaves like a particle.

$$\text{momentum of the photon} = \frac{\text{Planck's constant}}{\text{wavelength}}$$

$$p = \frac{h}{\lambda}$$

De Broglie Wavelength

In 1924, a young doctoral candidate named Louis de Broglie reasoned that, if photons could display momentum, particles should be able to display wave characteristics. If the momentum of a photon = hf/c, then

$$\lambda = \frac{h}{m\mathbf{v}}$$

It is indeed possible to calculate the wavelength, sometimes called a *matter wave,* for any known mass whose velocity is specified.

Davisson and Germer

De Broglie wavelengths were soon verified in electron diffraction experiments by Davisson and Germer in the United States, and Thompson in England.

Atoms and Radioactivity

There are over 100 different chemical elements. One way in which they differ from each other is in the charge of the nucleus of the individual atom. The *atomic number* (Z) of an element is the number of protons in the nucleus of the atom; since the neutral atom is electrically uncharged, the atomic number is also the number of electrons surrounding the nucleus. (Hydrogen has an atomic number of 1; oxygen, an atomic number of 8; uranium, an atomic number of 92.) All elements whose atomic numbers are greater than 92 are man-made; they are known as the *transuranic* elements. Many of these were first made in the United States; one is nobelium ($Z = 102$). Lawrencium has an atomic number of 103.

All atoms, except those of common hydrogen, contain neutrons in the nucleus in addition to protons. The *mass number* is the total number of protons and neutrons in the nucleus. Most chemically pure elements contain atoms that are identical in atomic number but different in mass number because they differ in the number of neutrons; these are called *isotopes* of the element.

The actual mass of the proton is extremely small; the mass of the neutron is slightly greater. Each of these is about 1,840 times as massive as the electron. The mass of the nucleus is less than the sum of the masses of the *nucleons* (protons and neutrons) that compose it. Einstein had predicted that mass and energy might be converted into one another.

$$E = mc^2$$

The equivalent amounts are given by the relation in which m is the mass converted, **c** the speed of light, and E the equivalent amount of energy. If m is in kilograms, and **c** in meters per second, E is in joules. The *binding energy* of a nucleus is the energy equivalence of the difference in mass between the nucleus and the sum of the masses of the nucleons; it is the energy that would have to be supplied to break the nucleus into its component protons and neutrons. The nature of the nuclear force that holds the nucleus together is not fully understood. It is greater than electromagnetic and gravitational forces, but acts over short distances only.

Henri Becquerel discovered nuclei that broke up spontaneously, and in this process were found to give off some "radiation" that ionized the chemicals in a photographic emulsion. Such nuclei are said to be *radioactive.* This radioactivity is not affected by temperature or any other treatment we can give. Natural radiation was found to consist of three different types of particles; they are called alpha, beta, and gamma. *Alpha particles* are helium nuclei; *beta particles* are electrons; *gamma rays* are electromagnetic waves similar to X rays, but of higher frequency.

An element can change into a different element; in this process, known as *transmutation,* a charged particle is emitted or absorbed by a nucleus. The *half-life* of a radioactive element is the length of time required for one-half of a given mass of an element to transmute into a different element. For example, an ejected particle may be a proton, an alpha particle, or an electron. Although not normally present in the nucleus, electrons may be ejected, since a neutron can change into a proton and an electron upon disintegration.

Nuclear Changes

Atomic Mass Unit

When dealing with nuclear changes we often find it convenient to use a small mass unit. The *atomic mass unit* (*amu* or *u*) is defined as one-twelfth of the mass of an atom of carbon-12, that is, of the isotope of carbon that has a mass number of 12. One atomic mass unit then turns out to be approximately equal to 1.66×10^{-27} kilograms. Often conventional energy units are used instead of mass units:

$$\begin{aligned} E &= mc^2 \\ &= (1.66 \times 10^{-27}\ \text{kg})(3 \times 10^{6}\ \text{m/s})^2 \\ &= 1.49 \times 10^{-10}\ \text{J} \\ &= \frac{1.49 \times 10^{-10}}{1.6 \times 10^{-19}}\ \text{eV} \\ &= 0.931 \times 10^{9}\ \text{eV} \end{aligned}$$

In other words, 1 atomic mass unit is approximately equal to 931 million electron volts.

Some Nuclear Reactions

We mentioned above that there are naturally radioactive elements and that their nuclei break up. Changes in nuclei also occur in the laboratory as a result of bombardment with fast-moving particles. In his experiments Rutherford used alpha particles obtained from radioactive substances. Now bombardment is frequently done by particles speeded up with particle accelerators. Let us look at some equations that represent the reactions.

In all of these nuclear reactions, atomic number and mass number are conserved; that is, the algebraic sum of the atomic numbers on the left side of the equation must equal the sum on the right; and the sum of the mass numbers on the left side must equal the sum on the right. Atomic number is written as a subscript of the chemical symbol of the element, and mass number as a superscript.

Natural Radioactivity–Alpha Decay

$$^{226}_{88}\text{Ra} \rightarrow {}^{222}_{86}\text{Rn} + {}^{4}_{2}\text{He}$$

Note that the atomic numbers on each side add up to 88 and the mass numbers to 226. You will recall that the alpha particle is the nucleus of the common helium atom, which escapes this reaction with great speed. This is an example of *alpha decay.*

Natural Radioactivity–Beta Decay

$$^{24}_{11}\text{Na} \rightarrow {}^{24}_{12}\text{Mg} + {}^{0}_{-1}\text{e}$$

In beta decay an electron (the negative beta particle) is given off. Note that its mass number is zero and its charge is –1. For the electron and neutron, atomic number is replaced by charge. Algebraically the subscripts on the right add up to 11, which is the starting number on the left side. In beta decay neutrinos are also given off, a fact not always shown. The charge and mass of the neutrino are zero. Here we do not distinguish between neutrino and antineutrino.

$$^{99}_{43}\text{Tc} \rightarrow {}^{99}_{43}\text{Tc} + \gamma$$

Natural Radioactivity–Gamma Radiation

Gamma radiation occurs during a reorganization of nuclear energies. Atomic numbers are unaffected; that is, the element stays the same.

First Artificial Transmutation

$$^{14}_{7}\text{N} + {}^{4}_{2}\text{He} \rightarrow {}^{17}_{8}\text{O} + {}^{1}_{1}\text{H}$$

Bombarding nitrogen with alpha particles in 1919, Rutherford became the first person to convert one element into another in a laboratory.

Discovery of the Neutron

$$^{9}_{4}\text{Be} + {}^{4}_{2}\text{He} \rightarrow {}^{12}_{6}\text{C} + {}^{1}_{0}\text{n}$$

In 1932, Chadwick noted that, when beryllium is bombarded with alpha particles, a free neutron is emitted along with the production of an isotope of carbon. Note that the mass number of the neutron (like that of the proton) is 1, but its charge is zero.

Note also that both of the above artificial transmutations require only the input energies of alpha particles resulting from natural radioactivity. Now we shall consider some machines used in physics for higher energy experimentation.

Particles and Particle Accelerators

Particles

As indicated above, it is believed that the nuclei of all atoms except common hydrogen contain protons and neutrons (the *nucleons*). The nucleus is surrounded by electrons. In the neutral atom, the number of these electrons is equal to the number of protons in the nucleus (the atomic number). Other particles have been discovered, some after their existence was predicted by theory. Many more than 100 of these particles, referred to as *elementary particles,* are now known. Some appear to fit the ancient Greek concept of indivisible matter. As yet, there is no theory available that accounts completely for the existence, characteristics, and behavior of these particles. Most are now believed to be composed of a small number of *quarks,* whose charge is thought to be only one-third of an electron's. For example, it is believed that the proton and neutron contain three quarks each. Particle accelerators have been built to obtain more facts and examine theories on the nature of matter.

In 1934, Hideki Yukawa predicted the existence of a particle with mass intermediate between that of the electron and that of the proton. Later several different ones were actually discovered and are called *mesons.* Neutral mesons as well as positive and negative mesons have been found. The negative pi-meson fits Yukawa's description. The pi-meson (also known as the *pion*) has a mass about 270 times that of the electron and is unstable. The breakdown of the pi-meson results in the production of a lighter particle, the *muon.*

Particles heavier than the neutron have been found. An example is the upsilon particle, which has a mass about 10 times that of the proton.

Antiparticles have also been discovered. The antiparticle of a charged particle has the same mass as the corresponding particle but an opposite charge. The *positron* is the antiparticle of the electron.

These various particles are often referred to as elementary particles, but this term no longer implies that they are indivisible. For example, the neutron can last indefinitely inside the nucleus, but outside the nucleus it is unstable and decays into a proton and electron. In this and similar reactions conservation of energy and conservation of momentum suggested the existence of the *neutrino*—a neutral particle with practically zero rest mass. The concept of such a particle was introduced by Pauli in 1931 and advanced by Fermi, who suggested the name. The neutrino was detected experimentally in 1956.

The list of elementary particles usually includes the *photon.* It always moves with the speed of light, has zero rest mass, and has an amount of energy depending on the frequency of the electromagnetic wave it represents.

Like the electron, the muon has a neutrino associated with it. The *tau* particle, still heavier than the muon, also has a nuetrino associated with it. All have antiparticles.

Particle Accelerators

Particle accelerators are helpful for examination of the atomic nucleus. The greater the energy of the particles used to bombard or "smash" the nucleus, the more detail the physicist expects to find about the structure of the nucleus and its particles. This bombardment has led to the discovery of some of the particles mentioned above.

The energy possessed by the bombarding particles is usually described in terms of million electron volts (MeV) or billion electron volts (GeV: G is for *giga-;* it equals 10^9). One *electron volt* is the energy that one electron acquires on being accelerated through a difference of potential of 1 volt. Electrons or protons are frequently used as the particles or bullets of the atom smashers. These, like other charged particles, can be accelerated by being placed in an electric field. In the *Van de Graaff* machine and in the *linear accelerator* the particle gains speed along a straight path. In the linear accelerator a series of cylindrical electrodes is used, and the particles acquire more and more energy as they pass between successive pairs of electrodes. To achieve high energies a very long evacuated pipe is required.

In order to avoid the need for such long pipes, circular particle accelerators have been constructed. The equivalent length of path is achieved by having the particles go around the circular path many times. The charged particles are forced to move in a circular path by a magnetic field directed at right angles to the path of the particles. The first of these accelerators was the *cyclotron,* invented by E. O. Lawrence in 1932. In the cyclotron, a flat, evacuated cylindrical box is placed between the poles of a strong electromagnet. Inside the box are two hollow, D-shaped electrodes. The protons or deuterons to be accelerated are fed into the space between the electrodes. An ac voltage of high frequency (about 10^7 hertz) is applied to the electrodes, and the particles are accelerated across the gap. (See Figure 20.4.)

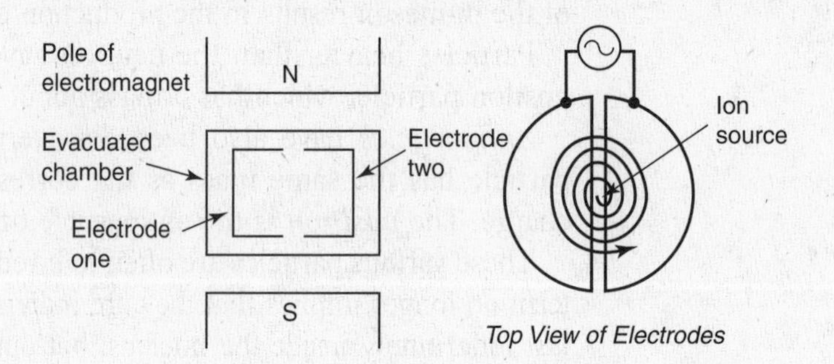

FIGURE 20.4

The magnetic field bends the moving particles into a circular path. The voltage reverses in time to accelerate the particles emerging from the electrode. After 100 trips around the cyclotron, the particles may acquire an energy of several million electron volts. Successive semicircular paths in the electrodes are of larger radius, but each of these is covered in the same length of time because the greater speed of the particles compensates for the increased length of path.

When a proton has acquired a speed corresponding to an energy of about 10 million electron volts, a relativity effect sets in; the particle's mass starts to increase noticeably with further increases in energy. Therefore, the particle takes longer to cover each successive semicircle. The frequency of the ac voltage applied to the dees must therefore change to compensate for this increase in mass. This is done in the *synchrocyclotron,* which uses a constant magnetic field but an electric field of changing frequency. (It uses a frequency-modulated oscillator.) In the *synchrotron* both the magnetic field and the frequency of the electric field are changed. The results are a circular path of almost constant radius for the particles and also a saving of material for the magnet. The largest accelerator currently operational in the United States is at Fermilab in Batavia, Illinois.

Fission and Fusion

Fission

Fission is the splitting of a massive nucleus into two large fragments with the simultaneous release of energy and some particles that can also produce fission. For example, the massive uranium nucleus (mass number 235) fissions on capture of a neutron and releases a large amount of energy and two or more neutrons. The total mass of all the particles produced is less than the mass of the original uranium atom plus neutron. The difference in mass accounts for the energy produced in the form of kinetic energy (fast-moving particles) and gamma rays. If the uranium is arranged so that the released neutrons will, in turn, fission other uranium nuclei, we have a *chain reaction.* This is done in a *nuclear reactor* or pile, in which the fissionable material is arranged so that the release of energy can be controlled. Boron rods are used as *control rods:* boron absorbs neutrons readily; and if the pile gets too active, the rods are inserted to lower the rate. Cadmium may also be used on the control rod. Fission of uranium-235 is most likely to proceed with slow-moving neutrons; *moderators* serve to slow down the neutrons. Moderator materials are graphite and deuterium (used in heavy water). In the so-called A-bomb or fission bomb, relatively large amounts of fissionable material are split in a short time with the consequent release of tremendous amounts of energy and radioactive material. Radiation hazards in the case of a nuclear reactor breakdown and in the disposal of nuclear wastes are a serious concern.

Fusion

In *fusion* the nuclei of some light elements such as lithium and hydrogen combine; again there is a loss of some mass with the consequent release of energy. The fusion reaction requires a temperature of millions of degrees. This is, therefore, called a *thermonuclear reaction* and has been achieved in an uncontrollable way in the H-bomb; here an A-bomb is first used to produce the required high temperature. Scientists are working toward achieving a controllable fusion reaction, which will make available tremendous amounts of energy everywhere by the use of the hydrogen in water.

Questions
Chapter 20

In each case, select the choice that best answers the question or completes the statement.

1. Of the following, the particle whose mass is closest to that of the electron is the

(A) positron
(B) proton
(C) neutron
(D) neutrino
(E) deuteron

2. When a beta particle is emitted from the nucleus of an atom, the effect is to

(A) decrease the atomic number by 1
(B) decrease the mass number by 1
(C) increase the atomic number by 1
(D) increase the mass number by 1
(E) decrease the atomic number by 2

3. Gamma rays consists of

(A) helium nuclei
(B) hydrogen nuclei
(C) neutrons
(D) high-speed neutrinos
(E) radiation similar to X rays

4. Neutrons penetrate matter readily chiefly because they

(A) occupy no more than one-tenth of the volume of electrons
(B) occupy no more than one-tenth of the volume of protons
(C) have a smaller mass than protons
(D) are electrically neutral
(E) are needlelike in shape

5. It is characteristic of alpha particles emitted from radioactive nuclei that they

(A) are sometimes negatively charged
(B) usually consist of electrons
(C) are helium nuclei
(D) are hydrogen nuclei
(E) are the ultimate unit of positive electricity

Questions 6 and 7

In the accompanying diagram, *R* is a natural radioactive material. *F* is a fluorescent screen. A magnetic field is to be imagined perpendicular down into the paper. Bright spots appear on the screen at *T, S,* and *P.* The space between *R* and the screen is evacuated.

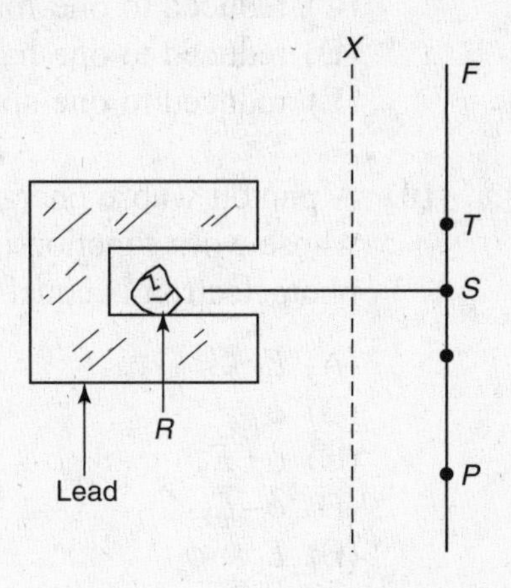

6. The emanation from *R* that produced spot *P* consists of

(A) electrons
(B) protons
(C) alpha particles
(D) north poles
(E) south poles

7. If paper is inserted at *X,* which of the following is (are) most likely to be blocked by the paper?

(A) north pole
(B) south pole
(C) electrons
(D) alpha particles
(E) gamma rays

---

8. When describing isotopes of the same element, the most accurate statement is that they have

(A) the same spin as a top
(B) the same atomic mass but different atomic numbers
(C) the same atomic number but different atomic masses
(D) the same chemical properties and therefore cannot be separated
(E) a coexistence limit; that is, no element can have more than three isotopes

9. A metal plate is illuminated with monochromatic light, and photoelectrons are observed to come off. If the intensity of this light is reduced to one-fourth of its original value, the maximum kinetic energy of the photoelectrons will be

(A) zero
(B) unchanged
(C) reduced to one-fourth of the original value
(D) reduced to one-half of the original value
(E) reduced to one-sixteenth of the original value

10. A photon whose energy is E_P joules strikes a photosensitive surface whose work function is ϕ joules. The maximum energy of the ejected photoelectron is equal to

(A) E_P
(B) ϕ
(C) $\phi + E_P$
(D) $\phi - E_P$
(E) $E_P - \phi$

11. In the nuclear reaction

$$^{2}_{1}\mathrm{H} + {}^{3}_{1}\mathrm{H} \rightarrow {}^{4}_{2}\mathrm{He} + {}^{1}_{0}\mathrm{n} + Q$$

Q represents the energy released. The masses of the nuclei, in atomic mass units, are as follows: $^{2}_{1}\mathrm{H} = 2.01472$, $^{3}_{1}\mathrm{H} = 3.01697$, $^{4}_{2}\mathrm{He} = 4.00391$, $^{1}_{0}\mathrm{n} = 1.00897$. This reaction is primarily an example of

(A) fission
(B) fusion
(C) ionization
(D) alpha decay
(E) neutralization

12. In the reaction shown in question 11, the value of Q, in atomic mass units, is closest to

(A) 5.03169
(B) 5.01288
(C) 0.01881
(D) 5.01288
(E) 2.01472

13. In the nuclear reaction shown below, what is the value of the coefficient y?

$$^{235}_{92}\mathrm{U} + {}^{1}_{0}\mathrm{n} \rightarrow {}^{144}_{56}\mathrm{Ba} + {}^{89}_{36}\mathrm{Kr} + y\,{}^{1}_{0}\mathrm{n}$$

(A) 0
(B) 1
(C) 2
(D) 3
(E) 4

Explanations to Questions Chapter 20

Answers

1. (A)	**5.** (C)	**9.** (B)	**13.** (D)
2. (C)	**6.** (A)	**10.** (E)	
3. (E)	**7.** (D)	**11.** (B)	
4. (D)	**8.** (C)	**12.** (C)	

Explanations

1. A The positron is sometimes referred to as a positive electron to suggest that it is similar to the negative electron in all respects except electric charge. The mass of the neutrino is zero or practically zero; the masses of the proton and neutron are more than 1,800 times as great as the mass of the electron. The deuteron is a combination of a proton and neutron and forms the nucleus of an isotope of hydrogen. The triton is the nucleus of the heavy hydrogen isotope containing one proton and two neutrons.

2. C The emission of an electron from the nucleus is the result of the break-up of a neutron in the nucleus into an electron and a proton. The electron is emitted, leaving the proton in the nucleus and thus raising the positive charge in the nucleus by 1. Since atomic number is defined as the number of protons in the nucleus, the atomic number went up by 1.

3. E Gamma rays and X rays are electromagnetic waves. The wavelength of a gamma or X ray depends on the method of production, but is roughly about 1 Å or 10^{-8} cm.

4. D Neutrons are electrically neutral. The volume of a neutron is about the same as that of an electron or proton. The mass of a neutron is slightly greater than that of a proton. There is a great deal of empty space in all atoms; because neutrons are electrically neutral, they can go through this space near electrons and nuclei without experiencing electric forces.

5. C Alpha particles emitted from radioactive sources are helium nuclei; that is, an alpha particle has a positive charge and consists of two protons and two neutrons. Hydrogen nuclei are protons. Both protons and positrons are smaller units of positive charge, each having one-half of the charge of the alpha particle.

6. A The magnetic field at right angles to the path of the emanation from the radioactive material deflects positive charges in one direction, negative charges in the opposite direction. Gamma rays are not deflected at all and produce a spot at *S*. Electrons are usually deflected more than alpha particles because they have a much smaller mass. Based on the orientation of the magnetic field, electrons will be deflected to position *P*.

7. D No north or south poles are emitted. (Furthermore, they occur only in pairs.) In natural radioactivity, gamma rays are most penetrating. Beta particles are next, and then come alpha particles, which are most likely to be blocked by the paper.

8. C Isotopes of the same element have the same atomic number but differ in atomic mass because their nuclei differ in the number of neutrons. They are similar in chemical properties, but many methods have been devised for their separation. One of these is roughly similar to the set-up in question 6. Ions of the element are produced. They are projected into a magnetic field. The ions of the isotopes have the same charge and are deflected in the same direction, but because they differ in mass, the amount of deflection will differ for the different isotopes.

9. B Reducing the intensity of monochromatic light merely reduces the number of photons in the light; its frequency is not charged. Therefore, each photon that interacts with an electron in the metal has the same amount of energy as before, and so each electron that is emitted can have the same maximum kinetic energy as before, $hf - \phi$. (However, the number of photoelectrons will be reduced.)

10. E In the photoelectric effect, the energy of the photon is used in two ways when it interacts with an electron in the metal. Part is used to do the necessary work to overcome the force that holds the electron in the metal; the rest, to give the electron kinetic energy: $E_P = \phi + E_K$. This, of course, is equivalent to $E_K = E_P - \phi$.

11. B Fusion is the combining of two nuclei to produce a more massive nucleus. Here the nuclei of two isotopes of hydrogen combine to produce helium. It is true that the alpha particle is a helium nucleus, but in alpha decay the alpha particle is emitted from a single unstable, heavy nucleus.

12. C You can save time if you recognize that this is a mental problem. The value of Q is the difference between the sum of the atomic mass units on the left side of the equation and the sum of those on the right side. The sum on the left side is a little over 5, and the sum on the right side is also a little over 5. The difference, therefore, has to be less than 1. The only possible choice is (C): 0.01881.

13. D The sum of the mass numbers (superscripts) on the left side is 236. The sum on the right side has to be the same. For the first two products, Ba and Kr, the sum is only 233. We need 3 neutrons to provide the additional 3.

CHAPTER 21

Special Relativity

The Postulates of Einstein's Special Theory of Relativity

Physicists who believed in the electromagnetic wave theory of light guessed that a material in space known as "ether" was responsible for transmitting light through space. After all, they thought, water vibrates up and down to transmit water waves. They wanted something physical in space to vibrate up and down also. The ether was supposed to move in a similar manner to carry light.

The velocity of light is different when measured across a stream of moving water as compared to when the light is moving with the stream of water. Because Earth revolves about the Sun in its orbit, it was expected that the velocity of light might be different in different circumstances.

In 1887, an experiment was performed by two American physicists, Albert Michelson and Edward Morley. They were attempting to measure the difference between the velocity of light in the direction of Earth's revolution in its orbit about the Sun and the velocity of light in a direction perpendicular to Earth's revolution. To their surprise, the "race" turned out to be a tie! The velocity of light was found to be the same in all directions. In fact, the velocity of light is an *invariant* quantity. This means that all observers, no matter where they are or what they are doing, measure the velocity of light in vacuum to be the same value.

In 1905, Albert Einstein published his theory under the title "On the Electrodynamics of Moving Bodies." Two other physicists, Hendrick Lorentz and George Fitzgerald, had independently predicted a few years before that, in order to explain the Michelson-Morley experiment, there might be a con traction of length in the direction of motion by a factor of $\sqrt{1-v^2/c^2}$. However, they still thought that the velocity of light varied depending on the motion of the observer.

Einstein disagreed and stated the following postulates in his *theory of special relativity*:

1. The laws of physics are the same for all frames of reference moving at constant velocity with respect to one another.
2. The velocity of light in a vacuum is the same for all observers regardless of their states of motion or the motion of the source of light.

The term *special relativity* means that we do not consider the effects of gravity on the motion of light and time.

Length Contraction

In special relativity, the measurement of length is based on the observer's frame of reference; therefore, this measurement differs. The proper length, L_0, is the length of a rod as measured by someone moving along with it in an *inertial frame of reference* (i.e., a frame moving with constant velocity).

To a *fixed* observer, the length of the rod appears to be contracted according to the formula

$$L = L_0\sqrt{1-(\mathbf{v}^2/\mathbf{c}^2)}$$

This phenomenon is called *length contraction.*

Example: Imagine a meter stick moving relative to Earth with a velocity of $0.85c$. By what factor will its length appear to be contracted?

Solution: Using the length contraction formula, we have

$$\frac{L}{L_0} = \sqrt{1 - \frac{(0.85\mathbf{c})^2}{\mathbf{c}^2}} = 0.53$$

which is approximately one-half the rest length.

Simultaneity and Time Dilation

Two events are simultaneous to an observer who measures that the two events occur at the same time in a given frame of reference. In classical physics, if one observer records simultaneous events, then all observers will likewise record simultaneous events. This fact is not true in special relativity, however, since the velocity of light must be the same for all observers. An interesting consequence of this fact is the apparent difference in the passage of time.

If t_0 is the proper time measured by a clock moving with an observer O (in a frame of reference), the observed time interval in a fixed frame of reference is given by

$$t = \frac{t_0}{\sqrt{1-(\mathbf{v}^2/\mathbf{c}^2)}}$$

Since the denominator is less than 1, the time interval, as measured in the fixed frame of reference, is apparently larger. This phenomenon is known as *time dilation.* In other words, if an astronaut were to travel at close to the speed of light, we would observe a slowing down time!

Confirmation of these predictions can be found in cosmic ray particles called *muons.* These particles are created in the nuclear collisions of particle accelerators on Earth or in the ionosphere and magnetosphere surrounding Earth. In a laboratory, muons have a lifetime of approximately

2.2 microseconds (2.2. × 10^{-6} s). At velocities of about 0.998**c**, a particle would travel only 0.66 kilometer before decaying. How is it possible for muons to be detected on Earth if they are created high in the atmosphere? The answer is time dilation.

Using Einstein's formula and the velocity of muons as measured on Earth, we find that while the frame of reference moving with a muon would observe a lifetime of 2.2 microseconds, we on Earth would observe a lifetime of

$$t = \frac{2.2 \times 10^{-6}}{\sqrt{1-(0.998c)^2/\mathbf{c}^2}} = 34.8 \times 10^{-6}\ \text{s} = 34.8\ \mu\text{s}$$

Relativistic Mass and Energy

Inertia is the property of matter that resists a net force changing its state of motion. In Newtonian mechanics, the inertia of an object is measured as its *mass.* In all Newtonian interactions, the mass of an object remained the same unless something physical was happening to it (e.g., an explosion or a rocket's use of fuel).

In his paper on special relativity, Einstein demonstrated that inertial mass varies with velocity. The experimental evidence for this phenomenon is observed in the mass of high-speed electrons accelerated by magnetic fields. According to special relativity, the mass, m, varies according to the formula

$$m = \frac{m_0}{\sqrt{1-\mathbf{v}^2/\mathbf{c}^2}}$$

where m_o is the rest mass (a constant).

We now see why it is impossible for a mass to be accelerated to the velocity of light in a vacuum. As the velocity **v** approaches the velocity of light, **c**, the mass of the object increases toward infinity and is undefined if **v** = **c**. The relativistic momentum, **p** = m**v**, can now be expressed as

$$\mathbf{p} = m\mathbf{v} = \frac{m_0\mathbf{v}}{\sqrt{1-\mathbf{v}^2/\mathbf{c}^2}}$$

In his theory of special relativity, Einstein also introduced the most famous equation of the twentieth century. Any object with a rest mass of m_o has an equivalent rest energy given by

$$E_o = m_o\mathbf{c}^2$$

This equation accounts for the binding energy factor of 931.5 million electron volts per atomic mass unit.

Example: What is the rest energy for a 60-kilogram student?

Solution: Use the formula

$$E_o = m_o\mathbf{c}^2 = (60)(9 \times 10^{16}) = 5.4 \times 10^{18}\ \text{J}$$

Questions
Chapter 21

In each case, select the choice that best answers the question or completes the statement.

1. Which of the following statements is a postulate of special relativity?

(A) All motion is relative.
(B) Objects can never go faster than the velocity of light in any medium.
(C) Accelerated frames of reference produce fictitious forces.
(D) The laws of physics are the same for all inertial frames of reference.
(E) All laws of physics reduce to Newton's laws at low velocity.

2. What is the perceived length of a meter stick if it is moving with a velocity of 0.5**c** relative to Earth?

(A) 0.87 m
(B) 0.75 m
(C) 0.50 m
(D) 0.25 m
(E) 0.35 m

3. What is the energy equivalent of an electron at rest?

(A) 931.5 MeV
(B) 0.017 MeV
(C) 0.512 MeV
(D) 535 MeV
(E) 675 MeV

4. An observer is moving with a velocity of 0.95**c** in a direction perpendicular to a rod of length *L*. The observer will measure the length of the rod to be

(A) equal to *L*
(B) less than *L*
(C) greater than *L*
(D) zero
(E) none of these

5. According to the special theory of relativity, the density of a moving object should

(A) increase only
(B) decrease only
(C) increase and then decrease
(D) remain the same
(E) decrease and then increase

Explanations to Questions Chapter 21

Answers

1. (D)
2. (A)
3. (C)
4. (A)
5. (A)

Explanations

1. D "The laws of physics are the same for all inertial frames of reference" is one of Einstein's postulates of relativity.

2. A We use the length contraction formula with $L_o = 1.0$ m:

$$L = L_0\sqrt{1 - \mathbf{v}^2/\mathbf{c}^2} = (1.0)\sqrt{1-(0.5\mathbf{c})^2/\mathbf{c}^2} = 0.87 \text{ m}$$

3. C We use the mass-energy relationship for a stationary electron:

$$E_o = m_o\mathbf{c}^2 = (9.1 \times 10^{-31})(9 \times 10^{16}) = 8.14 \times 10^{-14} \text{ J} = 0.512 \text{ MeV}$$

4. A If the rod is laid along the x-axis and the motion is in the y-direction, there will be no observed change in length.

5. A Density is the ratio of mass to volume. Since the inertial mass of an object increases while the volume (because of length contraction) decreases, only the density of a moving object should increase.

Answer Sheet: Practice Test 1

1. A B C D E
2. A B C D E
3. A B C D E
4. A B C D E
5. A B C D E
6. A B C D E
7. A B C D E
8. A B C D E
9. A B C D E
10. A B C D E
11. A B C D E
12. A B C D E
13. A B C D E
14. A B C D E
15. A B C D E
16. A B C D E
17. A B C D E
18. A B C D E
19. A B C D E
20. A B C D E
21. A B C D E
22. A B C D E
23. A B C D E
24. A B C D E
25. A B C D E
26. A B C D E
27. A B C D E
28. A B C D E
29. A B C D E
30. A B C D E
31. A B C D E
32. A B C D E
33. A B C D E
34. A B C D E
35. A B C D E
36. A B C D E
37. A B C D E
38. A B C D E
39. A B C D E
40. A B C D E
41. A B C D E
42. A B C D E
43. A B C D E
44. A B C D E
45. A B C D E
46. A B C D E
47. A B C D E
48. A B C D E
49. A B C D E
50. A B C D E
51. A B C D E
52. A B C D E
53. A B C D E
54. A B C D E
55. A B C D E
56. A B C D E
57. A B C D E
58. A B C D E
59. A B C D E
60. A B C D E
61. A B C D E
62. A B C D E
63. A B C D E
64. A B C D E
65. A B C D E
66. A B C D E
67. A B C D E
68. A B C D E
69. A B C D E
70. A B C D E
71. A B C D E
72. A B C D E
73. A B C D E
74. A B C D E
75. A B C D E

CHAPTER 22

Practice Tests

PRACTICE TEST 1

Part A

Directions: For each group of questions below, there is a set of five lettered choices, followed by numbered questions. For each question select the one choice in the set that best answers the question and fill in the corresponding circle on the answer sheet. You may use a lettered choice once, more than once, or not at all in each set. Do not use a calculator.

Questions 1–6 refer to the following laws and principles:

(A) Conservation of momentum
(B) Conservation of kinetic energy
(C) Boyle's law
(D) Charles' law
(E) None of these

KEY for Questions 1–6
m = mass
$\mathbf{v}$ = velocity
$\mathbf{p}$ = pressure
V = volume
T = absolute temperature
h = vertical height
$\mathbf{g}$ = acceleration due to gravity

To which choice is each of the following equations most closely related?

1. $m_1\mathbf{v}_1^2 = m_2\mathbf{v}_2^2$

2. $m_1\mathbf{v}_1 = m_2\mathbf{v}_2$

3. $V_1T_1 = V_2T_2$

4. $V_1T_2 = V_2T_1$

5. $\mathbf{p}_1V_1 = \mathbf{p}_2V_2$

6. $m\mathbf{g}h = m\mathbf{v}^2$

Questions 7–13 refer to the five sketched graphs below:

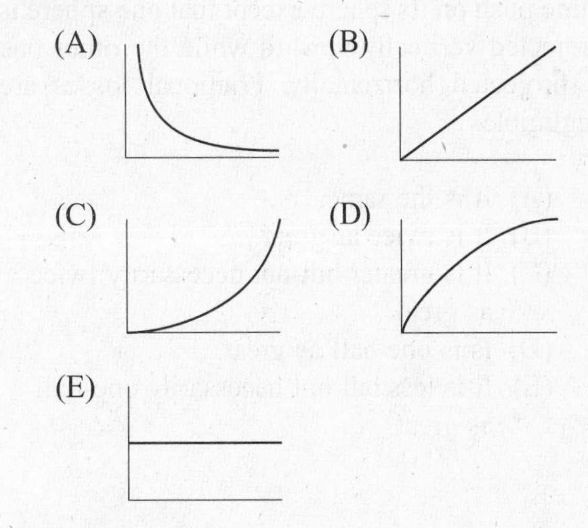

For each question, select the graph that best represents the relationship described.

7. The relationship between frequency and wavelength for a mechanical wave moving through a uniform medium at constant velocity

8. The relationship between mass and acceleration for a mass acted upon by a constant net force

9. The relationship between velocity and time for an object accelerating uniformly from rest.

10. The relationship between the length of a pendulum and the corresponding period of oscillation

GO ON TO THE NEXT PAGE ►

11. The relationship between the kinetic energy of a moving object and velocity as its velocity changes

12. The relationship between voltage and current for a simple circuit at constant temperature

13. The relationship between distance and time for an object at rest

Questions 14 and 15 refer to the following laboratory experiment.

Two small, identical metal spheres are projected at the same time from the same height by two identical spring guns. Each gun provides the same push on its sphere except that one sphere is projected vertically upward while the other one is projected horizontally. Frictional losses are negligible.

(A) It is the same.
(B) It is twice as great.
(C) It is greater but not necessarily twice as great.
(D) Is is one-half as great.
(E) It is less but not necessarily one-half as great.

14. How does the time required for the vertically projected sphere to hit the level floor compare with that for the horizontally projected sphere?

15. How does the kinetic energy with which the vertically projected sphere hits the floor compare with that for the horizontally projected sphere?

GO ON TO THE NEXT PAGE ➤

Part B

Directions: Each of the questions or incomplete statements is followed by five suggested answers or completions. Select the choice that best answers the question or completes the statement and fill in the corresponding circle on the answer sheet. Do not use a calculator.

16. Of the following, the smallest quantity is

(A) 0.635 km
(B) 0.635×10^4 cm
(C) 6.35×10^4 m
(D) 0.635×10^6 mm
(E) 0.635×10^3 m

17. A length of 55 millimeters is approximately equivalent to a length of

(A) 0.055 m
(B) 55 cm
(C) 0.55 m
(D) 5.5 cm
(E) 0.0055 m

18. Two forces of 10 newtons and 7 newtons, respectively, are applied simultaneously to an object. The maximum value of their resultant is, in newtons,

(A) 7
(B) 10
(C) 17
(D) $17\sqrt{3}$
(E) 70

Questions 19 and 20

In the graphs below, v stands for speed, s stands for distance, t for time. The acceleration of a certain object starting from rest and moving with constant acceleration along a straight line is represented by graph X.

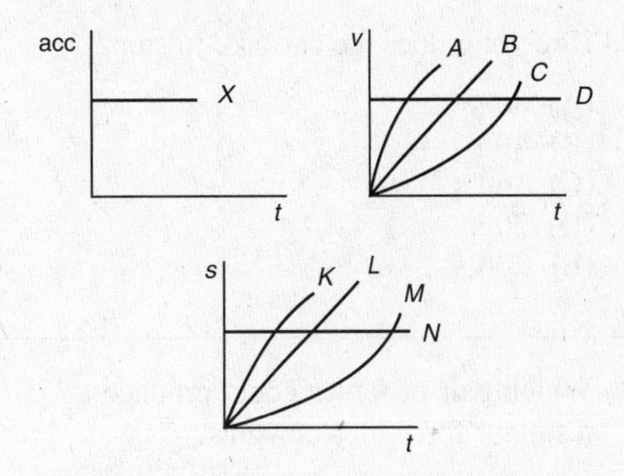

19. The speed of the object is represented by

(A) graph A
(B) graph B
(C) graph C
(D) graph D
(E) None of these

20. The distance covered by the object is represented by

(A) graph K
(B) graph L
(C) graph M
(D) graph N
(E) None of these

GO ON TO THE NEXT PAGE ►

Questions 21–22

A car is traveling at 20 meters per second and stops over a distance of 800 meters.

21. What was the average acceleration of the car?

(A) -0.25 m/s^2
(B) $+0.25$ m/s^2
(C) -0.5 m/s^2
(D) $+0.5$ m/s^2
(E) $+2.0$ m/s^2

22. How long does the car take to stop?

(A) 80 s
(B) 40 s
(C) 160 s
(D) 20 s
(E) 200 s

---

23. Which pair of forces could produce a resultant force of 7 newtons?

(A) 2 N, 6 N
(B) 3 N, 11 N
(C) 1 N, 5 N
(D) 4 N, 2 N
(E) 3 N, 3 N

24. An insulated tank contains 2.5 cubic meters of an ideal gas under a pressure of 0.5 atmosphere. If the pressure is raised to 3 atmospheres, while the temperature is held constant, what is the new volume, in cubic meters of the gas?

(A) 2.5
(B) 5
(C) 1.25
(D) 0.42
(E) Not enough information is given.

Questions 25 and 26

Two small masses, *X* and *Y*, are *d* meters apart. The mass of *X* is 4 times as great as that of *Y*, and *X* attracts *Y* with a force of 16 newtons.

25. *Y* attracts *X* with a force of

(A) 1 N
(B) 4 N
(C) 16 N
(D) 32 N
(E) 64 N

26. If the distance between *X* and *Y* is changed to 2*d* meters, *X* will attract *Y* with a force of

(A) 1 N
(B) 4 N
(C) 8 N
(D) 16 N
(E) 32 N

---

27. The period of a simple pendulum in a laboratory does NOT depend on

(A) the altitude of the laboratory
(B) the acceleration due to gravity in the laboratory
(C) the length of the string
(D) the vibration in the laboratory
(E) the mass of the bob

28. A car is traveling on a level highway at a speed of 15 meters per second. A braking force of 3,000 newtons brings the car to a stop in 10 seconds. The mass of the car is

(A) 1,500 kg
(B) 2,000 kg
(C) 2,500 kg
(D) 3,000 kg
(E) 45,000 kg

29. An elevator weighing 2.5×10^4 newtons is raised to a height of 10 meters. If friction is neglected, the work done is

(A) 2.5×10^4 J
(B) 2.5×10^5 J
(C) 2.5×10^3 J
(D) 7.5×10^4 J
(E) 98 J

GO ON TO THE NEXT PAGE ►

30. A 10-kilogram rocket fragment falling toward Earth has a net downward acceleration of 5 meters per second squared. The net downward force acting on the fragment is

(A) 5 N
(B) 10 N
(C) 50 N
(D) 98 N
(E) 320 N

Questions 31 and 32

A racing car is speeding around a flat, unbanked circular track whose radius is 250 meters. The car's speed is a constant 50.0 meters per second. The mass of the car is 2.00×10^3 kilograms.

31. The centripetal force necessary to keep the car in its circular path is provided by

(A) the engine
(B) the brakes
(C) friction
(D) the steering wheel
(E) the stability of the car

32. The magnitude of the centripetal force on the car is, in newtons,

(A) 1.00×10^1
(B) 4.00×10^2
(C) 4.00×10^3
(D) 2.00×10^4
(E) 4.00×10^4

Questions 33 and 34

Three wires of the same length and cross-sectional area are connected in series to a battery. The wires are made of copper, iron, and nichrome, respectively.

33. The current through the copper is

(A) the same as that through the iron or nichrome
(B) greater than that through the iron or nichrome
(C) less than that through the iron or nichrome
(D) greater than that through the iron but less than that through the nichrome
(E) greater than that through the nichrome but less than that through the iron

34. The potential difference across the copper is

(A) the same as that across the iron or nichrome
(B) greater than that across the iron or nichrome
(C) less than that across the iron or nichrome
(D) greater than that across the iron but less than that across the nichrome
(E) greater than that across the nichrome but less than that across the iron

---

35. A projectile is launched at an angle θ to the horizontal with an initial velocity **v**. In the absence of any air resistance, which of the following statements is correct?

(A) The horizontal velocity increases during flight.
(B) The horizontal velocity remains constant during flight.
(C) The horizontal velocity decreases during flight.
(D) The vertical acceleration increases during flight.
(E) The vertical acceleration decreases during flight.

36. To make an ammeter, you need a galvanometer and a resistor. Specifically, you should

(A) not use the resistor at all
(B) place a high resistance in series with the galvanometer coil
(C) place a low resistance in series with the galvanometer coil
(D) place a high resistance in parallel with the galvanometer coil
(E) place a low resistance in parallel with the galvanometer coil

GO ON TO THE NEXT PAGE ►

Questions 37 and 38

In the following circuit, *V* stands for a good voltmeter and *A* for a good ammeter.

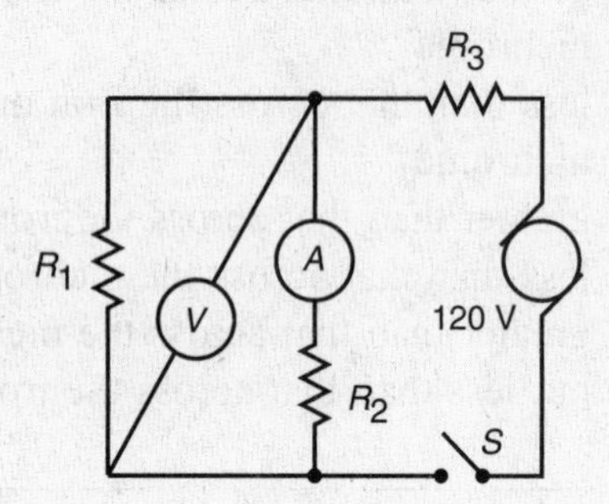

37. The voltmeter should measure the potential difference across

(A) the ammeter
(B) R_1
(C) R_3
(D) the generator
(E) the switch

38. The ammeter should measure the current through

(A) the voltmeter
(B) R_1
(C) R_2
(D) R_3
(E) the generator

---

39. Which of the following charges is NOT possible?

(A) 8.0×10^{-19} C
(B) 3.2×10^{-18} C
(C) 1.2×10^{-18} C
(D) 6.4×10^{-18} C
(E) 4.8×10^{-19} C

40. Which of the following electromagnetic waves has the shortest wavelength?

(A) Light rays
(B) Radio waves
(C) X rays
(D) Ultraviolet rays
(E) Infrared rays

41. The sound of a siren to the west of you is transmitted to your ear by air that is

(A) vibrating in a north-south direction
(B) vibrating in a west-east direction
(C) vibrating in a vertical direction
(D) moving continuously westward
(E) moving continuously eastward

42. Which of the following statements is (are) correct about transverse waves?

(A) Transverse waves can be polarized.
(B) Transverse waves can be diffracted.
(C) Microwaves are transverse waves.
(D) Vibrations are perpendicular to the direction of propagation.
(E) All of the above

43. Total internal reflection can occur between two transparent media if their relative index of refraction

(A) is less than 1.0
(B) is greater than 1.0
(C) is equal to 1.0
(D) is sometimes greater than 1.0
(E) varies depending on the wavelength of light used

44. Each of the following particles is traveling with the same velocity. Which one will have the smallest de Broglie wavelength?

(A) An electron
(B) A proton
(C) A positron
(D) A neutron
(E) An alpha particle

45. In the Rutherford scattering experiments, the trajectory of the alpha particles was in the shape of a(n)

(A) hyperbola
(B) parabola
(C) circle
(D) ellipse
(E) straight line

GO ON TO THE NEXT PAGE ►

Questions 46 and 47

Two heating coils, *X* and *Y*, are each connected to a 120-volt direct current. The resistance of *X* is twice that of *Y*.

46. The current through *X* is

(A) twice that through *Y*
(B) equal to that through *Y*
(C) one-half that through *Y*
(D) one-fourth that through *Y*
(E) one-eighth that through *Y*

47. The rate at which *X* produces heat is

(A) 4 times that of *Y*
(B) 2 times that of *Y*
(C) the same as that of *Y*
(D) one-half that of *Y*
(E) one-fourth that of *Y*

---

48. An atom of an element differs from an atom of one of its isotopes in the number of

(A) neutrons in the nucleus
(B) protons in the nucleus
(C) valence electrons
(D) electrons outside the nucleus
(E) protons outside the nucleus

49. A woman is 10 meters away from a plane mirror. Her distance from her image appears to be

(A) 5 m
(B) 10 m
(C) 15 m
(D) 20 m
(E) 25 m

50. When a person uses a convex lens as a magnifying glass, the distance that the object must be from the lens is

(A) less than 1 focal length
(B) more than 1 but less than 2 focal lengths
(C) 2 focal lengths
(D) more than 2 but less than 4 focal lengths
(E) at least 4 focal lengths

51. Which of the following properties does NOT apply to sound?

(A) reflection
(B) refraction
(C) diffraction
(D) interference
(E) polarization

52. A 3-centimeter-tall candle is placed in front of a concave spherical mirror. An inverted image, half the size of the original, is observed on a screen. Where, relative to the mirror, must the candle have been placed?

(A) at 2f
(B) beyond 2f
(C) between f and 2f
(D) between f and the mirror
(E) This is not a realistic question for this type of mirror.

53. The pages of this book are visible because they

(A) absorb light
(B) emit light
(C) reflect light
(D) refract light
(E) polarize light

54. In monochromatic red light, a blue book will probably appear to be

(A) blue
(B) black
(C) purple
(D) yellow
(E) green

55. When light emerges from water and enters air,

(A) the light will be refracted
(B) the frequency of the light will increase
(C) the frequency of the light will decrease
(D) the speed of the light will decrease
(E) the speed of the light will increase

GO ON TO THE NEXT PAGE 

56. A girl holds a nickel in a space ship flying at a speed close to the speed of light. To a fixed observer viewing the face of the coin as it passes by, the nickel may appear to be the width of a

(A) dime
(B) nickel
(C) quarter
(D) half-dollar
(E) silver dollar

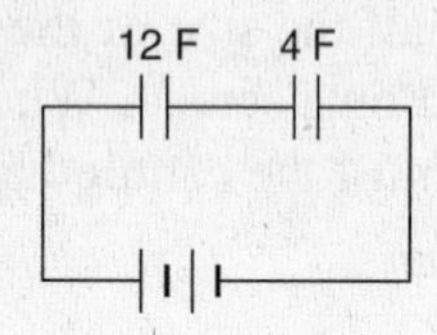

57. What is the total capacitance for the circuit shown above?

(A) 16 F
(B) 1/3 F
(C) 3 F
(D) 48 F
(E) 1 F

Questions 58 and 59

The diagram below shows a conducting loop rotating clockwise in a uniform magnetic field

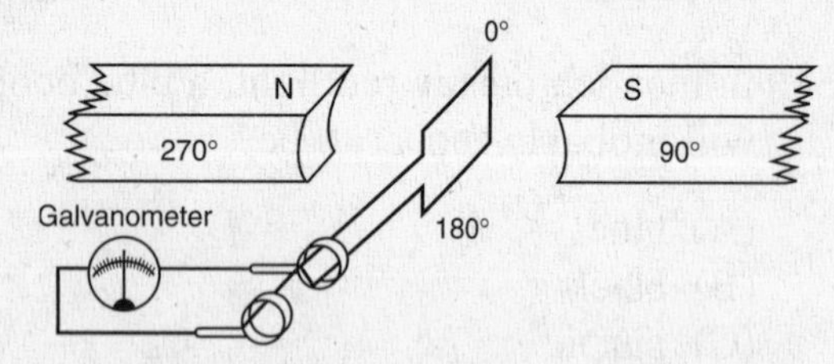

58. As the loop rotates, the induced voltage will be maximum

(A) at 0° and 90°
(B) at 0° and 180°
(C) at 90° and 270°
(D) at 180° and 270°
(E) and constant throughout the rotation

59. The induced voltage will NOT be decreased by doing each of the following EXCEPT

(A) increasing the speed of rotation
(B) using stronger magnets
(C) using a rotating loop with an iron core
(D) using a commutator
(E) moving the north and south poles further apart

---

60. Which of the following is (are) present at the red end of the visible spectrum?

I. Slower waves than at the violet end
II. Lower energy quanta than at the violet end
III. Longitudinal waves
IV. Longer wavelengths than at the violet end

(A) I, II, and III only
(B) I and III only
(C) II and IV only
(D) IV only
(E) None of the above

61. The principle of interference is necessary to explain adequately which of the following?

I. The production of beats
II. The production of the green color of leaves
III. The appearance of colors in thin oil films
IV. The appearance of a virtual image, rather than a real image, in a plane mirror

(A) I, II, and III only
(B) I and III only
(C) II and IV only
(D) IV only
(E) None of the above

GO ON TO THE NEXT PAGE ►

62. Which of the following represent(s) the same quantity as 6.50×10^{-3} ampere?

I. 6.50 mA
II. 65.0×10^{-4} A
III. 0.00650 A
IV. 65.0×10^{-2} A

(A) I, II, and III only
(B) I and III only
(C) II and IV only
(D) IV only
(E) none of the above

63. Which of the following properties of the light is (are) NOT changed when a beam of light goes obliquely from a rarer to a denser medium?

I. Amplitude
II. Direction
III. Wavelength
IV. Frequency

(A) I, II, and III only
(B) I and III only
(C) II and IV only
(D) IV only
(E) None of the above

64. Which of the following is (are) true of an object starting from rest and accelerating uniformly?

I. Its kinetic energy is proportional to its displacement.
II. Its speed is proportional to the square root of its displacement.
III. Its kinetic energy is proportional to the square of its speed.
IV. Its velocity is proportional to the square of the elapsed time.

(A) I, II, and III only
(B) I and III only
(C) II and IV only
(D) IV only
(E) None of the above

65. A man pulls an object up an inclined plane with a force **F** and notes that the object's acceleration is 3 meters per second squared. When he triples the force without changing its direction, what will be the effect(s) on the acceleration?

I. It decreases.
II. It increases.
III. It remains the same.
IV. It triples.

(A) I, II, and III only
(B) I and III only
(C) II and IV only
(D) IV only
(E) None of the above

66. A woman is standing in an elevator that rises with constant positive acceleration. Which of the following is (are) true of the push she exerts on the floor of the elevator?

I. It is equal to her weight.
II. Its value depends on the elevator's acceleration.
III. It is equal to less than her weight.
IV. It is equal to more than her weight.

(A) I, II, and III only
(B) I and III only
(C) II and IV only
(D) IV only
(E) None of the above

67. The lowest note that can be produced by a vibrating object is known as which of the following?

I. Fundamental frequency
II. First overtone
III. First harmonic
IV. First basso

(A) I, II, and III only
(B) I and III only
(C) II and IV only
(D) IV only
(E) None of the above

GO ON TO THE NEXT PAGE ►

68. Assume that you have a coil of copper wire whose resistance is 100 ohms. To obtain a coil of copper wire with less resistance, which of the following may you use?

I. Copper wire that has the same length but is thicker.
II. Copper wire that has the same thickness but is shorter.
III. Copper wire that is thicker and shorter.
IV. Copper wire that is thinner and longer.

(A) I, II, and III only
(B) I and III only
(C) II and IV only
(D) IV only
(E) None of the above

69. Which of the following metals has the highest permeability?

(A) Copper
(B) Gold
(C) Silver
(D) Nickel
(E) Iron

70. Monochromatic light with a wavelength of 6.0×10^{-7} meter is incident upon two slits that are 2.0×10^{-5} meter apart. As a result, an interference pattern appears on a screen 2.0 meters away from the slits and parallel to the slits. What is the expected distance between the central maximum and the next bright line on the screen?

(A) 1.0×10^{-2} m
(B) 3.0×10^{-2} m
(C) 1.7×10^{-2} m
(D) 6.0×10^{-2} m
(E) 1.2×10^{-1} m

71. A metal is illuminated by light above its threshold frequency. Which of the following properties of the light determines the number of electrons emitted by the metal?

(A) Color
(B) Frequency
(C) Intensity
(D) Speed
(E) Wavelength

Questions 72 and 73 refer to the information and diagram below.

Some of the energy levels of hydrogen are shown on the diagram (not to scale).

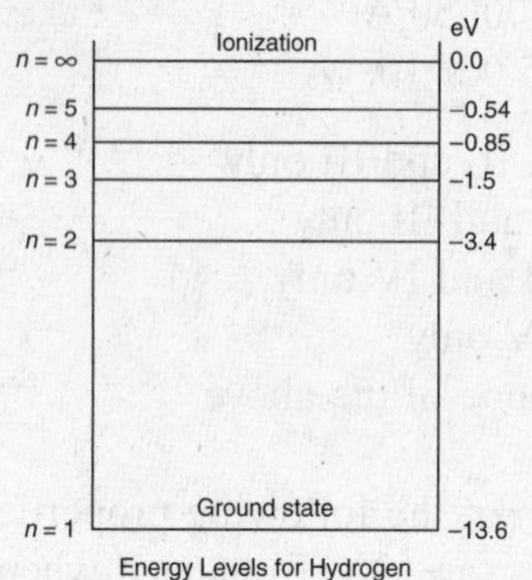

72. Which transition will result in the emission of the photon with the greatest energy?

(A) $n = 5$ to $n = 4$
(B) $n = 5$ to $n = 3$
(C) $n = 5$ to $n = 2$
(D) $n = 3$ to $n = 2$
(E) $n = 2$ to $n = 1$

73. Visible light may be produced as electrons in the hydrogen atom fall from higher energy states to level

(A) 5
(B) 4
(C) 2
(D) 3
(E) 1

---

74. If the half-life of ${}^{234}_{90}\text{Th}$ is 24 days, the amount of a 12-gram sample remaining after 96 days is

(A) 1 g
(B) 0.75 g
(C) 6 g
(D) 1.6 g
(E) 3 g

75. Given the equation ${}^{27}_{13}\text{Al} + {}^{4}_{2}\text{He} \rightarrow {}^{30}_{15}\text{P} + X$; the correct representation for X is

(A) ${}^{0}_{+1}\text{e}$
(B) ${}^{0}_{-1}\text{e}$
(C) ${}^{4}_{2}\text{He}$
(D) ${}^{1}_{0}\text{n}$
(E) ${}^{1}_{+1}\text{p}$

STOP

Answer Key for Practice Test 1

1. **B**	16. **B**	31. **C**	46. **C**	61. **B**
2. **A**	17. **A**	32. **D**	47. **D**	62. **A**
3. **E**	18. **C**	33. **A**	48. **A**	63. **D**
4. **D**	19. **B**	34. **D**	49. **D**	64. **A**
5. **C**	20. **C**	35. **B**	50. **A**	65. **E**
6. **E**	21. **A**	36. **E**	51. **E**	66. **C**
7. **A**	22. **A**	37. **B**	52. **B**	67. **B**
8. **A**	23. **A**	38. **C**	53. **C**	68. **A**
9. **B**	24. **D**	39. **C**	54. **B**	69. **E**
10. **D**	25. **C**	40. **C**	55. **E**	70. **D**
11. **C**	26. **B**	41. **B**	56. **A**	71. **C**
12. **B**	27. **E**	42. **E**	57. **C**	72. **E**
13. **E**	28. **B**	43. **A**	58. **C**	73. **C**
14. **C**	29. **B**	44. **E**	59. **E**	74. **B**
15. **A**	30. **C**	45. **A**	60. **C**	75. **D**

Answer Explanations for Practice Test 1

1. B The kinetic energy of an object with mass m equals $\frac{1}{2}m\mathbf{v}^2$. If, as in an elastic condition, kinetic energy is conserved, the kinetic energy before collision is equal to the kinetic energy of the masses after collision: $\frac{1}{2}m_1\mathbf{v}_1^2 = \frac{1}{2}m_2\mathbf{v}_2^2$. If we divide both sides by ½, we get the equality of question 1: $m_1\mathbf{v}_1^2 = m_2\mathbf{v}_2^2$.

2. A The momentum of an object is the product of its mass and velocity: $m\mathbf{v}$. In the collision of two objects or in an explosion, momentum is conserved; the momentum of the masses before the event equals the momentum of the masses after the event.

3. E See explanation for question 4.

4. D Write the relationship as a proportion:

$$\frac{V_1}{V_2} = \frac{T_1}{T_2} \text{ (check by cross-multiplying).}$$

This expresses Charles' law: The volume of a gas is proportional to its absolute temperature, if the pressure is constant.

5. C This is one way of expressing Boyle's law: If the temperature remains constant, the product of the pressure and volume of a gas remains constant.

6. E The formula mgh gives the potential energy of an object; $m\mathbf{v}^2$ is *not* the kinetic energy of the object. (Kinetic energy, of course, is $\frac{1}{2}m\mathbf{v}^2$.)

7. A The formula for wave velocity is $\mathbf{v} = f\lambda$. If velocity is constant, then frequency and wavelength vary inversely with one another.

8. A The formula for net force is $\mathbf{F} = m\mathbf{a}$. If force is constant, then mass and acceleration vary inversely with one another.

9. B The formula is $\mathbf{v} = at$, which is a straight diagonal line from the origin.

10. D The period of oscillation of a pendulum varies directly with the square root of the length of the pendulum.

11. C The formula is $KE = \frac{1}{2} m\mathbf{v}^2$. This is a parabolic relationship between kinetic energy and velocity.

12. B Ohm's law governs this circuit: $V = IR$. This is a direct relationship.

13. E An object at rest does not change distance with time.

14. C The time required for the horizontally projected sphere to hit the floor depends only on the vertical height from which it starts. The horizontal velocity does not affect the vertical velocity; the initial vertical velocity of this sphere is zero. The initial vertical velocity of the vertically projected sphere is not zero. This sphere is gradually slowed by gravity as it rises. At some greater height its velocity will be momentarily zero, and its time of fall from that greater height is greater than the time of fall for the other sphere. In addition, of course, some time was required to reach that greater height. Not enough information is provided to calculate actual times.

15. A Since the guns are identical, the springs do the same work on the spheres. As a result, the two spheres leave the guns with the same kinetic energy. In addition, the spheres initially have the same gravitational potential energy because they leave the guns from the same height. As the spheres fall, this potential energy is converted to kinetic energy. In the case of the vertically projected sphere another conversion occurs first. While it rises during the first part of its motion, its initial kinetic energy is converted to gravitational potential energy. This adds to the initial potential energy.

16. B One way to find the smallest quantity is to first write each quantity in standard scientific notation using the same unit, such as the meter. This converts the choices to: **(A)** 6.35×10^2 m **(B)** 6.35×10 m **(C)** 6.35×10^4 m **(D)** 6.35×10^2 m **(E)** 6.35×10^2 m. Of these, 6.35×10 m is the smallest.

17. A 55 mm = 55×10^{-3} m = 0.055 m.

18. C The resultant of two forces is greatest when they act in the same direction. Their resultant is then equal to their sum: 10 N + 7 N = 17 N.

19. B For an object starting from rest and moving with constant acceleration, the speed is given by $\mathbf{v} = \mathbf{a}t$. The graph of $\mathbf{v}$ versus t for this relationship is a straight line going through the origin.

20. C For such an object the distance covered is given by $\mathbf{d} = \frac{1}{2}\mathbf{a}t^2$. The graph of $\mathbf{d}$ versus t is a parabola symmetrical about the x-axis.

21. A Use the formula $\mathbf{v}_f^2 - \mathbf{v}_i^2 = 2\mathbf{ad}$. The final velocity is zero, so the acceleration will be negative. Simple substitutions yield $\mathbf{a} = -0.25$ m/s^2.

22. A Since $\Delta\mathbf{v} = \mathbf{a}t$ and $\mathbf{v}_f = 0$ m/s, the time to stop will be 80 s.

23. A In order to produce a possible resultant of 7 N, the pair of forces must have a maximum possible resultant that is more than 7 N and a minimum possible resultant that is less than 7 N. Only choice (A) meets those criteria (max = 8 N; min = 4 N).

24. D Use Boyle's law: (0.5 atm)(2.5 m^3) = (3 atm)V. Calculation shows that $V = 0.42$ m^3.

25. C In accordance with Newton's third law: If X exerts a force on Y, then Y exerts an equal and opposite force on X.

26. B Gravitational, magnetic, and electrostatic forces vary inversely as the square of the distance between the small objects. Since the distance is doubled, the force must become one-fourth as great, that is, 4 N.

27. E Recall the formula for the period of a simple pendulum: $T = 2\pi\sqrt{L/\mathbf{g}}$. This should remind you that the period depends on the length and on the acceleration due to gravity. The greater the altitude, the less the acceleration due to gravity. Vibration in the laboratory may interfere with the point of suspension and affect the period. Changing the mass of the bob of a simple pendulum does not change its period.

28. B The car is decelerated by an unbalanced force of 3,000 N. We apply Newton's second law of motion, which can be stated as: The unbalanced force acting on an object is equal to the product of the object's mass and the resulting acceleration:

$$\mathbf{F} = m\mathbf{a}.$$

First we calculate the acceleration. Acceleration is defined as the ratio of the change in velocity divided by the time in which the change takes place:

$$\begin{aligned}\mathbf{a} &= \Delta\mathbf{v}/t \\ &= (15\ \text{m/s})/(10\ \text{s}) \\ &= 1.5\ \text{m/s}^2\end{aligned}$$

Substituting in $\mathbf{F} = m\mathbf{a}$ gives

$$\begin{aligned}3{,}000\ \text{N} &= m(1.5\ \text{m/s}^2) \\ m &= 2{,}000\ \text{kg}\end{aligned}$$

Another method is to use a different form of the second law: The impulse acting on an object is equal to the resulting change in momentum. The *impulse* is equal to the product of the unbalanced force and the time during which it acts. Therefore,

$$\begin{aligned}\mathbf{F}t &= m\mathbf{v} \\ (3{,}000\ \text{N})(10\ \text{s}) &= m(15\ \text{m/s}) \\ m &= 2{,}000\ \text{kg}\end{aligned}$$

Note that the object started with a velocity of 15 m/s and then came to rest; therefore, the change in velocity is 15 m/s.

29. B The work done in lifting an object is equal to the product of the object's weight and the vertical height through which it is lifted.

$$\begin{aligned}\text{work} &= \text{weight} \times \text{height} \\ &= (2.5 \times 10^4\ \text{N})(10\ \text{m}) \\ &= 2.5 \times 10^5\ \text{J}\end{aligned}$$

30. C According to Newton's second law of motion, the net or unbalanced force acting on an object is equal to the product of the object's mass and the resulting acceleration:

$$\begin{aligned}\mathbf{F} &= m\mathbf{a} \\ &= (10\ \text{kg})(5\ \text{m/s}^2) \\ &= 50\ \text{N}\end{aligned}$$

31. C When an object moves at constant speed around a circle, the centripetal acceleration is always directed toward the center of the circle. An unbalanced force acting on the object from the outside is needed to provide this acceleration. Here it is the friction between the wheels and the road that provides the centripetal force acting on the car. The engine turns the wheels to make the motion possible; the steering wheel turns the wheels in the desired direction of motion, but unless there is friction the car will merely continue sliding in a straight line.

32. D The centripetal force is equal to the product of the mass of the car and its centripetal acceleration:

$$\mathbf{F} = m\mathbf{v}^2/r$$
$$= (2 \times 10^3 \text{ kg})(50 \text{ m/s})^2/(250 \text{ m})$$
$$= 2 \times 10^4 \text{ N}$$

33. A In a series circuit the current is the same in every part.

34. D The potential difference across each wire is equal to the IR product. Since the current I is the same in each of the wires, the potential difference is proportional to the resistance. You should know that, of all metals, iron has the highest conductivity (or lowest resistivity); copper and aluminum are close behind. Therefore, the resistance R of the iron wire is the least, that of the copper next, and that of the nichrome highest. The potential differences follow in the same sequence.

35. B In the absence of air resistance, a projectile's horizontal velocity remains constant.

36. E An ammeter is a galvanometer with a low resistance ("shunt") placed in parallel across its coil.

37. B The voltmeter measures the potential difference between its terminals. The voltmeter is in parallel with R_1; therefore, the voltage across them is the same. Remember: in such a diagram the straight line implies zero resistance. The fact that the voltmeter is drawn along a diagonal means nothing special.

38. C The ammeter is in series with R_2; the current through these two is the same.

39. C All charges must be (in magnitude) whole-integer multiples of one elementary charge, which is 1.6×10^{-19}C. Of the five choice, only 1.2×10^{-18} C is not a whole multiple of this value.

40. C In the electromagnetic spectrum, X rays have the shortest wavelength of the choices given.

41. B The energy must travel from west to east; it is transmitted by a sound wave. Sound waves are longitudinal waves: waves in which the medium vibrates back and forth in a direction parallel to the direction in which the energy travels.

42. E All of the statements presented about transverse waves are correct.

43. A In order for light to undergo total internal reflection, it must first go from a substance of higher index of refraction to one of lower index of refraction. It must then have an angle of incidence greater than the critical angle for the first medium. This means that the relative index of refraction, which is the ratio of the two indices (n_2/n_1) must be less than 1.0.

44. E The de Broglie wavelength of a particle is inversely proportional to its linear momentum. Since all the particles travel with the same velocity, the most massive particle will have the smallest wavelength. Alpha particles are the most massive of the choices.

45. A Rutherford discovered that the scattered alpha particles followed hyperbolic trajectories.

46. C Apply Ohm's law: $I = V/R$. Since the voltage V is the same for X and Y, the currents through them compare inversely as the resistance. Then X has a resistance that is twice as large; therefore, it has a current half as large as that through Y.

47. D Since the voltage is the same for X and Y, it is convenient to compare their rates of heat production (or power consumption) by $P = V^2/R$. This shows that their rates of heat production vary inversely as their resistances. (You may recall that in a parallel circuit the device with the lesser resistance consumes the greatest power.) Then X has twice as much resistance as Y; therefore, it consumes ½ the power and produces heat at ½ the rate of Y.

48. A This is a simple fact. Of course, in ordinary neutral atoms all the protons are in the nucleus and the number of electrons outside the nucleus equals the number of protons in the nucleus.

49. D The woman is 10 m in front of the mirror. In a plane mirror her image is 10 m behind the mirror. The distance between her and her image is 20 m.

50. A This is almost pure recall of a fact. You should also think of the fact that, when you use a magnifying glass, you are using a convex lens to get a virtual, enlarged, erect image. A virtual image is obtained with a convex lens only if the object is less than one focal length away from the lens.

51. E Sound is a longitudinal wave, which cannot be polarized.

52. B According to ray diagrams from geometrical optics, if an object is located more than 2 focal lengths in front of a concave, spherical mirror, a smaller, real, inverted image will form. Recall: for a concave, spherical mirror, the location of twice the focal length is approximately equal to the center of curvature.

53. C Nonluminous, opaque objects, such as the pages of this book, are seen by the light they reflect. Luminous objects are seen by the light they emit.

54. B Monochromatic red light contains only red light. An ideal blue object will reflect only blue light; when illuminated by monochromatic red light, there is no blue light for the object to reflect. When no light enters our eye we "see" blackness. (An actual book may reflect other colors, such as green, in addition to the blue. However, such colors will not be present in monochromatic red light.)

55. E You should know as a fact that light travels faster in air than in water or glass. In addition, the light will be refracted if it reaches the surface obliquely; it will not be refracted if it is normal to the surface.

56. A Einstein's special theory of relativity tells us that, at speeds near the velocity of light, lengths are shortened according to

$$l_{\text{relativistic}} = l_{\text{rest}}\sqrt{1-(\mathbf{v}^2/\mathbf{c}^2)}$$

where **v** = velocity of spaceship and **c** = velocity of light. Of the choices, only a dime has a shorter width than a nickel.

57. C The total capacitance for capacitors in series, as in the circuit shown, is found by adding the capacitances reciprocally. Note that this is the opposite of what is done for resistors! In this case,

$$C_{eq} = (C_1C_2) / (C_1 + C_2) = 3 \text{ F}$$

58. C At the instant shown in the diagram, the magnetic flux from the N-pole to the S-pole is perpendicular to the face of the loop. At this instant, the voltage induced in the loop is zero. One way to look at this is to think of the top and bottom wires of the loop. At the instant shown, they are moving parallel to the magnetic flux and therefore are not cutting any lines of force; thus no voltage is induced. One-quarter of a rotation later, the top wire is at 90° and the bottom wire is at 270°. At that instant a maximum voltage is induced in them since they are moving perpendicularly across the lines of force. Another half of a rotation later, the top wire is at 270° and the bottom wire is at 90°. Again a maximum voltage is induced, but this time in the opposite direction around the loop because the wire that moved down across the field before is now moving up, and vice versa.

59. E The voltage induced in the coil is proportional to the flux density provided by the external magnets and the speed of rotation of the loop. The flux density decreases between the two poles when the poles are moved further apart. This decreases the induced voltage. When a rotating loop is wound around an iron core, the flux density is increased. Using a commutator merely changes the current in the external circuit to direct current.

60. C In a vacuum, all electromagnetic waves travel at the same speed. In materials such as glass and water, violet light travels more slowly than red light. The wavelengths of red light are greater than those of violet light. Therefore, the frequencies of red light are lower than those of violet light ($\mathbf{c} = f\lambda$). The energy of a quantum equals Planck's constant times the frequency ($E = hf$). Hence each quantum of red light has less energy than a quantum of violet light.

61. B Beats are produced by the constructive and destructive interference of two different frequencies. Where the waves are in phase, they reinforce each other; where they are out of phase, they tend to annul each other. Reinforcement leads to comparative loudness; annulment, to comparative quiet.

A color spectrum is produced with thin oil films and soap bubbles because light is reflected by both surfaces of the film; the beams from the two surfaces interfere with each other. At different angles of vision different colors are annulled.

Green leaves appear green in ordinary light because the leaves absorb most of the light. Green predominates in the reflected light.

62. A You know that 1 mA = 10^{-3} A. Therefore, 6.50 mA = 6.50×10^{-3} A = 65.0×10^{-4} A = 0.00650 A. But, choice (IV), 65.0×10^{-2} A = 6.50×10^{-1} A.

63. D The amplitude changes because there is always some reflection when a wave goes from one medium to another with a different index of refraction. When the energy of a given wave decreases, the amplitude of the wave decreases.

The direction changes: when a ray goes obliquely from a rarer to a denser medium, it is bent toward the normal. The frequency of the wave, choice (IV), is the same in the two media, but the speed is lower in the denser medium. Therefore, the wavelength is smaller in the denser medium ($\mathbf{v} = f\lambda$).

64. A The kinetic energy of the object equals $\frac{1}{2}m\mathbf{v}^2$ (proportional to the square of its speed). For an object starting from rest, $\mathbf{v}^2 = 2\mathbf{a}d$. Therefore, kinetic energy = $\frac{1}{2}m \times 2\mathbf{a}d = mad$ (kinetic energy is proportional to its displacement d). Since $\mathbf{v}^2 = 2\mathbf{a}d$, $\mathbf{v} = \sqrt{2\mathbf{a}d}$ (speed proportional to the square root of its displacement).

Its velocity is proportional to time; $v = at$, and a is constant.

65. E If the net force were tripled, the acceleration would be tripled. However, the force the man exerts is not the net force. Suppose the force he exerts is 5 N. Part of this is needed to overcome friction, and another part is needed to overcome a component of the weight. Suppose these two parts add up to 1 N. The net force is only 4 N. Tripling the original force results in an applied force of 15 N; the net force will be 14 N. In this example, the net force, and therefore the acceleration, are more than tripled (14/4 = 3.5).

66. C The woman's push on the floor of the elevator equals the floor's push on her. This push moves her up with the same acceleration as the elevator. The push that would be required merely to keep the woman in equilibrium is equal to her weight. An additional force is required to accelerate her. According to Newton's second law, this force is proportional to the acceleration.

67. B First harmonic is another name for fundamental frequency. In addition to the fundamental frequency, many vibrating objects produce higher frequencies called overtones. (First basso is not a physics term.)

68. A The resistance of a wire = kL/A. If you make the length smaller, the resistance becomes less. If you make the cross-section area A larger, the resistance becomes less. Doing the opposite will increase the resistance.

69. E Iron, cobalt, and nickel have the three highest magnetic permeabilities, forming what are called *ferromagnets*.

70. D In the interference pattern produced by two slits, the distance between two adjacent bright lines is given by a fairly simple relationship. If x is this distance,

$$x = \lambda L/d$$

where λ = wavelength of the light, L = distance between the double slit and the screen, and d = distance between the two slits. Substituting gives

$$x = (6.0 \times 10^{-7}\text{ m})(2.0\text{ m})/(2.0 \times 10^{-5}\text{ m})$$
$$= 6.0 \times 10^{-2}\text{ m}$$

71. C The intensity of the arriving light determines how many photons arrive each second. If the frequency of the light is above the threshold frequency, the greater are the number of arriving photons and the number of emitted electrons.

72. E This question requires only rough mental calculation. Quick inspection shows that a transition from $n = 2$ to $n = 1$ results in a release of a photon with an energy of about 10 eV. Any drop to a level other than the ground state will involve at most a release of 3.4 eV.

73. C According to observations and the Bohr theory of the hydrogen atom, visible light is produced when electrons fall from higher energy states to level 2. All visible spectral lines formed for hydrogen are called the Balmer series.

74. B The *half-life* of an isotope of an element is the time required for one-half of the nuclei of a sample of the isotope to break up. This example starts with 12 g of thorium (Th). In 24 days only 6 g are left. In another 24 days (a total of 48 days) one-half of the 6 g disintegrates, and only 3 g of thorium are left. After still another 24 days (a total of 72 days) one-half of the remaining 3 g is lost, leaving 1.5 g of thorium. Finally, after another 24 days (a total of 96 days) one-half of the 1.5 g has disintegrated, leaving 0.75 g of thorium.

75. D When the symbol of an element is written with a superscript and a subscript, the superscript is the mass number and the subscript is the atomic number. In the case of the given isotope of aluminum (Al), the mass number is 27 and the atomic number is 13. The equation describes a nuclear change. In all nuclear changes mass number and atomic number are conserved. This means that the sum of the mass numbers on the left side of the equation must equal the sum of the mass numbers on the right. The sum of the mass numbers on the left is 27 + 4, or 31. On the right side the mass number of phosphorus (P) is 30. Therefore, the mass number of X must be 1, because 30 + 1 = 31. Since atomic number is also conserved, the algebraic sum of the subscripts on the left side must equal the algebraic sum on the right. The sum on the left is 13 + 2, or 15. The subscript for X must be zero, because 15 + 0 = 15. The neutron is the particle that has a mass number of 1 and an atomic number (or charge) of zero.

Answer Sheet: Practice Test 2

1. (A) (B) (C) (D) (E)	26. (A) (B) (C) (D) (E)	51. (A) (B) (C) (D) (E)
2. (A) (B) (C) (D) (E)	27. (A) (B) (C) (D) (E)	52. (A) (B) (C) (D) (E)
3. (A) (B) (C) (D) (E)	28. (A) (B) (C) (D) (E)	53. (A) (B) (C) (D) (E)
4. (A) (B) (C) (D) (E)	29. (A) (B) (C) (D) (E)	54. (A) (B) (C) (D) (E)
5. (A) (B) (C) (D) (E)	30. (A) (B) (C) (D) (E)	55. (A) (B) (C) (D) (E)
6. (A) (B) (C) (D) (E)	31. (A) (B) (C) (D) (E)	56. (A) (B) (C) (D) (E)
7. (A) (B) (C) (D) (E)	32. (A) (B) (C) (D) (E)	57. (A) (B) (C) (D) (E)
8. (A) (B) (C) (D) (E)	33. (A) (B) (C) (D) (E)	58. (A) (B) (C) (D) (E)
9. (A) (B) (C) (D) (E)	34. (A) (B) (C) (D) (E)	59. (A) (B) (C) (D) (E)
10. (A) (B) (C) (D) (E)	35. (A) (B) (C) (D) (E)	60. (A) (B) (C) (D) (E)
11. (A) (B) (C) (D) (E)	36. (A) (B) (C) (D) (E)	61. (A) (B) (C) (D) (E)
12. (A) (B) (C) (D) (E)	37. (A) (B) (C) (D) (E)	62. (A) (B) (C) (D) (E)
13. (A) (B) (C) (D) (E)	38. (A) (B) (C) (D) (E)	63. (A) (B) (C) (D) (E)
14. (A) (B) (C) (D) (E)	39. (A) (B) (C) (D) (E)	64. (A) (B) (C) (D) (E)
15. (A) (B) (C) (D) (E)	40. (A) (B) (C) (D) (E)	65. (A) (B) (C) (D) (E)
16. (A) (B) (C) (D) (E)	41. (A) (B) (C) (D) (E)	66. (A) (B) (C) (D) (E)
17. (A) (B) (C) (D) (E)	42. (A) (B) (C) (D) (E)	67. (A) (B) (C) (D) (E)
18. (A) (B) (C) (D) (E)	43. (A) (B) (C) (D) (E)	68. (A) (B) (C) (D) (E)
19. (A) (B) (C) (D) (E)	44. (A) (B) (C) (D) (E)	69. (A) (B) (C) (D) (E)
20. (A) (B) (C) (D) (E)	45. (A) (B) (C) (D) (E)	70. (A) (B) (C) (D) (E)
21. (A) (B) (C) (D) (E)	46. (A) (B) (C) (D) (E)	71. (A) (B) (C) (D) (E)
22. (A) (B) (C) (D) (E)	47. (A) (B) (C) (D) (E)	72. (A) (B) (C) (D) (E)
23. (A) (B) (C) (D) (E)	48. (A) (B) (C) (D) (E)	73. (A) (B) (C) (D) (E)
24. (A) (B) (C) (D) (E)	49. (A) (B) (C) (D) (E)	74. (A) (B) (C) (D) (E)
25. (A) (B) (C) (D) (E)	50. (A) (B) (C) (D) (E)	75. (A) (B) (C) (D) (E)

PRACTICE TEST 2

Part A

Directions: For each group of questions below, there is a set of five lettered choices, followed by numbered questions. For each question select the one choice in the set that best answers the question and fill in the corresponding circle on the answer sheet. You may use a lettered choice once, more than once, or not at all in each set. Do not use a calculator.

Questions 1–7 refer to the following units:

(A) hertz
(B) watt
(C) joule
(D) volt
(E) half-life

1. What unit is used to express a quantity of energy?

2. To express a measure of electromotive force (emf), which unit is used?

3. In which unit would you express a measure of radioactivity?

4. What is the unit for power?

5. If you want to express a quantity of heat, what unit would you use?

6. What unit is used for the same property as cycle per second?

7. What unit expresses work per unit charge?

Questions 8–13 refer to the following scientists. For each question, choose the letter that corresponds to the correct scientist.

(A) Galileo
(B) Halley
(C) Kepler
(D) Copernicus
(E) Newton

8. He predicted the periodic return of comets due to the force of gravity.

9. He first used a refracting telescope in 1610 to see the craters on the Moon.

10. He invented the reflecting telescope and also worked on the spectrum of sunlight.

11. He discovered the law of falling bodies.

12. He developed a heliocentric model.

13. He stated as one of his three laws that planets sweep out equal areas in equal time.

Questions 14 and 15 refer to the formation of images using an object, a screen, and one or more of these:

I. a plane mirror
II. a concave lens (diverging)
III. a convex lens (converging)

(A) I only
(B) II only
(C) III only
(D) I and II only
(E) None of these

14. Which of items I–III can be used to produce a virtual image larger than the object?

15. Which of items I–III can be used to produce a real image the same size as the object?

GO ON TO THE NEXT PAGE ➤

Part B

Directions: Each of the questions or incomplete statements is followed by five suggested answers or completions. Select the choice that best answers the question or completes the statement and fill in the corresponding circle on the answer sheet. Do not use a calculator.

16. Of the following, the one that is most like a gamma ray in its properties is the

(A) alpha ray
(B) beta ray
(C) X ray
(D) electron
(E) proton

17. The Sun's energy is believed to be a result of

(A) oxidation of carbon
(B) oxidation of hydrogen
(C) oxidation of helium
(D) nuclear fission
(E) nuclear fusion

18. The emission of one alpha particle from the nucleus of an atom produces a change of

(A) –1 in atomic number
(B) –1 in atomic mass
(C) –2 in atomic number
(D) –2 in atomic mass
(E) +2 in atomic mass

19. Two forces of 10 newtons and 7 newtons, respectively, are applied simultaneously to an object. The minimum value of their resultant is, in newtons,

(A) 0
(B) 3
(C) 7
(D) $7\sqrt{3}$
(E) 10

20. Two forces act together on an object. The magnitude of their resultant is greatest when the angle between the forces is

(A) 0°
(B) 45°
(C) 60°
(D) 90°
(E) 180°

21. Of the following, the largest quantity is

(A) 0.047 cm
(B) 47×10^{-4} cm
(C) 4.7×10^{-2} cm
(D) 0.00047×10^{2} cm
(E) 0.000047×10^{4} cm

22. The Celsius temperature of an object is increased by 90°. Its increase on the Kelvin scale will be

(A) 50 K
(B) 90 K
(C) 162 K
(D) 180 K
(E) 273 K

23. The work done in holding a weight of 40 newtons at a height of 3 meters above the floor for 2 seconds is, in joules,

(A) 0
(B) 40
(C) 80
(D) 120
(E) 240

24. When alpha particles are directed at a thin metallic foil, it can be observed that most of the particles

(A) pass through with virtually no deflection
(B) bounce backward
(C) are absorbed
(D) are widely scattered
(E) change to hydrogen

GO ON TO THE NEXT PAGE ➤

25. A lithium nucleus contains 3 protons and 4 neutrons. What is its atomic number?

(A) 1
(B) 7
(C) 3
(D) 4
(E) 12

26. Which of the following statements concerning simple harmonic motion described below is correct?

(A) The maximum velocity and maximum acceleration occur at the same time.
(B) The maximum velocity occurs when the acceleration is a minimum.
(C) The velocity is always directly proportional to the displacement.
(D) The maximum velocity occurs when the displacement is a maximum.
(E) None of the above statements is correct.

Questions 27 and 28

A sonar depth finder in a boat uses sound signals to determine the depth of water. Four seconds after the sound leaves the boat it returns to the boat because of reflection from the bottom. Assume the speed of the sound in water is 1,460 meters per second.

27. The speed of the sound in water, compared to the speed in air, is

(A) less
(B) the same
(C) greater
(D) sometimes less
(E) none of these

28. The depth of the water is, in meters,

(A) 2,200
(B) 4,400
(C) 4,800
(D) 2,920
(E) 19,200

Questions 29 and 30 refer to the following diagram and description.

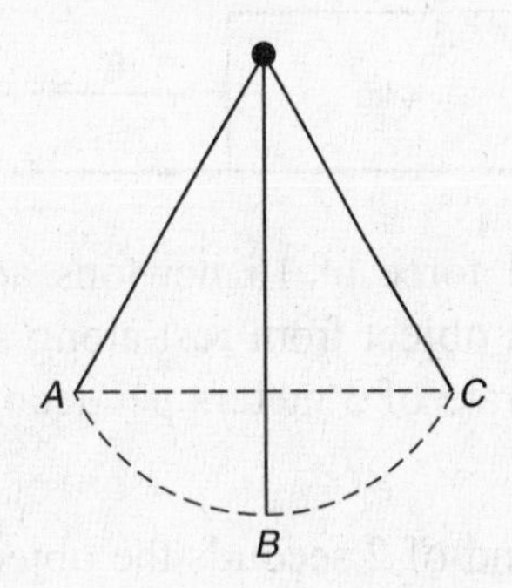

A pendulum, consisting of a string of length L and a small metal ball of mass m at the free end, is allowed to swing freely after being released from point A. The highest point it reaches on the other side is C and the lowest point of swing is B.

29. At B the mass m has

(A) zero potential energy and zero kinetic energy
(B) maximum potential and maximum kinetic energy
(C) minimum potential and maximum kinetic energy
(D) maximum potential and zero kinetic energy
(E) zero potential and the same kinetic energy as at C

30. If we want the period of the pendulum to be doubled, we should change the length of the string to

(A) $\frac{1}{4}L$
(B) $\frac{1}{2}L$
(C) $2L$
(D) $4L$
(E) $2L$ and the mass to $2m$

GO ON TO THE NEXT PAGE ➤

Questions 31 and 32 refer to the following diagram and description.

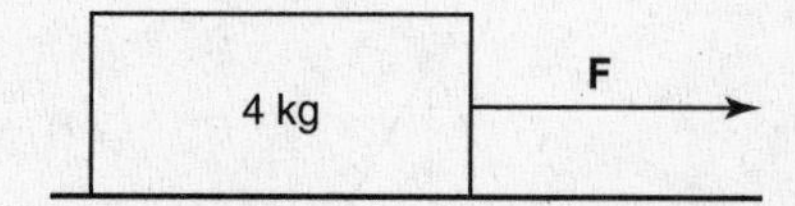

A horizontal force of 15 newtons accelerates a 4.0-kilogram object from rest along a horizontal surface at a rate of 3 meters per second squared.

31. At the end of 2 seconds the object's momentum will be, in kilogram • meters per second,

(A) 12
(B) 24
(C) 30
(D) 45
(E) 60

32. What is the frictional force, in newtons, that is retarding the forward motion of the object?

(A) 3
(B) 4
(C) 12
(D) 40
(E) 60

Questions 33 and 34

A 40-newton object is lifted 10 meters above the ground and then released.

33. After the object has fallen 4 meters and is 6 meters above the ground, its kinetic energy will be, in joules,

(A) 5
(B) 160
(C) 240
(D) 400
(E) 5,120

34. Just before the object hits the ground, its kinetic energy will be, in joules,

(A) 12
(B) 160
(C) 240
(D) 400
(E) 12,800

---

35. Ball *X* is projected horizontally at the same time as ball *Y* is released and allowed to fall. If air resistance is negligible and both start from the same height,

(A) *X* and *Y* will reach the level ground at the same time
(B) *X* will reach the level ground first
(C) *Y* will reach the ground first
(D) we must know the mass of *X* and *Y* to tell which will hit the ground first
(E) we must know the density to tell which will hit the ground first

36. Ball *X* is projected vertically up at the same time as ball *Y* is projected horizontally. If air resistance is negligible and both start from the same height,

(A) *X* and *Y* will reach the level ground at the same time
(B) *X* will reach the level ground first
(C) *Y* will reach the level ground first
(D) we must know the mass of *X* and *Y* to tell which will hit the ground first
(E) we must know the density to tell which will hit the ground first

37. Object X is dropped at the same time as object Y is projected vertically upward and object Z is projected horizontally. Air resistance is negligible. Which of the following is true?

(A) Objects X and Y have the same downward acceleration, and this is greater than Z's.
(B) Objects X and Z have the same downward acceleration, and this is less than Y's.
(C) Objects Y and Z have the same downward acceleration, and this is greater than X's.
(D) Objects Y and Z have the same downward acceleration, and this is less than X's.
(E) Objects X, Y, and Z have the same downward acceleration.

38. Isobaric changes in an ideal gas imply that there is no change in

(A) volume
(B) temperature
(C) pressure
(D) internal energy
(E) potential energy

Questions 39–41 relate to the following set of graphs.

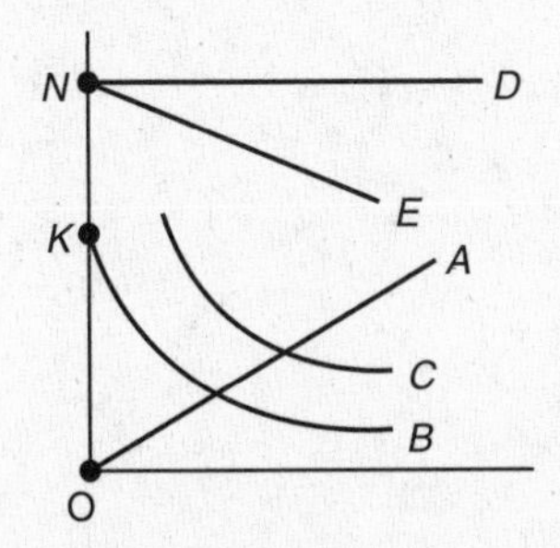

39. The graph that shows most nearly the variation of the pressure of a gas versus its volume, at constant temperature, is

(A) *A*
(B) *B*
(C) *C*
(D) *D*
(E) *E*

40. The graph that most nearly shows the variation of current through a given resistor versus the voltage applied to it is

(A) *A*
(B) *B*
(C) *C*
(D) *D*
(E) *E*

41. Which graph represents a car decelerating uniformly?

(A) A
(B) B
(C) C
(D) D
(E) E

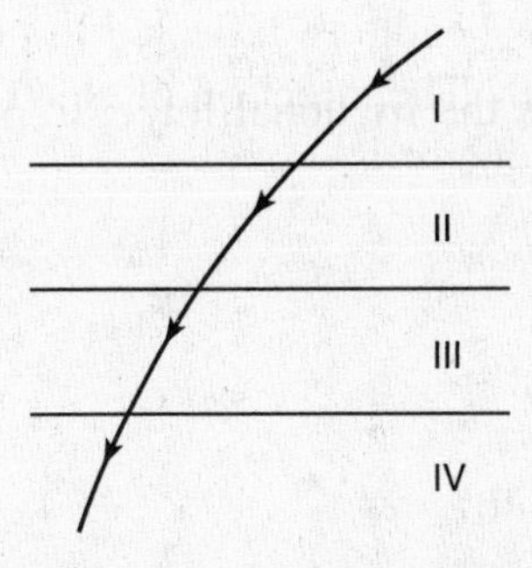

42. The diagram above shows the path of a ray of light through four media separated from each other by horizontal surfaces. The medium in which the speed of the light is greatest is

(A) I
(B) II
(C) III
(D) IV
(E) indeterminate on the basis of the given information

GO ON TO THE NEXT PAGE ►

43. A bright-line spectrum may be obtained by using light from

(A) the Sun
(B) a sodium vapor lamp
(C) a tungsten filament bulb
(D) a clean, glowing platinum wire
(E) a clean, glowing nichrome wire

44. Colors in a soap bubble are caused

(A) solely by pigments
(B) solely by reflection
(C) solely by absorption
(D) by polarization
(E) by interference

45. A compass is placed to the west of a vertical conductor. When electrons pass through the conductor, the north pole of the compass needle is deflected to point toward the south. The direction of electron flow in the conductor is

(A) down
(B) up
(C) east
(D) west
(E) south

Questions 46 and 47 refer to the combination circuit shown below.

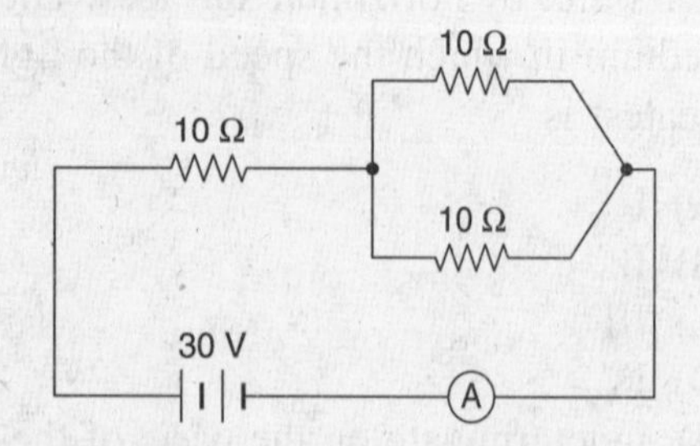

46. What is the total (equivalent) resistance for this circuit?

(A) 10 Ω
(B) 30 Ω
(C) 15 Ω
(D) 5 Ω
(E) 20 Ω

47. What is the circuit current as indicated by the ammeter?

(A) 2 A
(B) 5 A
(C) 1 A
(D) 3 A
(E) 6 A

---

48. If pith ball *X* attracts pith ball *Y* but repels pith ball *Z*,

(A) *Y* must be positively charged
(B) *Y* must be negatively charged
(C) *Y* may be positive or negative but not neutral
(D) *Y* may be neutral or charged
(E) *Y* must be neutral

49. A pendulum of length *L* has a period *T*. If the period is to be doubled, the length of the pendulum should be

(A) increased by $\sqrt{2}$
(B) quartered
(C) quadrupled
(D) halved
(E) doubled

GO ON TO THE NEXT PAGE ►

Questions 50–53 refer to the following diagram and description.

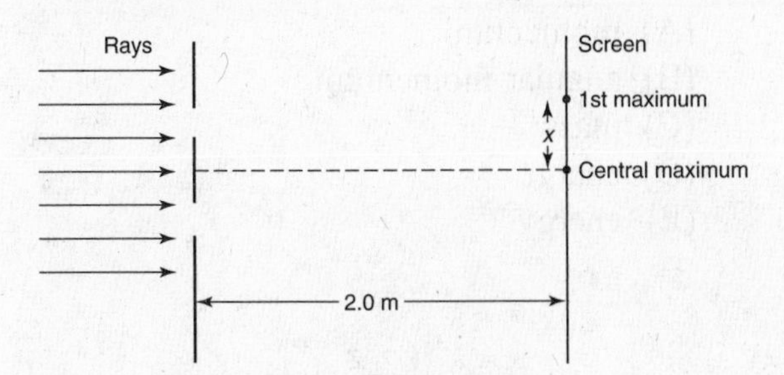

Two parallel slits 2.0×10^{-6} meter apart are illuminated by parallel rays of monochromatic light of wavelength 6.0×10^{-7} meter, as shown. The interference pattern is formed on a screen 2.0 meters from the slits.

50. Distance x is

(A) 6.0×10^{-1} m
(B) 6.0×10^{-7} m
(C) 3.0×10^{-1} m
(D) 3.3 m
(E) 330 m

51. The difference in path length for the rays of light from the two slits to the first maximum is

(A) λ
(B) 2λ
(C) $\frac{\lambda}{2}$
(D) 0
(E) 3λ

52. If the wavelength of the rays of light passing through the slits is doubled, the distance from the central maximum to the first maximum

(A) is halved
(B) is doubled
(C) remains the same
(D) is quadrupled
(E) is tripled

53. If the screen is moved to 1.0 meter from the slits, the distance between the central maximum and the first maximum

(A) is halved
(B) is doubled
(C) remains the same
(D) is quadrupled
(E) is tripled

---

54. Two hundred cubic centimeters of a gas are heated from 30°C to 60°C without letting the pressure change. The volume of the gas will

(A) be halved
(B) be decreased by about 10%
(C) remain the same
(D) be doubled
(E) be increased by about 10%

55. Which of the following is equivalent to 1 pascal of gas pressure?

(A) $1 \text{ kg} \cdot \text{m}^2/\text{s}^2$
(B) $1 \text{ kg/m} \cdot \text{s}^2$
(C) $1 \text{ kg} \cdot \text{m}^3/\text{s}^2$
(D) $1 \text{ kg} \cdot \text{m/s}$
(E) $1 \text{ kg} \cdot \text{m}^2/\text{s}^3$

56. If yellow light emerges from a box, which of the following will be true in regard to the light?

I. It may contain red light.
II. It must contain a wavelength shorter than 8,000 angströms ($1 \text{ Å} = 10^{-10}$ m).
III. It may contain only a single wavelength.
IV. It must contain a wavelength that by itself would appear yellow.

(A) I, II, and III only
(B) I and III only
(C) II and IV only
(D) IV only
(E) None of the above

GO ON TO THE NEXT PAGE ➤

57. Light, with a frequency *f*, is incident on a photoemissive metal. It is observed that electrons are emitted with both minimal and maximum kinetic energies. Which of the following can be done in order to increase the maximum kinetic energy of the emitted electrons?

I Increase the intensity of the incident light
II Increase the frequency of the incident light
III Decrease the frequency of the incident light
IV Decrease the work function by using a different metal

(A) I only
(B) II and IV
(C) I and III
(D) III and IV
(E) I and IV

58. A gas is allowed to expand at constant temperature from a volume of 100 cubic centimeters to a volume of 400 cubic centimeters. The original pressure of the gas was 4 atmospheres. What will the new pressure be?

I. 1 atmosphere
II. 16 atmospheres
III. 76 centimeters of mercury
IV. It cannot be calculated without knowing the temperature.

(A) I, II, and III only
(B) I and III only
(C) II and IV only
(D) IV only
(E) None of the above

59. Lenz's law is an electrical restatement of which conservation law?

(A) momentum
(B) angular momentum
(C) mass
(D) charge
(E) energy

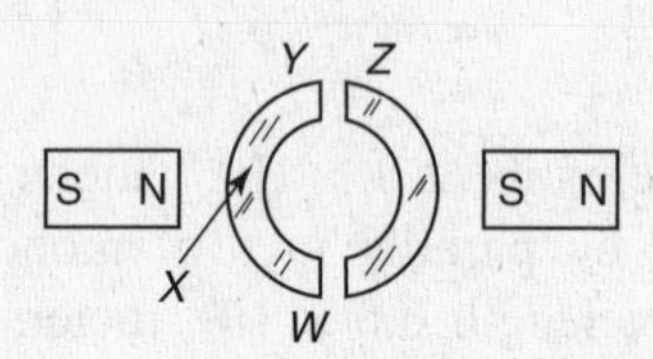

60. Two halves of an iron washer are placed between two magnets as shown in the above diagram. What will be the result(s)?

I. *X* will be an S-pole and *Y* an N-pole.
II. *Y* will be an N-pole and *Z* an S-pole.
III. *Y* and *W* will be S-poles.
IV. *X* and *Z* will be S-poles.

(A) I, II, and III only
(B) I and III only
(C) II and IV only
(D) IV only
(E) None of the above

Questions 61 and 62 relate to the following information.

An electromotive force (emf) of 0.003 volt is induced in a wire when it moves at right angles to a uniform magnetic field with a speed of 4.0 meters per second.

61. If the length of wire in the field is 15 centimeters, what is the flux density, in teslas?

(A) 0.003
(B) 0.005
(C) 6
(D) 12
(E) 2,000

GO ON TO THE NEXT PAGE ►

62. If the wire were to move at an angle of 30° to the field instead of at right angles, the value of the induced emf would be

(A) unchanged
(B) two times as great
(C) three times as great
(D) four times as great
(E) one-half as great

63. What is the weight of a 5.0-kilogram object at the surface of Earth?

(A) 5.0 kg
(B) 25 N
(C) 49 N
(D) 49 kg
(E) 98 N

64. A wool cloth becomes positively charged as it

(A) gains protons
(B) gains electrons
(C) loses protons
(D) loses electrons
(E) gains neutrons

65. If the frequency of a sound wave in air at standard temperature and pressure remains constant, the energy of the wave can be varied by changing its

(A) amplitude
(B) speed
(C) wavelength
(D) period
(E) direction

66. A magnetic field is produced by

(A) moving electrons
(B) moving neutrons
(C) stationary protons
(D) stationary ions
(E) stationary electrons

67. A charged particle is placed in an electric field *E*. If the charge on the particle is doubled, the force exerted on the particle by field *E* is

(A) unchanged
(B) doubled
(C) halved
(D) quadrupled
(E) quartered

68. An electron is observed to pass undeflected through a region in which an electric field and a magnetic field are at right angles to one another. If **E** represents the strength of the electric field in units of newtons per concomb and **B** represents the strength of the magnetic field in units of newtons per (amperes × meter), then the ratio **E/B** is expressed simply in units of

(A) C/s
(B) m/s
(C) m/s^2
(D) m/C
(E) N

Questions 69 and 70

In the following graph, the velocity of an object as it moves along a horizontal straight line is plotted against time.

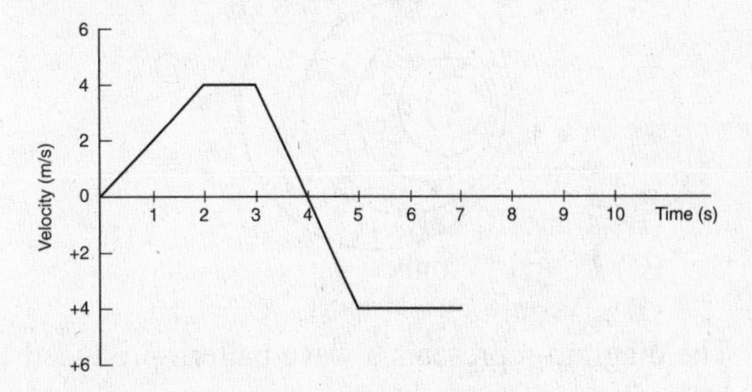

69. The distance traveled by the object during the first 2 seconds is

(A) 2 m
(B) 4 m
(C) 8 m
(D) 16 m
(E) 24 m

GO ON TO THE NEXT PAGE ➤

70. The average speed of the object during the first 4 seconds is, in meters per second,

(A) zero
(B) 1.2
(C) 2.5
(D) 4
(E) 8

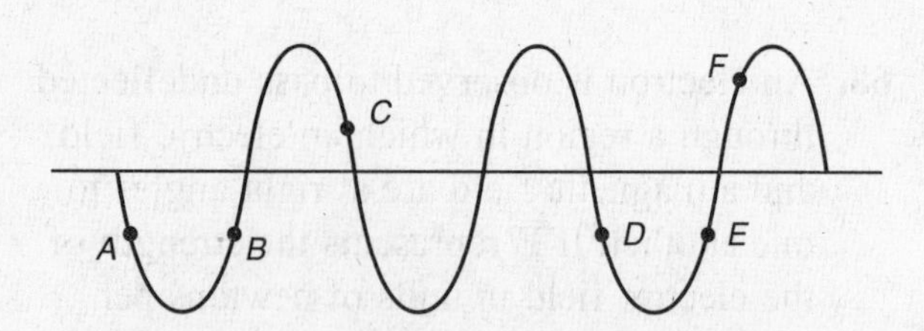

71. Which point in the wave shown in the above diagram is in phase with point *A*?

(A) *F*
(B) *B*
(C) *C*
(D) *D*
(E) *E*

Questions 72 and 73 refer to the following diagram and description.

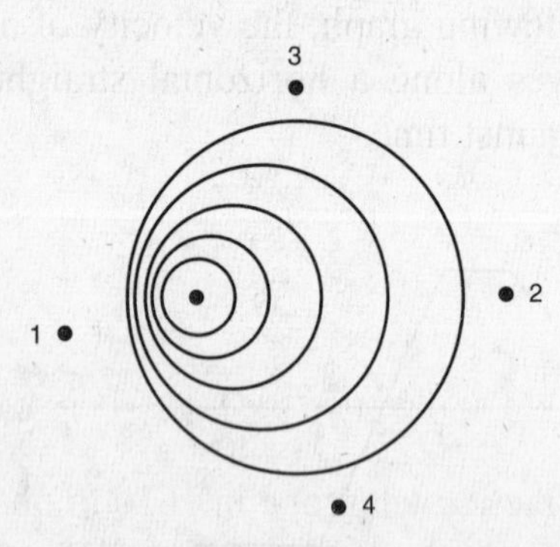

The diagram represents a wave pattern produced by a vibrating source moving at constant speed.

72. Toward which point is the source moving?

(A) 1
(B) 2
(C) 3
(D) 4
(E) It is moving perpendicularly into the page.

73. If the source were to accelerate, the wavelength immediately in front of the source would

(A) increase only
(B) decrease only
(C) first increase and then decrease
(D) first decrease and then increase
(E) remain the same

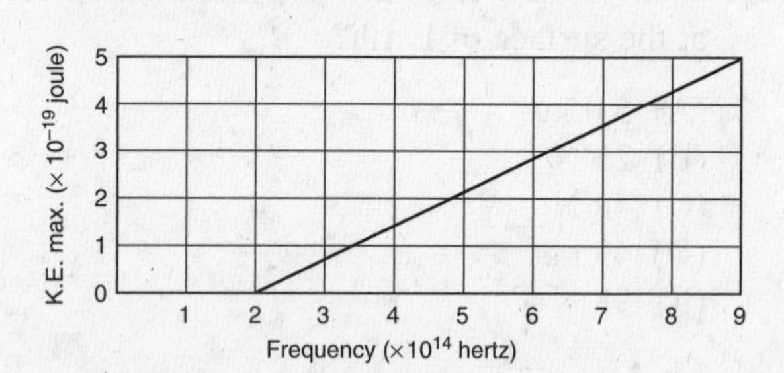

74. The graph above shows the maximum kinetic energy of the photoelectrons ejected when photons of different frequencies strike a metal surface. The slope of the graph is Planck's constant, *h*. The work function of this metal is, in joules,

(A) $h \times 10^{14}$
(B) $2h \times 10^{14}$
(C) $3h \times 10^{14}$
(D) $4h \times 10^{14}$
(E) $8h \times 10^{14}$

75. An electron moves with constant speed at right angles to a uniform magnetic flux. If the flux density were doubled, the force on the electron would be

(A) unaffected
(B) reduced to one-fourth
(C) halved
(D) doubled
(E) quadrupled

Answer Key for Practice Test 2

1. **C**	16. **C**	31. **B**	46. **C**	61. **B**
2. **D**	17. **E**	32. **A**	47. **A**	62. **E**
3. **E**	18. **C**	33. **B**	48. **D**	63. **C**
4. **B**	19. **B**	34. **D**	49. **C**	64. **D**
5. **C**	20. **A**	35. **A**	50. **A**	65. **A**
6. **A**	21. **E**	36. **C**	51. **A**	66. **A**
7. **D**	22. **B**	37. **E**	52. **B**	67. **B**
8. **B**	23. **A**	38. **C**	53. **A**	68. **B**
9. **A**	24. **A**	39. **C**	54. **E**	69. **B**
10. **E**	25. **C**	40. **A**	55. **B**	70. **C**
11. **A**	26. **B**	41. **E**	56. **A**	71. **D**
12. **D**	27. **C**	42. **A**	57. **B**	72. **A**
13. **C**	28. **D**	43. **B**	58. **B**	73. **B**
14. **C**	29. **C**	44. **E**	59. **E**	74. **B**
15. **C**	30. **D**	45. **B**	60. **E**	75. **D**

Answer Explanations for Practice Test 2

1. C Three units of energy are the joule, watt-second, and newton-meter.

2. D The emf is the difference of potential produced by devices such as batteries and generators. It is the voltage between the terminals when the current is zero.

3. E The half-life of a radioactive substance is the length of time required for one-half of its mass to decay into something else.

4. B Power is energy per unit of time. In SI units, 1 J/1 s = 1 W.

5. C Heat is a form of energy. See explanation 1.

6. A The hertz or cycle per second is the unit of frequency.

7. D The volt is the unit of work per unit charge: $V = W/q$.

8. B The periodic return of comets due to gravity was predicted by Edmund Halley.

9. A. Galileo Galilei first used a refracting telescope in 1610 to observe the Moon.

10. E Isaac Newton invented the reflecting telescope and identified the colored spectrum of the Sun.

11. A The law of falling bodies states that all objects near Earth's surface fall with the same rate of acceleration. This was discovered by Galileo.

12. D The modern heliocentric (sun-centered) model was developed by Nicholas Copernicus.

13. C Kepler's second law states that planets sweep out equal areas in equal time.

14. C A *plane mirror* always produces a virtual image that is the same size as the object. (An image in a plane mirror will *appear* to get smaller with increasing distance from it for the same reason that a real person will look smaller as the distance between the observer and the person increases: it is a visual effect.) A *concave lens* produces a virtual image whose size is smaller than the object and decreases as the distance between the object and the lens increases. A *convex lens* can produce both real and virtual images.

15. C A convex lens produces a real image the same size as the object when the object's distance from the lens is two focal lengths. (See also explanation 14.)

16. C Both X rays and gamma rays are usually classified as electromagnetic waves. Their wavelengths are of the order of 10^{-8} cm. The other choices are classified as parts of ordinary matter. Alpha rays are streams of helium nuclei; beta rays are streams of electrons.

17. E Oxidation reactions yield very little energy compared with nuclear fission or fusion. The Sun's vast energy is believed to be the result of a cycle of reactions involving the fusion of hydrogen nuclei.

18. C Alpha particles are helium nuclei. They have a mass number of 4 and charge of +2 (they consist of 2 protons and 2 neutrons). Therefore, when an alpha particle emerges from a nucleus, the positive charge of the remaining portion has decreased by 2; that is, its atomic number, which is the charge of the nucleus, is reduced by 2.

19. B The resultant of two forces applied to an object at the same time depends on their directions with respect to each other. The resultant is largest when the two forces act in the same direction—at an angle of 0°. Then the resultant is equal to the sum of the two forces. As the angle between the two forces increases, the value of the resultant decreases until the angle between the forces is 180°. The resultant then has a minimum value and is equal to the difference between the two forces.

20. A See explanation 19.

21. E You may be able to get the answer quickly by eliminating some of the choices after a quick inspection. Systematic comparison can be made by writing all the choices in standard form (scientific notation) or as decimal fractions. We'll use the former method. The choices then become: **(A)** 4.7×10^{-2} **(B)** 4.7×10^{-5} **(C)** 4.7×10^{-2} **(D)** 4.7×10^{-2} **(E)** 4.7×10^{-1}. Therefore, of the choices given, 0.000047×10^{4} cm is the largest quantity.

22. B Each Kelvin unit represents the same *change* in temperature as 1 Celsius degree. Therefore, an increase of 90° on the Celsius scale is an increase of 90 kelvins on the Kelvin scale.

23. A Considering the choices, it is clear that you are asked to find the work done on the object. Although a force is exerted, the force doesn't move and the object doesn't move. Hence no work is done; work equals force times distance moved in the direction of the force.

24. A When Rutherford directed a stream of alpha particles at a thin metallic foil, he noticed that most of the alpha particles went right through without being deflected. This convinced him that most of the atom is empty space.

A very small fraction of the alpha particles bounce backward. They are reflected from the nucleus by the strong electrostatic repulsion between the positive nucleus and the positive alpha particle. Other alpha particles are scattered through various angles as they approach the nucleus.

25. C The *atomic number* of an atom is the number of protons in its nucleus.

26. B In simple harmonic motion, the maximum velocity occurs when the acceleration is a minimum.

27. C The speed of sound in water is always greater than in air.

28. D If the sound takes 4 s to go down and up again, it takes only 2 seconds just to go to the bottom.

$$\text{distance} = \text{speed} \times \text{time}$$
$$= 1{,}460 \text{ m/s} \times 2 \text{ s} = 2{,}920 \text{ m}$$

29. C As the pendulum bob swings back and forth, at the highest point of its swing, at *A* and *C,* all of its energy is potential. At those points it is momentarily at rest. As it swings through its arc from its highest point, the potential energy is converted to kinetic. Therefore, at the lowest point of the swing, at *B,* the energy is kinetic. Then, as the bob swings up toward *C,* the kinetic energy is converted to potential. If air friction is negligible, *C* is at the same height as *A,* and the bob has the same potential energy there as at *A*. Also, in that case, throughout the swing the sum of the potential and kinetic energies remains constant.

30. D The period of a simple pendulum is proportional to the square root of its length. If the length is multiplied by 4, the period is multiplied by the square root of 4, which is 2. The complete formula for the period of the pendulum is $T = 2\pi\sqrt{L/\mathbf{g}}$, where $\mathbf{g}$ is the acceleration due to gravity at that location.

31. B The momentum of an object is equal to the product of its mass and its velocity:

$$\text{momentum} = \text{mass} \times \text{velocity}$$
$$= (4.0 \text{ kg})(3 \text{ m/s}^2)(2 \text{ s})$$
$$= 24 \text{ kg} \cdot \text{m/s}$$

We calculated the final velocity at the end of 2 s by multiplying the constant acceleration by the time. Note also that the question does not say that the 15-N force is a net force. Therefore, quick use of the relationship that impulse is equal to change in momentum would not be well advised. (Unfortunately, in some questions the absence of friction has to be assumed.) This will be more apparent from question 32.

32. A The horizontal force of 15 N was used to give the object acceleration and also to overcome friction. According to Newton's second law of motion, the force that gives an object acceleration is equal to the product of the object's mass and its acceleration:

$$\mathbf{F} = m\mathbf{a}$$
$$= (4 \text{ kg})(3 \text{ m/s}^2) = 12 \text{ N}$$

The force of friction is equal to the difference between the total force and the net force: 15 N − 12 N = 3 N.

33. B Potential energy = weight × height. At the top, 10 m above the ground, the object's potential energy = 40 N × 10 m = 400 J. When the object is 6 m above the ground, it's potential energy = 40 N × 6 m = 240 J.

The difference in potential energy is the kinetic energy that the object acquired:

$$\text{kinetic energy} = (400 - 240)\text{ J} = 160\text{ J}$$

34. D Just before the object hits the ground, all of its potential energy (400 V) has been converted to kinetic energy.

35. A This is one of the classic experiments. It shows that horizontal motion does not affect vertical motion (that is, vertical motion is affected only by a vertical force or by a force having a vertical component).

36. C This follows from the result of the experiment described in question 35. Ball *X* must first rise to a greater height than ball *Y;* we can think of it as then falling freely from the highest point it reaches. In the meantime, ball *Y* has already started on its downward trip. Of course, mass and density have nothing to do with the time ball *Y* will take to reach ground, if air resistance is negligible.

37. E Any change in velocity of the three objects is due to gravity only; the acceleration due to gravity is the same for all objects near the surface of Earth.

38. C Isobaric changes occur in systems with constant pressure.

39. C This refers to Boyle's law: The pressure of a gas varies inversely as its volume. The graph is a hyperbola, as is (C). Choice (*B*) is not satisfactory, since this graph would indicate that the volume of the gas could be reduced to zero.

40. A This refers to Ohm's law: $V = IR$. As the applied voltage increases, the current increases. The only graph that shows such a variation is *A*.

41. E A car undergoing uniform deceleration would show a downward-sloping, diagonal line such as *E* in a graph of velocity versus time.

42. A An oblique ray is bent toward the normal when it enters a medium in which the wave is slowed. In the diagram, the ray is shown bent more and more toward the normal as it goes from medium I toward medium IV. The light travels more slowly in each successive medium.

43. B A bright-line spectrum is usually obtained from a glowing gas or vapor. In the laboratory, sodium vapor is often used as the source of such a spectrum. A special sodium vapor lamp may be used; frequently, ordinary salt (sodium chloride) is sprinkled into the gas flame, giving the necessary light. Glowing tungsten, platinum, or nichrome provides a continuous spectrum. The Sun's spectrum contains the dark absorption lines known as *Fraunhofer lines.*

44. E The rainbow colors of soap bubbles and oil films are produced by interference of light.

45. B Hold a pencil in a vertical position in front of you. Imagine that you are facing north. Then to the left of the pencil is west. Imagine the compass placed to the left of the pencil. According to the question, the compass needle points south. Apply the right-hand rule: Pretend to grasp a wire with your right hand so that your fingers, when on the left side of the wire, will point south. To do this, your thumb will have to point up; this indicates the direction of conventional current.

46. C This is a combination series-parallel circuit. To solve, first reduce the parallel branch. Since both resistors are 10 Ω, the equivalent resistance is 5 Ω, using the formula $\mathbf{R}_{eq} = \mathbf{R}_1\mathbf{R}_2/(\mathbf{R}_1 + \mathbf{R}_2)$. Next, this

5-Ω equivalent resistance is in series with the remaining 10-Ω resistor. Therefore, the total (equivalent) resistance is equal to 15 Ω.

47. A Use Ohn'slaw: $V = IR$. Substituting into the equation 30 V and 15 Ω yields 2 A.

48. D If two nonmagnetic objects repel each other, both objects must be electrified, and the charge on the two must be of the same kind, both positive or both negative. Hence, since *X* repels *Z*, *X* must be charged, but it may be positive or negative. A charged object will attract neutral objects and objects oppositely charged. Since charged object *X* attracts *Y*, *Y* may be neutral or charged.

49. C The period of a pendulum is proportional to the square root of its length. To double the period, the length must be quadrupled.

50. A This question refers to Young's double-slit experiment. When monochromatic light passes through two small slits, a series of bright lines is seen on a distant screen. The distance, *x*, of the first maximum or bright line away from the central maximum can be calculated from the following relationship:

$$\frac{x}{L} = \frac{\lambda}{d}$$

where *L* is the distance from the slits to the screen, λ is the wavelength of the light, *d* is the distance between the slits. Substituting gives:

$$\frac{x}{2.0 \text{ m}} = \frac{6 \times 10^{-7} \text{ m}}{2 \times 10^{-6} \text{ m}}$$

$$x = 6 \times 10^{-1} \text{ m}$$

51. A In the interference pattern, wherever there is a bright region or maximum, the waves from the two slits must arrive in phase. Therefore, they must arrive a whole number of wavelengths apart, meaning that the path lengths from the two slits to the bright regions must differ by a whole number of wavelengths. The path difference to the first maximum is one wavelength (λ).

52. B From the formula in explanation 50, we see that *x* and λ are proportional to each other. (If we increase the numerator on the left, we must also increase the one on the right to keep the two fractions equal.) Therefore, if the wavelength is doubled, the distance, *x*, between the two maxima also is doubled.

53. A From the formula in explanation 50, we see that *x* is proportional to *L*, the distance between the slits and the screen. (If the numerator of the fraction increases, the denominator must also increase if the value of the fraction doesn't change.) Therefore, when the screen is moved closer to the slits (from 2 m to 1 m), the distance between the two maxima is also halved.

54. E At constant pressure, the volume of a gas is proportional to the absolute temperature. The absolute temperature increases from (273 + 30) K to (273 + 60) K. An increase from 303 to 333 is an increase of about 10%. The volume should also increase by about 10%.

55. B One pascal of gas pressure is given by:

$$1 \text{ N/m}^2 = 1 \text{ kg/m} \cdot \text{s}^2$$

56. A Yellow light may be light of a single wavelength or it may be the result of a mixture of two or more wavelengths. For example, you should know that, if red and green light are mixed, yellow light is produced. The visible range of light is approximately 3,900–7,800 Å.

57. B In the photoelectric effect, we have the equation

$$KE_{max} = hf - \phi,$$

where ϕ is the work function. The maximum kinetic energy is independent of the intensity. To increase the maximum kinetic energy, one needs to increase the frequency and/or lower the work function by using a different metal.

58. B Boyle's law applies. Since the volume is quadrupled, the pressure is reduced to one-fourth of its original value: from 4 atm to 1 atm. The pressure of 1 atm is the same as the pressure of a column of mercury 76 cm high.

59. E Lenz's law is an electrical restatement of the law of conservation of energy.

60. E We may use the molecular theory of magnetism and the law of magnets to determine what will happen. The molecular magnets in the washer halves will orient themselves so that, near the poles of the permanent magnet, opposite poles will predominate: at *X*, mostly S-poles. The opposite poles of the molecular magnets have to point in the opposite direction from *X*. Hence, at *Y* and *W*, N-poles will predominate. The reverse happens in the other half of the washer, which is nearer to the S-pole. Therefore, *Z* will be an S-pole.

61. B An emf is induced in a wire when it cuts across magnetic flux lines. This emf is maximum when the wire cuts across at right angles to the flux, and it is then equal to the product of the magnetic flux density, the length of wire in the field, and the speed of cutting.

$$\text{emf} = BL\mathbf{v}$$
$$0.003 \text{ V} = B\,(0.15 \text{ m})(4 \text{ m/s})$$
$$= 0.005 \text{ T}$$

62. E As stated in question 61, the emf is maximum when the angle of cutting is 90°. It is less for any other angle. Only choice (E) satisfies this. (Actually the emf is proportional to the sine of the angle. The sine of 30° is 0.5.)

63. C The weight of an object on Earth is equal to the product of the object's mass and the free-fall acceleration at the place on Earth:

$$\mathbf{w} = mg$$
$$= (5.0 \text{ kg})(9.8 \text{ m/s}^2)$$
$$= 49 \text{ N}$$

Note that $1 \text{ N} = 1 \text{ kg} \cdot \text{m/s}^2$.

64. D Usually objects become electrically charged by gaining or losing electrons. Since electrons are negatively charged, the wool cloth had to lose electrons to become positively charged.

65. A When the amplitude of a wave is increased, the energy of the wave increases. Since temperature and pressure of the air are constant at standard conditions, the speed of sound will remain constant. Since frequency is being kept constant, the wavelength must also remain constant because the speed of a wave equals the product of frequency and wavelength. The period is the reciprocal of frequency, and so it is also constant. Direction is not involved.

66. A A magnetic field is produced by an electric current; an electric current in a wire consists of moving electrons. Neutrons are electrically neutral; magnetic fields are not produced by a stream of neutral particles. Stationary particles do not produce a magnetic field.

67. B The electric field is defined as the force exerted on a unit charge placed in that field:

$$\mathbf{E} = \frac{\mathbf{F}}{q}$$

This shows that the force on any charge placed in the field is proportional to the field and the magnitude of the charge:

$$\mathbf{F} = \mathbf{E}q$$

If the charge is doubled, the force exerted by the field is also doubled.

68. B From electromagnetic theory, if an electron (charge q) is passing through crossed fields without being deflected, then the electrical and magnetic forces balance:

$$\mathbf{F}_e = \mathbf{E}q = \mathbf{B}q\mathbf{v} = \mathbf{F}_m$$

From this we see that $\mathbf{v} = \mathbf{E}/\mathbf{B}$, so the units must be m/s.

69. B On a graph of velocity against time, the distance traveled (actually the displacement) is given by the area under the graph. Another method, which we shall use here, is to multiply the average speed by the time traveled. The graph for the first 2 s is an oblique straight line. This represents motion with constant acceleration. For such motion the average speed is the sum of the initial and final speeds divided by 2. The initial speed is zero. The average speed is (final speed/2) = (4 m/s)/2 = 2 m/s.

$$\begin{aligned} \text{distance} &= (\text{average speed})(\text{time}) \\ &= (2 \text{ m/s})(2 \text{ s}) \\ &= 4 \text{ m} \end{aligned}$$

70. C We first find the total distance traveled during the first 4 s. We already know the distance from question 69 traveled during the first 2 s. During the third second, the speed remains constant at 4 m/s.

$$\begin{aligned} \text{distance} &= (4 \text{ m/s})(1 \text{ s}) \\ &= 4 \text{ m} \end{aligned}$$

During the fourth second the average speed is again equal to (4 m/s)/2, or 2 m/s.

$$\text{distance} = (2 \text{ m/s})(1 \text{ s}) = 2 \text{ m}$$

Total distance = (4 m + 4 m + 2 m) = 10 m. Then

$$\begin{aligned}\text{average speed} &= (\text{total distance})/\text{time} \\ &= (10 \text{ m})/4 \text{ s} \\ &= 2.5 \text{ m/s}\end{aligned}$$

71. D Two points in a wave are in phase if they are going through the same part of the vibration at the same time. We can think of the diagram as

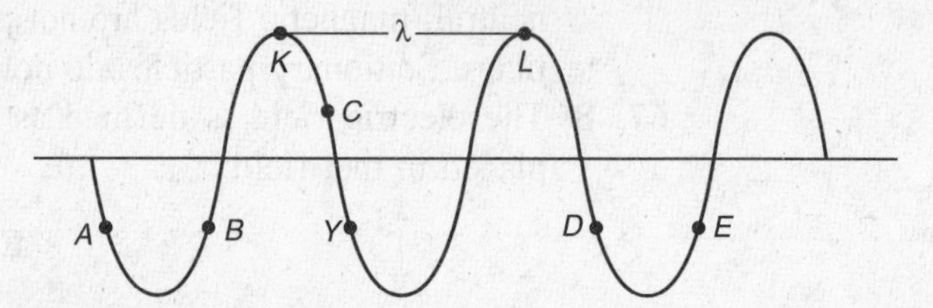

representing a periodic wave that is continuously transmitting energy to the right. *Two points in phase are then a whole number of wavelengths apart.* In the above diagram, one wavelength is the distance between *K* and *L,* two adjacent peaks on the curve. This is equal to the distance from *A* to *Y.* We can think of curve *ABKCY* as completing one cycle; then the same wave pattern is shown again by curve *YLD.* The distance between *Y* and *D* is another wavelength. *D* is 2 wavelengths from *A.* Points *A, Y,* and *D* are going through the same part of the vibration at the same time. Point *D* is in phase with point *A*.

72. A The *Doppler effect* deals with the apparent change in the frequency of a wave when there is relative motion between the source of the wave and an observer. Here we are told that the vibrating source is moving. The circles represent the spherical wave fronts produced by the source. The wavelength is the distance between two adjacent wave fronts—in this diagram, the distance between two adjacent circles. If the source were stationary, the distance between wave fronts would be the same in all directions. We notice in the diagram that the circles are much closer to each other near point 1 than anywhere else.

This shows that the source keeps moving closer to point 1 as it sends out compressions. At the same time it keeps moving away from point 2. As the wave reaches an observer at point 1, the length, and therefore whose frequency, of the wave are different from what they would be if the source were stationary. The frequency would be greater than the actual frequency of the source.

73. B Students are often confused about this part of the Doppler effect. In question 72 the source was moving with constant velocity toward point 1. An observer at point 1 would receive a wave of constant frequency and wavelength, but different from what it would be if the source were stationary. If the source now started to move faster, it would tend to catch up with the wave front it had just sent out toward point 1. The new wave front would therefore be closer to the preceding wave front than if the source were moving with constant velocity. The wavelength, therefore, would decrease.

74. B As we lower the frequency of the radiation incident on a photoemissive material, the maximum kinetic energy of the photoelectrons becomes less and less. The energy of the photon that would just give it zero kinetic energy is the work function of this material. Its frequency would be the threshold frequency, f_0. On the graph this is the point where the line intersects the axis, 2×10^{14} Hz. The value of the work function is equal to the product of Planck's constant and the threshold frequency:

$$\phi = hf_0$$

$$= h\,(2 \times 10^{14}\text{ Hz}) = 2h \times 10^{14}$$

75. D The force on an electron moving at right angles to a uniform magnetic flux is equal to the product of the flux density, the charge of the electron, and the speed of the electron ($\mathbf{F} = q\mathbf{v}\mathbf{B}$). If the flux density is doubled, the force on the electron is doubled.

Answer Sheet: Practice Test 3

1. (A) (B) (C) (D) (E)
2. (A) (B) (C) (D) (E)
3. (A) (B) (C) (D) (E)
4. (A) (B) (C) (D) (E)
5. (A) (B) (C) (D) (E)
6. (A) (B) (C) (D) (E)
7. (A) (B) (C) (D) (E)
8. (A) (B) (C) (D) (E)
9. (A) (B) (C) (D) (E)
10. (A) (B) (C) (D) (E)
11. (A) (B) (C) (D) (E)
12. (A) (B) (C) (D) (E)
13. (A) (B) (C) (D) (E)
14. (A) (B) (C) (D) (E)
15. (A) (B) (C) (D) (E)
16. (A) (B) (C) (D) (E)
17. (A) (B) (C) (D) (E)
18. (A) (B) (C) (D) (E)
19. (A) (B) (C) (D) (E)
20. (A) (B) (C) (D) (E)
21. (A) (B) (C) (D) (E)
22. (A) (B) (C) (D) (E)
23. (A) (B) (C) (D) (E)
24. (A) (B) (C) (D) (E)
25. (A) (B) (C) (D) (E)
26. (A) (B) (C) (D) (E)
27. (A) (B) (C) (D) (E)
28. (A) (B) (C) (D) (E)
29. (A) (B) (C) (D) (E)
30. (A) (B) (C) (D) (E)
31. (A) (B) (C) (D) (E)
32. (A) (B) (C) (D) (E)
33. (A) (B) (C) (D) (E)
34. (A) (B) (C) (D) (E)
35. (A) (B) (C) (D) (E)
36. (A) (B) (C) (D) (E)
37. (A) (B) (C) (D) (E)
38. (A) (B) (C) (D) (E)
39. (A) (B) (C) (D) (E)
40. (A) (B) (C) (D) (E)
41. (A) (B) (C) (D) (E)
42. (A) (B) (C) (D) (E)
43. (A) (B) (C) (D) (E)
44. (A) (B) (C) (D) (E)
45. (A) (B) (C) (D) (E)
46. (A) (B) (C) (D) (E)
47. (A) (B) (C) (D) (E)
48. (A) (B) (C) (D) (E)
49. (A) (B) (C) (D) (E)
50. (A) (B) (C) (D) (E)
51. (A) (B) (C) (D) (E)
52. (A) (B) (C) (D) (E)
53. (A) (B) (C) (D) (E)
54. (A) (B) (C) (D) (E)
55. (A) (B) (C) (D) (E)
56. (A) (B) (C) (D) (E)
57. (A) (B) (C) (D) (E)
58. (A) (B) (C) (D) (E)
59. (A) (B) (C) (D) (E)
60. (A) (B) (C) (D) (E)
61. (A) (B) (C) (D) (E)
62. (A) (B) (C) (D) (E)
63. (A) (B) (C) (D) (E)
64. (A) (B) (C) (D) (E)
65. (A) (B) (C) (D) (E)
66. (A) (B) (C) (D) (E)
67. (A) (B) (C) (D) (E)
68. (A) (B) (C) (D) (E)
69. (A) (B) (C) (D) (E)
70. (A) (B) (C) (D) (E)
71. (A) (B) (C) (D) (E)
72. (A) (B) (C) (D) (E)
73. (A) (B) (C) (D) (E)
74. (A) (B) (C) (D) (E)
75. (A) (B) (C) (D) (E)

PRACTICE TEST 3

Part A

Directions: For each group of questions below, there is a set of five lettered choices, followed by numbered questions. For each question select the one choice in the set that best answers the question and fill in the corresponding circle on the answer sheet. You may use a lettered choice once, more than once, or not at all in each set. Do not use a calculator.

Questions 1–3 refer to the following situation.

A 70-newton object rests on the middle of a rough horizontal board. Then one end of the board is raised slowly, rotating the board through an angle of 50° without causing the object to slide.

(A) It remains zero.
(B) It remains 70 newtons.
(C) It first increases and then decreases.
(D) It steadily decreases.
(E) It steadily increases.

1. While one end of the board is being raised, what is the effect on the force of friction between the board and the object?
2. What is the effect on the normal force between the board and the object?
3. What is the effect on the gravitational pull on the object?

Questions 4 and 5 refer to the following measurements:

(A) 353.0 mm
(B) 55.0 cm
(C) 0.056 m
(D) 0.731 m
(E) 0.034 cm

4. Which of the measurements listed above is the smallest?
5. Which of the measurements listed above is the largest?

Questions 6 and 7 deal with the following subatomic particles:

(A) Photon
(B) Electron
(C) Newton
(D) Alpha particle
(E) Proton

6. Which particle always has the same charge as the positron?
7. The mass of which particle is the same as that of the positron?

GO ON TO THE NEXT PAGE ➤

Questions 8–10 refer to the following experiment. A ball is projected vertically upward from the surface of Earth and reaches its maximum height in 4.0 seconds.

(A) Zero
(B) 20
(C) 40
(D) 80
(E) 100

8. Approximately what was the ball's initial speed, in meters per second?

9. Approximately what was the maximum height, in meters, reached by the ball?

10. What is the total displacement, in meters, of the ball from the time it is thrown until it returns to the point from which it was thrown?

Questions 11–13 refer to the diagram below. A 6.0-kilogram object is being moved at constant speed along a straight line on a level surface by a force of 10.0 newtons.

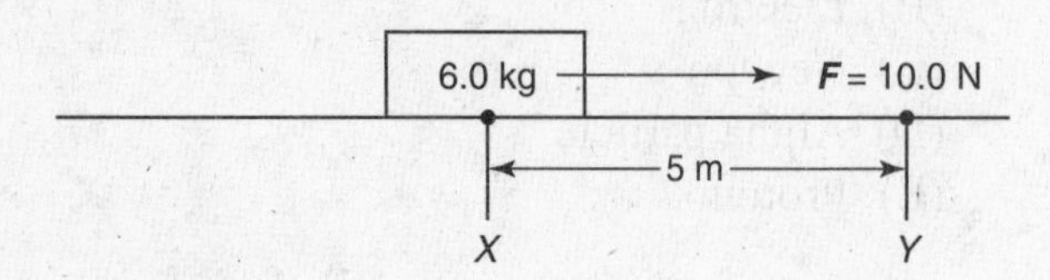

(A) Zero
(B) Less than 2, but not zero
(C) 2
(D) 10
(E) 20

11. What is the change in kinetic energy, in joules, of the object as it moves from point *X* to point *Y*?

12. If energy is supplied at the rate of 10 watts, how much energy, in joules, is supplied in 2 seconds?

13. If the object's momentum is 12 kilogram • meters per second, what is its velocity, in meters per second?

Questions 14 and 15 refer to the concept of conductivity.

(A) Free electrons only
(B) Positive ions only
(C) Negative ions only
(D) Positive and negative ions only
(E) Positive ions, negative ions, and electrons

14. On which particle(s) does the conductivity in metallic wires depend?

15. On which particle(s) does the conductivity of salt solutions depend?

GO ON TO THE NEXT PAGE ►

Part B

Directions: Each of the questions or incomplete statements is followed by five suggested answers or completions. Select the choice that best answers the question or completes the statement and fill in the corresponding circle on the answer sheet. Do not use a calculator.

16. Three forces are acting on a particle in equilibrium. The three forces are 6, 8, and 10 newtons, respectively. The resultant of the 8-newton and 10-newton forces has a magnitude of

(A) 2 N
(B) 6 N
(C) 9 N
(D) 13 N
(E) 18 N

17. Which of the following combinations of fundamental units represents a joule?

(A) kg × m/s^2
(B) kg × m/s
(C) kg × m^2/s^2
(D) kg × s/m^2
(E) kg × m^2/s

Questions 18–21 relate to the graph below. Cars *A* and *B* are moving on straight, parallel tracks. Car *B* passes Car *A* at the same instant that Car *A* starts from rest at t = 0 second.

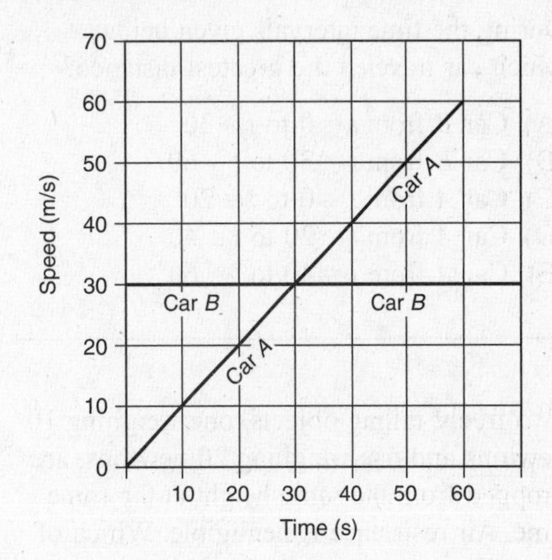

18. What is the acceleration, in meters per second squared, of Car *A* during the interval between t = 0 and t = 60 seconds?

(A) 0
(B) 1
(C) 20
(D) 30
(E) 40

19. How far did Car *A* travel in the interval between t = 0 and t = 60?

(A) 15 m
(B) 30 m
(C) 360 m
(D) 1,800 m
(E) 3,600 m

PRACTICE TEST 3

GO ON TO THE NEXT PAGE ➤

20. How long, in seconds, after $t = 0$ did Car *A* take to catch up to Car *B*?

(A) 0
(B) 10
(C) 20
(D) 30
(E) 60

21. During the time intervals given below, which car traveled the greatest distance?

(A) Car *B* from $t = 0$ to $t = 30$
(B) Car *B* from $t = 30$ to $t = 60$
(C) Car *A* from $t = 0$ to $t = 20$
(D) Car *A* from $t = 20$ to $t = 40$
(E) Car *A* from $t = 40$ to $t = 60$

---

22. Two freely falling objects, one weighing 10 newtons and one weighing 20 newtons, are dropped from the same height at the same time. Air resistance is negligible. Which of the following statements is (are) true?

I. Both objects have the same potential energy at the top.
II. Both objects fall with the same acceleration.
III. Both objects have the same speed just before hitting the ground.

(A) I only
(B) III only
(C) I and II only
(D) II and III only
(E) I, II, and III

23. Which of the following units may be used to express the amount of heat absorbed by a body?

I. Calories
II. Joules
III. Ergs

(A) I only
(B) III only
(C) I and II only
(D) II and III only
(E) I, II, and III

24. A steel ball is projected horizontally from a certain height. It hits the ground in 10 seconds. If air resistance is negligible, and the initial speed of the object is 30 meters per second, its acceleration during the third second is

(A) 9.8 m/s^2
(B) 3.2 m/s^2
(C) 9.6 m/s^2
(D) 32.7 m/s^2
(E) greater than during the first second

25. The horizontal distance traveled by the steel ball in question 24 during the first 5 seconds is

(A) 150 m
(B) 160 m
(C) 80 m
(D) 20.5 m
(E) 4,900 m

26. If a stone at the end of a string is whirled in a circle, the inward pull of the string on the stone

(A) is known as the centrifugal force
(B) is inversely proportional to the speed of the object
(C) is inversely proportional to the square of the speed
(D) is proportional to the speed
(E) is proportional to the square of the speed

27. Two carts are connected together, at rest, by a massless spring on a frictionless surface. One cart has a mass of 0.4 kilogram, while the other cart has a mass of 1.2 kilograms. When released, the speed of the 0.4-kilogram cart is observed to be 0.6 meter per second. What is the speed of the 1.2-kilogram cart?

(A) 0.6 m/s
(B) 1.8 m/s
(C) 1.2 m/s
(D) 0.3 m/s
(E) 0.2 m/s

GO ON TO THE NEXT PAGE ►

28. A man weighing 600 newtons pushes against a box with a force of 20 newtons, but the box does not move. The box weighs 3,000 newtons. The force that the box exerts on the man is

(A) zero
(B) 20 N
(C) 100 N
(D) 150 N
(E) 3,000 N

29. A change in temperature of 45°C corresponds to a change in kelvins of

(A) 25
(B) 45
(C) 81
(D) 113
(E) 7

30. A temperature of 300°C corresponds to an absolute temperature in Kelvin units of

(A) 27
(B) –27
(C) 523
(D) 573
(E) 572

Questions 31 and 32

A spring with a force constant k is stretched a distance x.

31. By what factor must the spring's elongation be changed so that the elastic potential energy in the spring is doubled?

(A) ¼
(B) ½
(C) 2
(D) 4
(E) $\sqrt{2}$

32. Which of the following are equivalent units for the force constant k?

(A) $kg \cdot m^2/s^2$
(B) $kg \cdot m/s$
(C) $kg \cdot s^2$
(D) kg/s^2
(E) $kg \cdot s/m$

---

33. What average braking force is applied to a 2,500-kilogram car having a velocity of 30 meters per second if the car is to stop in 15 seconds?

(A) 5,000 N
(B) 6,000 N
(C) 8,000 N
(D) 10,000 N
(E) 12,000 N

34. An object weighing 40 newtons is dragged a distance of 20 meters by a force of 5 newtons in the direction of motion. The work done by the 5-newton force

(A) is 5 J
(B) is 100 J
(C) is 200 J
(D) is 800 J
(E) cannot be calculated without knowing the force of friction

35. A man holds two solid steel objects above the ground; one weighs 10 newtons and the other 20 newtons. As compared with the potential energy of the 10-newton object, the potential energy of the 20-newton object is

(A) greater even if air is present
(B) greater only in vacuum
(C) the same even if air is present
(D) the same only in absence of air
(E) less

36. If a gas is heated at constant pressure, which of the following statements will apply?

I. Its volume increase is proportional to the temperature.
II. The kinetic energy of the molecules decreases.
III. The kinetic energy of the molecules increases.

(A) I only
(B) III only
(C) I and II only
(D) I and III only
(E) II and III only

37. What is the total capacitance of the circuit shown below?

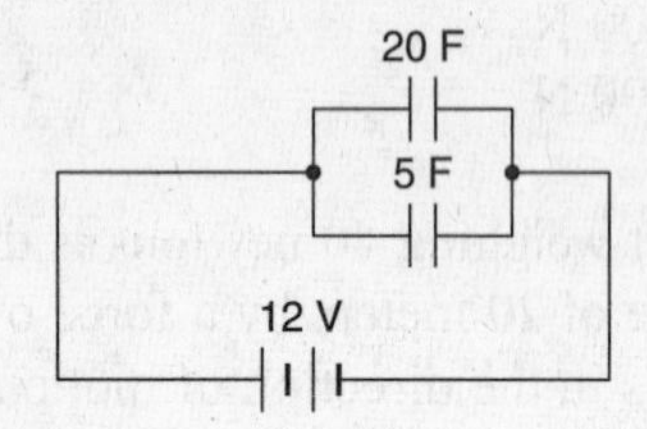

(A) 25 F
(B) 20 F
(C) 15 F
(D) 5 F
(E) 4 F

38. A motor operates from a 120-volt electrical source and uses 4 amperes of current. What is the power generated by the motor?

(A) 30 W
(B) 48 W
(C) 120 W
(D) 480 W
(E) 640 W

39. A 2-ohm and a 60-ohm resistor are connected in series to a direct-current generator. The voltage across the 20-ohm resistor is 80 volts. The current through the 60-ohm resistor

(A) cannot be calculated with the given information
(B) is about 1.3 A
(C) is 4.0 A
(D) is 1.0 A
(E) is 5.0 A

40. A stone is thrown vertically upward from sea level to a height of 20 meters. Assume that air resistance is negligible. After the stone leaves the hand, its acceleration is

(A) downward at a decreasing value until it reaches its highest point
(B) upward at a decreasing value until it reaches its highest point
(C) upward at 9.8 m/s^2
(D) not known
(E) None of the above

41. An object is placed 10 centimeters from a concave spherical mirror whose radius of curvature is 12 centimeters. What is the distance of the image from the mirror?

(A) 5 cm
(B) 10 cm
(C) 15 cm
(D) 20 cm
(E) 24 cm

42. If two sounds have the same wavelength in air at the same temperature, what other property must they also have in common?

I. Intensity
II. Amplitude
III. Frequency

(A) I only
(B) III only
(C) I and II only
(D) II and III only
(E) I, II, and III

43. Two frequencies sounded together produce 3 beats per second. If one of the frequencies is 400 vibrations per second, what may the other frequency be?

I. 1,200 vib/s
II. 403 vib/s
III. 397 vib/s

(A) I only
(B) III only
(C) I and II only
(D) II and III only
(E) I, II, and III

44. Two conducting spheres are negatively charged. The distance between them is 20 centimeters. The force that these two spheres exert on each other is 4.0×10^{-4} newton. If the distance between them is increased to 40 centimeters, the electrostatic force between them becomes, in newtons,

(A) 1.0×10^{-4}
(B) 2.0
(C) 2.0×10^{-4}
(D) 8.0×10^{-4}
(E) 1.6

45. A 3.0-centimeter length of wire is moved at right angles across a magnetic field with a speed of 2.0 meters per second. If the flux density is 10 teslas, what is the magnitude of the induced electromotive force (emf)?

(A) 0.06 V
(B) 0.6 V
(C) 1.2 V
(D) 10 V
(E) 20 V

46. With which of the following may a virtual image be obtained?

I. Plane mirror
II. Convex lens
III. Concave lens

(A) I only
(B) III only
(C) I and II only
(D) II and III only
(E) I, II, and III

47. An electric toaster-oven is rated 120 volts, 1,500 watts. What is the resistance of the heater coil when it operates under these conditions, and what is the current going through it?

(A) 12.5 Ω, 9.6 A
(B) 1,190 Ω, 8.0 A
(C) 8.0 Ω, 15.0 A
(D) 18.7 Ω, 8.0 A
(E) 9.6 Ω, 12.5 A

48. A ^{238}U nucleus differs from a ^{235}U nucleus in that it contains three more

(A) neutrons
(B) alpha particles
(C) protons
(D) electrons
(E) positrons

49. The mass of an ^4_2He nucleus is

(A) 2 g
(B) 4 g
(C) equal to the mass of two protons plus two electrons
(D) equal to the mass of two protons plus two neutrons
(E) slightly less than the mass of two protons plus two neutrons

50. If graphite is used in a nuclear reactor, its function is to

(A) absorb alpha particles
(B) absorb neutrons
(C) slow down neutrons
(D) absorb electrons
(E) slow down electrons

51. A gamma ray photon makes a collision with an electron at rest. During the interaction the change in the photon will be

(A) a decrease in momentum only
(B) a decrease in kinetic energy only
(C) an increase in momentum only
(D) an increase in kinetic energy only
(E) a decrease in momentum and kinetic energy

52. X rays consist of

(A) a stream of neutrons
(B) a stream of Compton particles
(C) a stream of electrons
(D) radiation similar to radon
(E) radiation similar to gamma rays

53. If a lens produces an image the same size as the object when the object is 20 centimeters from the lens, then the largest real image of this object can be obtained when the distance of the object from the lens is approximately

(A) 5 cm
(B) 10 cm
(C) 15 cm
(D) 30 cm
(E) 40 cm

54. If two sounds have the same pitch but different loudness, the louder sound has

(A) the higher frequency
(B) the lower frequency
(C) the greater amplitude
(D) the smaller amplitude
(E) the greater repetition rate

55. For a given sound wave traveling through air at constant temperature, the speed of a compression

(A) is unrelated to the speed of the rarefaction
(B) is the same as the speed of the rarefaction
(C) is always less than the speed of the rarefaction
(D) is always greater than the speed of the rarefaction
(E) is greater than the speed of the rarefaction only when the temperature is greater than the critical temperature

56. During the time that sound travels 331 meters in air, light can travel in vacuum a distance of about

(A) 1,100 m
(B) 200,000 cm
(C) 200,000 m
(D) 300,000 km
(E) 200,000 km

57. Maximum destructive interference of two waves occurs at points where the phase difference between the two waves is

(A) 0°
(B) 45°
(C) 90°
(D) 180°
(E) 270°

GO ON TO THE NEXT PAGE ➤

Questions 58 and 59 refer to the diagram below. A metal sphere *M,* whose mass is 2.0 kilograms, is suspended by a string from point *P,* as in a pendulum. The sphere is released from a vertical height *h* and passes through its lowest position (sometimes referred to as its *rest position*) with a speed of 10 meters per second.

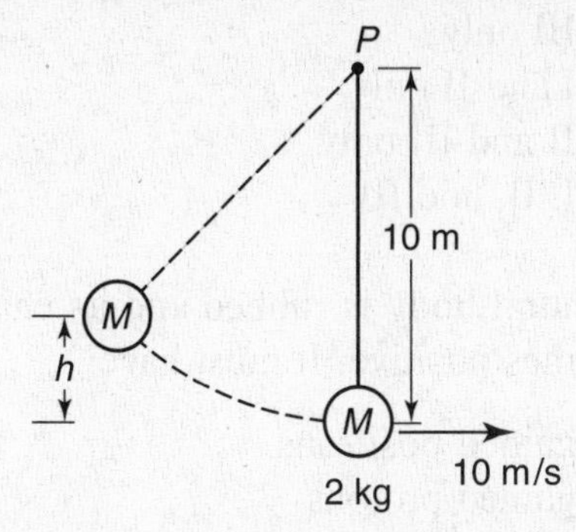

58. The height *h* from which the object was released is approximately

(A) 2.5 m
(B) 5.0 m
(C) 7.0 m
(D) 8.0 m
(E) 9.0 m

59. The centripetal force on object *M* as it passes through the rest position is approximately

(A) 10 N
(B) 20 N
(C) 50 N
(D) 200 N
(E) 2,000 N

60. It is desired to reduce the resistance of a coil of wire to one-half of its value. Which of the following methods will accomplish this?

I. Cut off one-half of the wire.
II. Reduce the coil's temperature to one-half of its value.
III. Double the current.

(A) I only
(B) III only
(C) I and II only
(D) II and III only
(E) I, II, and III

61. An object with a mass of 4.0 kilograms is kept motionless on an inclined plane by a force parallel to the plane. The angle of the incline is 30°; the object is at a vertical height of 10 meters above the bottom; friction is negligible. What is the value of the force, in newtons?

(A) 4.0
(B) 4.9
(C) 9.8
(D) 19.6
(E) 39.2

62. Since light rays are always diverged by concave lenses, such lenses

(A) cannot form images
(B) form only black and white images
(C) form only inverted images
(D) form only erect images
(E) form only magnified images

63. For measuring resistance quickly, the most useful device is the

(A) resistometer
(B) voltmeter
(C) ohmmeter
(D) potentiometer
(E) pyknometer

GO ON TO THE NEXT PAGE ►

64. All of the following pure elements are good electrical conductors EXCEPT

(A) copper
(B) aluminum
(C) silver
(D) silicon
(E) iron

65. All of the following can become good magnets EXCEPT

(A) iron
(B) cobalt
(C) nickel
(D) a solenoid
(E) copper

66. The most common isotope of each of the following elements is radioactive EXCEPT

(A) lead
(B) polonium
(C) radium
(D) radon
(E) uranium

67. All of the following may be used to express a quantity of energy EXCEPT the

(A) joule
(B) kilogram • meter squared per second squared
(C) newton per second
(D) watt-hour
(E) erg

68. All of the following refer to electromagnetic waves EXCEPT

(A) X rays
(B) beta rays
(C) gamma rays
(D) radio waves
(E) red light

69. By which of the following can cathode rays be appreciably deflected?

I. An electric field
II. A nearby stationary magnet
III. A magnetic field

(A) I only
(B) III only
(C) I and II only
(D) II and III only
(E) I, II, and III

70. A neutral body is rubbed and its charge becomes positive. It must have

(A) gained positrons
(B) gained protons
(C) gained neutrons
(D) lost protons
(E) lost electrons

Questions 71–73

Two parallel slits are 1.0×10^{-3} meter apart. The light passing through them has a wavelength of 5.4×10^{-7} meter. It falls on a viewing screen 2.0 meters away from the slits.

71. What is the wavelength of the light, in millimeters?

(A) 5.4×10^{-3}
(B) 5.4×10^{-4}
(C) 5.4×10^{-6}
(D) 5.4×10^{-10}
(E) 5.4×10^{4}

72. Approximately what is the distance, in meters, between the central bright band and the first band next to it?

(A) 1.1×10^{-3}
(B) 2.1×10^{-3}
(C) 2.6×10^{-3}
(D) 3.0×10^{-3}
(E) 3.6×10^{-3}

GO ON TO THE NEXT PAGE ➤

73. Approximately what is the distance between the first and second bright bands compared with the correct answer for question 72?

(A) the same
(B) half as much
(C) twice as much
(D) 2½ times as much
(E) three times as much

---

74. The approximate speed of light in water is, in meters per second,

(A) 5.8×10^8
(B) 1.0×10^8
(C) 2.3×10^8
(D) 2.8×10^8
(E) 3.0×10^8

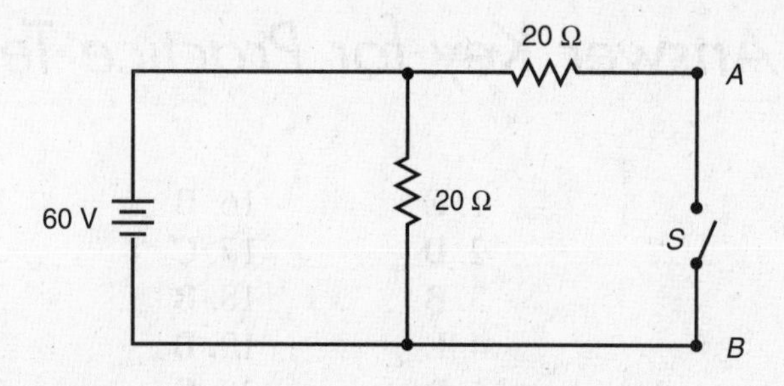

75. When switch *S* is open, as shown in the diagram, what is the difference of potential between points *A* and *B*?

(A) zero
(B) 20 V
(C) 30 V
(D) 40 V
(E) 60 V

Answer Key for Practice Test 3

1. **E**	16. **B**	31. **E**	46. **E**	61. **D**
2. **D**	17. **C**	32. **D**	47. **E**	62. **D**
3. **B**	18. **B**	33. **A**	48. **A**	63. **C**
4. **E**	19. **D**	34. **B**	49. **E**	64. **D**
5. **D**	20. **E**	35. **A**	50. **C**	65. **E**
6. **E**	21. **E**	36. **B**	51. **E**	66. **A**
7. **B**	22. **D**	37. **A**	52. **E**	67. **C**
8. **C**	23. **E**	38. **D**	53. **B**	68. **B**
9. **D**	24. **A**	39. **C**	54. **C**	69. **E**
10. **A**	25. **A**	40. **E**	55. **B**	70. **E**
11. **A**	26. **E**	41. **C**	56. **D**	71. **B**
12. **E**	27. **E**	42. **B**	57. **D**	72. **A**
13. **C**	28. **B**	43. **D**	58. **B**	73. **A**
14. **A**	29. **B**	44. **A**	59. **B**	74. **C**
15. **D**	30. **D**	45. **B**	60. **A**	75. **E**

Answer Explanations for Practice Test 3

1. E As long as there is no sliding, the force of friction is just equal and opposite to the component of the weight parallel to the plane. This component is zero when the plane is horizontal and steadily increases when the angle of incline increases: component parallel to the plane is equal to the weight times the height of the plane divided by the length of the plane.

2. D The normal force is the component perpendicular to the board. It is equal to the weight when the plane is horizontal and steadily decreases as the angle of incline increases: component perpendicular to the plane is equal to the weight multiplied by the base divided by the length of the plane.

3. B The gravitational pull on the object is practically Earth's pull on it toward the center of Earth. It is the weight of the object (70 N) even when it is supported.

4. E Convert all measurements to meters and compare. The smallest is 0.034 m.

5. D Since all measurements have been converted to meters, compare. The largest is 0.731 m.

6. E The proton has a single positive charge equal to that of the positron. The alpha particle has twice this charge. The charge of the electron is negative. Neutrons may be positive, negative, or neutral. Photons have no charge.

7. B The positron is the antiparticle of the electron. It has the same mass as the electron, but an equal and opposite charge.

8. C Throughout its upward motion the ball is slowed down by the acceleration due to gravity. Since the final speed at the top is momentarily zero, the initial speed is equal to the product of the acceleration due to gravity and the time to reach the top:

$$\mathbf{v} = \mathbf{g}t = (9.8 \text{ m/s}^2)(4.0 \text{ s}) = 39.2 \text{ m/s}$$

9. D One method we can use to calculate the distance traveled is to multiply the average speed by the time of travel:

$$\text{distance} = (\tfrac{1}{2} \times 39.2 \text{ m/s})(4 \text{ s}) = 78 \text{ m}$$

10. A The *displacement* of an object is the change in its position. It is a vector quantity and is the distance from the starting point to the end point (in the direction from the first point to the second one). Here the two points coincide; therefore, the displacement is zero.

11. A The *kinetic energy* of an object is equal to one-half of the product of its mass and the square of its speed. Here the mass and speed do not change. Therefore the kinetic energy does not change, either.

12. E One watt is equivalent to 1 J/s. If energy is supplied at the rate of 10 J/s, then, in 2 s, 20 J are supplied.

13. C The momentum of an object is equal to the product of its mass and its velocity:

$$\text{momentum} = \text{mass} \times \text{velocity}$$
$$12 \text{ kg} \bullet \text{m/s} = (6.0 \text{ kg})(\mathbf{v})$$
$$\mathbf{v} = 2 \text{ m/s}$$

14. A In metals, some electrons are only loosely held by the atoms. When a difference of potential is applied to the wire, these electrons are free to move in the direction opposite to the electric field. The rest of the atom is fixed in position.

15. D The solid salts have a crystalline structure consisting of positive and negative ions. When the salts are dissolved in a liquid such as water, they dissociate and the negative ions are free to move toward a positive electrode, and the positive ions toward a negative electrode.

16. B If three forces acting on a particle are in equilibrium, any one force is the equilibrant of the other two. Therefore, the 6-N force is the equilibrant of the 8-N and 10-N forces. The equilibrant is equal and opposite to the resultant, which is also 6 N.

17. C *Joules* are units of energy. Since PE = mgh, the units that represent a joule are $(\text{kg})(\text{m/s}^2)(\text{m}) = (\text{kg})(\text{m}^2/\text{s}^2)$.

18. B *Acceleration* is defined as the change in velocity divided by the time in which the change takes place:

$$\mathbf{a} = \Delta\mathbf{v}/\Delta t$$

Car *A* starts from rest: its initial speed is zero. At the end of the 60 s, its speed, as read from the graph, is 60 m/s. Therefore, its acceleration is

$$\mathbf{a} = (60 \text{ m/s})/(60 \text{ s}) = 1 \text{ m/s}^2$$

19. D For an object starting from rest and moving with constant acceleration, the distance traveled is equal to one-half the product of its acceleration and the square of the time traveled:

$$\text{distance} = \tfrac{1}{2}\mathbf{a}t^2$$
$$= \tfrac{1}{2}(1\ \text{m/s}^2)(60\ \text{s})^2 = 1{,}800\ \text{m}$$

20. E We see from the graph that Car *B* is traveling at a constant speed of 30 m/s. For objects moving at constant speed, the distance traveled is equal to the product of that speed and the length of time traveled:

$$\text{distance} = \text{speed} \times \text{time} = 30t$$

In this question, Car *A* covers the same distance as Car *B* in the same time. We set the two expressions for distance equal to each other.

$$\tfrac{1}{2}\mathbf{a}t^2 = 30t$$

We recall that $\mathbf{a} = 1\ \text{m/s}^2$. Making that substitution, and dividing both sides by *t*, we get $\tfrac{1}{2}t = 30$. Therefore $t = 60$ s.

21. E Probably the easiest way to answer this question is to examine the graph and to make use of the fact that, on a graph of speed against time, the distance traveled is given by the area under the graph. Let us consider the first two choices. Car *B* travels at a constant speed. In each 30-s interval, the area under the graph is 9 squares. For Car *A* we have to examine 20-s intervals. We notice quickly that for the first two intervals the area is small: two squares for the first one, and 5½ for the second. For the third interval the area adds up to 10 squares, representing the greatest distance traveled.

22. D Acceleration is the same and constant for all falling objects, if air resistance is negligible. The speed of a falling object starting from rest is proportional to the acceleration and the time of fall, $\mathbf{v} = \boldsymbol{a}t$, and both are independent of the mass of the object. The *potential energy* of the object is proportional to its mass. Only II and III are correct.

23. E Since heat is a form of energy, any unit of energy may be used to describe the quantity of heat.

24. A Horizontal motion does not affect vertical motion. The acceleration remains constant for vertical motion and has the value of *g,* which is $9.8\ \text{m/s}^2$.

25. A Vertical motion does not affect horizontal motion. The speed in the horizontal direction remains constant: $\mathbf{d} = \mathbf{v}t$; $\mathbf{d} = 30\ \text{m/s} \times 5\ \text{s} = 150$ m.

26. E The pull of the string on the stone is the centripetal force. This force is proportional to the square of the speed of the object moving in the circle.

27. E. Momentum is conserved. Since the 1.2-kg cart is three times the mass of the other one, the more massive cart must have a speed that is one-third that of the 0.4-kg cart. Then, the lighter cart has a speed of 0.6 m/s, the heavier cart's speed must be 0.2 m/s.

28. B This is an example of Newton's third law of motion: action equals reaction. The force exterted by the box is 20 N.

29. B The temperature changes are the same.

30. D Kelvin units = °C + 273° = 300° + 273° = 573 K.

31. E The elastic potential energy of a spring is proportional to the square of its elongation. To double the energy, the elongation must be increased by a factor of $\sqrt{2}$.

32. D The units for force constant are N/m. Since 1 N = 1 kg • m/s^2, the equivalent units for force constant must be kg/s^2.

33. A Use:

$$\mathbf{F} = m\mathbf{a} = m\frac{\Delta \mathbf{v}}{\Delta t} = 2{,}500 \text{ kg}\left(\frac{30 \text{ m/s}}{15 \text{ s}}\right) = 5{,}000 \text{ N}$$

34. B Work done by effort force = effort force × distance moved by effort force in direction of force: 5 N × 20 m = 100 J.

35. A Potential energy (with respect to ground) = weight × vertical distance (above ground). The vertical distance is the same for both objects; therefore the heavier object has the greater potential energy. The presence of air has nothing to do with it.

36. B As a gas is heated at constant pressure, the kinetic energy of the molecules increases and the temperature rises. Perhaps you were misled by choice (A). According to Charles' law, the volume (not the volume change) is proportional to the *absolute* temperature of the gas.

37. A When capacitors are connected in parallel, the total capacitance is equal to the numerical sum of the capacitances. In this case, the total capacitance equals 25 F. Note that this is the *opposite* of how resistors are combined algebraically!

38. D Use:

$$P = VI = (120 \text{ V})(4 \text{ A}) = 480 \text{ W}$$

39. C Across the 20-Ω resistor the voltage is 80 V. Since current is equal to the ratio of voltage to resistance, it can be calculated:

$$I = \frac{80 \text{ V}}{20 \ \Omega}$$

$$= 4 \text{ A}$$

In a series circuit, the current is the same everywhere. The 4 A also go through the 60-Ω resistor.

40. E Throughout the stone's motion, the only force acting on it is Earth's gravitational pull *downward,* since air resistance is negligible. According to Newton's second law of motion, the acceleration is in the direction of the unbalanced force.

41. C The calculation for a concave spherical mirror is similar to the calculation for a convex lens: The reciprocal of the object distance plus the reciprocal of the image distance is equal to the reciprocal of the focal length:

$$\frac{1}{d_o} + \frac{1}{d_i} = \frac{1}{f}$$

The focal length of a spherical mirror is one-half of its radius:

$$f = \frac{r}{2}$$

$$= \frac{12\,\text{cm}}{2}$$

$$= 6\ \text{cm}$$

Substituting in the above equation gives:

$$\frac{1}{10\ \text{cm}} = \frac{1}{6\ \text{cm}}$$

$$d_i = 15\ \text{cm}$$

42. B Wavelength = speed/frequency. At the same temperature, all sound waves have the same speed. Since the wavelength is the same, the frequency is also the same.

43. D The number of beats per second equals the difference between the two frequencies.

44. A The force between two point charges, or two charged conducting spheres, varies inversely as the square of the distance between them. When the distance between them increases, the electrostatic force between them decreases. When the distance between them becomes twice as great, the force between them becomes one-fourth as much:

$$(\tfrac{1}{2})^2\ (4.0 \times 10^{-4}\ \text{N}) = 1.0 \times 10^{-4}\ \text{N}$$

45. B When a wire is moved at right angles to a magnetic field, the voltage induced in the wire is equal to the product of the flux density, the length of wire in the field, and the speed with which the wire is moved:

$$V = \mathbf{B}l\mathbf{v}$$

$$= (10.\ \text{T})(0.03\ \text{m})(2.0\ \text{m/s})$$

$$= 0.6\ \text{V}$$

46. E All images produced by plane mirrors, concave lenses, and convex mirrors are virtual. A virtual image may be obtained with a convex lens if the object is placed less than a focal length from the lens.

47. E The *power* consumed by an electric device is given by the relationship

$$P = V^2/R$$

where V is the voltage applied to the device, and R is the resistance of the device. Substituting, we get

$$1{,}500\ \text{W} = (120\ \text{V})^2/R$$

$$R = 9.6\ \Omega$$

According to Ohm's law, the current is

$$I = V/R$$

$$= (120\ \text{V})/9.6\ \Omega$$

$$= 12.5\ \text{A}$$

48. A Isotopes of the same element differ from each other in the number of neutrons in the nucleus.

49. E The $^4_2$He nucleus contains two protons and two neutrons; the total mass is a tiny fraction of a gram. The reason that the mass of the nucleus is slightly less than the sum of the masses of the individual particles is that significant energy (binging energy) would have to be used to break up the nucleus into separate protons and neutrons. The mass equivalent of this energy accounts for the greater mass of the sum of the isolated particles.

50. C Graphite is a *moderator* material used in nuclear reactors. Moderators slow down the neutrons during fission.

51. E This problem refers to the Compton effect. When a high-energy photon, such as we get from X rays and gamma rays, interacts with an electron, an elastic collision such as occurs between two particles results. Momentum and energy are conserved. The photon loses part of its kinetic energy and momentum to the electron.

52. E Choices (A), (B), (C), and (D) refer to particle emissions. Radon is a radioactive chemical element that emits alpha particles. X rays and gamma rays are both types of electromagnetic wave radiations.

53. B To produce an image the same size as the object, the object must be two focal lengths away from a convex lens. Therefore, $2f = 20$ cm; $f = 10$ cm. As the object gets closer to the lens, the image on the screen becomes larger. However, when the object distance is less than 1 focal length, a real image is no longer obtained.

54. C Other things remaining the same, the greater the amplitude of a sound wave the greater is the loudness.

55. B A sound wave in air consists of compressions and rarefactions traveling in succession away from the source. They travel with the same speed, the speed of the wave.

56. D It takes sound approximately 1 s to travel 331 m in air. In 1 s, light travels 3.0×10^8 m, or 300,000 km, in vacuum.

57. D Two waves are 180° out of phase (or have a phase difference of 180°) when the peak of one coincides with the trough of the other. The two waves then interfere with each other destructively, and the resultant wave has an amplitude equal to the difference in amplitudes of the original waves. In the diagram below, waves IV and V are 180° out of phase. With this phase difference the amplitude of the resultant wave, wave VI, is minimum. If the amplitudes of IV and V were equal, there would be complete annulment of other waves.

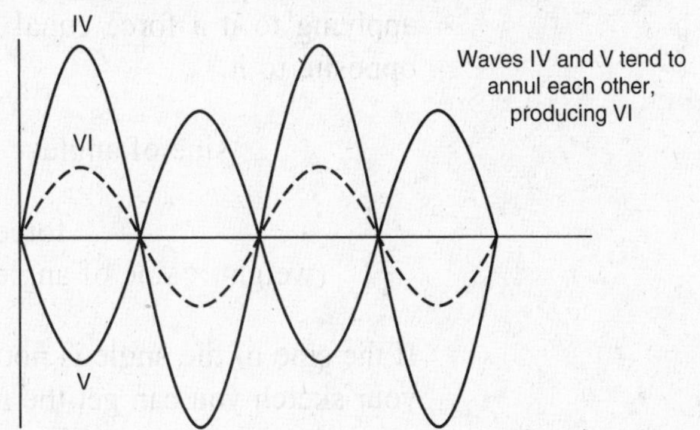

58. B The gain in gravitational potential energy of an object is equal to the product of its weight and the vertical height it is raised above the reference level. The weight of an object is equal to the product of its mass and the value of the acceleration due to gravity: **w** = *mg*. The reference level here is the lowest point during the swing. Therefore, when object *M* is at its highest point, it has a potential energy with respect to the lowest point equal to *mgh*. At the lowest point, all of this has been converted to kinetic energy:

$$E_p = mgh = \tfrac{1}{2}m\mathbf{v}^2$$
$$gh = \tfrac{1}{2}\mathbf{v}^2$$
$$(9.8\ \text{m/s}^2)h = \tfrac{1}{2}(10\ \text{m/s})^2$$
$$h = 5.0\ \text{m}$$

59. B The centripetal force acting on an object is given by the expression $\mathbf{F}_c = m\mathbf{v}^2/r$. Making the substitutions, we obtain

$$\mathbf{F}_c = \frac{(2\ \text{kg})(10\ \text{m/s})^2}{10\ \text{m}}$$
$$= 20\ \text{N}$$

60. A The resistance of a wire is practically independent of the applied voltage and the current through it (except for the change in temperature that may be produced). The resistance of a wire does depend on its temperature but is not proportional to it. The resistance of a wire is proportional to its length.

61. D A quick sketch is helpful. For an inclined plane, the sine of the angle of incline is equal to the ratio of the component of the object's weight parallel to the plane, to the weight of the object.

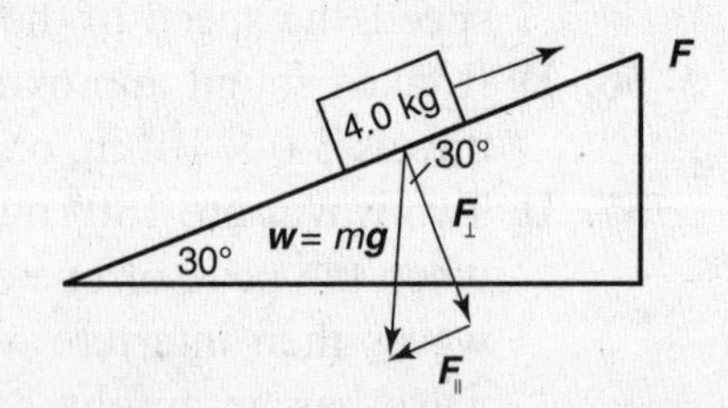

$$\text{sine of angle} = \frac{\text{parallel component}}{\text{weight}}$$

When this friction is negligible, the object can be kept from sliding by applying to it a force equal to this parallel component and directed opposite to it.

$$\text{sine of angle} = \frac{\text{force parallel to plane}}{\text{weight}}$$

force parallel to plane

$$(\text{weight} \times \text{sine of angle}) = (4.0\ \text{kg})(9.8\ \text{m/s}^2)(0.5) \doteq 20\ \text{N}$$

If the sine of the angle is not given, and you don't remember it, from your sketch you can get the ratio: height of plane/length.

62. D Diverging lenses always form erect, *virtual* images; that is, the light rays do not actually travel to the images.

63. C The ohmmeter is commonly used for quick measurements of resistance if high precision is not needed.

64. D Among the solids, metals and graphite are good conductors of electricity. Silicon and germanium are semiconductors.

65. E Iron, nickel, and cobalt are good magnetic elements. A *solenoid* is a metallic coil wound around a magnetic core; when electric current goes through the coil, it becomes a magnet. The greater the current, the stronger is the magnet. Copper is not a magnetic substance.

66. A Radium, radon, and all the isotopes of uranium are radioactive elements. You may not be sure about polonium (it *is* radioactive), but you should know that ordinary or common lead is not radioactive, although lead has some radioactive isotopes.

67. C You should recognize the joule, watt-hour, and erg as units of energy. You should be able to figure out that the unit (or dimension) of choice (B) represents the product of a unit of force ($1\ N = 1\ kg \cdot m/s^2$) and a unit of distance (m); that is, it is a unit of energy. Choice (C) does not represent any special quantity.

68. B Beta rays are streams of electrons. The electron is usually thought of as a particle. (Under certain conditions it shows wave properties.)

69. E Cathode rays consist of moving electrons. Moving electrons are deflected by a magnetic field (which may be provided by a magnet). Moving or stationary electrons are deflected by an electric field.

70. E Ordinary solids are made of neutrons, protons, and electrons. When two different solids are rubbed together, one solid may lose electrons to the other. For example, when a rubber rod is rubbed with a piece of wool, the wool loses electrons to the rod. The rod becomes negatively charged, and the wool becomes positive. The protons and neutrons are tightly held in the nucleus. Positrons are not present in ordinary matter.

71. B The wavelength is given as 5.4×10^{-7} m. One meter equals 10^3 mm.

72. A This question refers to the interference phenomenon produced when monochromatic light goes through two parallel slits and falls on a screen. On the screen is seen a central bright band with equally spaced bright bands on each side of the central band. The distance from the central bright maximum to the first bright band on either side is given by the following relationship:

$$\frac{\lambda}{d} = \frac{x}{L}$$

where λ is the wavelength of the light, d is the distance between the two slits, x is the distance between the central bright maximum and the first bright band next to it, and L is the distance from the slits to the screen. Substituting, we get:

$$\frac{5.4 \times 10^{-7}\ \text{m}}{1.0 \times 10^{-3}\ \text{m}} = \frac{x}{2.0\ \text{m}}$$

$$x = 1.1 \times 10^{-3}\ \text{m}$$

73. A It can be shown that for the above situation the distance between adjacent bright bands is approximately the same as the distance between the central bright band and its adjacent bright band.

74. C The absolute index of refraction of water is the smallest of the values for the common liquids and solids that we encounter. It is 1.33 and is worth remembering. The index of refraction of a substance is equal to the ratio of the speed of light in vacuum to the speed of light in the substance:

$$n = \frac{c}{v}$$

$$1.33 = \frac{3.0 \times 10^8 \text{ m/s}}{v}$$

$$v = 2.3 \times 10^8 \text{ m/s}$$

75. E When the switch is open, there is no current through the 20-Ω resistor. Therefore, there is no voltage drop across the resistor, and the difference of potential between A and B is the same (60 V) as that across the terminals of the battery.

APPENDIX I

Values of the Trigonometric Functions

Angle	Sine	Cosine	Tangent	Angle	Sine	Cosine	Tangent
1°	.0175	.9998	.0175	46°	.7193	.6947	1.0355
2°	.0349	.9994	.0349	47°	.7314	.6820	1.0724
3°	.0523	.9986	.0524	48°	.7431	.6691	1.1106
4°	.0698	.9976	.0699	49°	.7547	.6561	1.1504
5°	.0872	.9962	.0875	50°	.7660	.6428	1.1918
6°	.1045	.9945	.1051	51°	.7771	.6293	1.2349
7°	.1219	.9925	.1228	52°	.7880	.6157	1.2799
8°	.1392	.9903	.1405	53°	.7986	.6018	1.3270
9°	.1564	.9877	.1584	54°	.8090	.5878	1.3764
10°	.1736	.9848	.1763	55°	.8192	.5736	1.4281
11°	.1908	.9816	.1944	56°	.8290	.5592	1.4826
12°	.2079	.9781	.2126	57°	.8387	.5446	1.5399
13°	.2250	.9744	.2309	58°	.8480	.5299	1.6003
14°	.2419	.9703	.2493	59°	.8572	.5150	1.6643
15°	.2588	.9659	.2679	60°	.8660	.5000	1.7321
16°	.2756	.9613	.2867	61°	.8746	.4848	1.8040
17°	.2924	.9563	.3057	62°	.8829	.4695	1.8807
18°	.3090	.9511	.3249	63°	.8910	.4540	1.9626
19°	.3256	.9455	.3443	64°	.8988	.4384	2.0503
20°	.3420	.9397	.3640	65°	.9063	.4226	2.1445
21°	.3584	.9336	.3839	66°	.9135	.4067	2.2460
22°	.3746	.9272	.4040	67°	.9205	.3907	2.3559
23°	.3907	.9205	.4245	68°	.9272	.3746	2.4751
24°	.4067	.9135	.4452	69°	.9336	.3584	2.6051
25°	.4226	.9063	.4663	70°	.9397	.3420	2.7475
26°	.4384	.8988	.4877	71°	.9455	.3256	2.9042
27°	.4540	.8910	.5095	72°	.9511	.3090	3.0777
28°	.4695	.8829	.5317	73°	.9563	.2924	3.2709
29°	.4848	.8746	.5543	74°	.9613	.2756	3.4874
30°	.5000	.8660	.5774	75°	.9659	.2588	3.7321
31°	.5150	.8572	.6009	76°	.9703	.2419	4.0108
32°	.5299	.8480	.6249	77°	.9744	.2250	4.3315
33°	.5446	.8387	.6494	78°	.9781	.2079	4.7046
34°	.5592	.8290	.6745	79°	.9816	.1908	5.1446
35°	.5736	.8192	.7002	80°	.9848	.1736	5.6713
36°	.5878	.8090	.7265	81°	.9877	.1564	6.3138
37°	.6018	.7986	.7536	82°	.9903	.1392	7.1154
38°	.6157	.7880	.7813	83°	.9925	.1219	8.1443
39°	.6293	.7771	.8098	84°	.9945	.1045	9.5144
40°	.6428	.7660	.8391	85°	.9962	.0872	11.4301
41°	.6561	.7547	.8693	86°	.9976	.0698	14.3007
42°	.6691	.7431	.9004	87°	.9986	.0523	19.0811
43°	.6820	.7314	.9325	88°	.9994	.0349	28.6363
44°	.6947	.7193	.9657	89°	.9998	.0175	57.2900
45°	.7071	.7071	1.0000	90°	1.0000	.0000	

APPENDIX II

Summary of Formulas

torque = force × length of moment arm

sum of clockwise torques = sum of counterclockwise torques

$\mathbf{d} = \mathbf{v}_{av}t$

$\mathbf{a} = \frac{\mathbf{v}_f - \mathbf{v}_i}{t}$

$\mathbf{v}_{av} = \frac{\mathbf{v}_f}{2}$

$\mathbf{v}_f = \mathbf{a}t$

$\mathbf{d} = \frac{1}{2}\mathbf{a}t^2$

$\mathbf{v}_f^2 = 2\mathbf{ad}$

$\mathbf{F} = m\mathbf{a}$

$\mathbf{w} = m\mathbf{g}$

$\mathbf{F}t$ = change in momentum

momentum = mass × velocity

$\mathbf{F} = \frac{Gm_1m_2}{r^2}$

work = force × distance; $W = \mathbf{Fd}$

potential energy = $\mathbf{w}h$

kinetic energy = $\frac{1}{2}m\mathbf{v}^2$

energy produced = $m\mathbf{c}^2$

coefficient of friction = $\frac{\text{force of friction during motion}}{\text{normal force}}$

work against friction = friction × distance that object moves

$\frac{\mathbf{F}}{\mathbf{w}} = \frac{h}{l}$

$\mathbf{F} = \mathbf{w} \sin \theta$

$\mathbf{F} = kx$; $\text{P.E.}_s = \frac{1}{2}kx^2$

power = $\frac{\text{work}}{\text{time}}$

$$\text{power} = \frac{\text{force} \times \text{distance}}{\text{time}}$$

$$\text{mechanical advantage (MA)} = \frac{\text{resistance}}{\text{actual effort}}; \text{MA} = \frac{F_R}{F_E}$$

$$\text{work output} = \mathbf{F}_R D_R$$

$$\text{work input} = \mathbf{F}_E D_E$$

$$\frac{\mathbf{F}_R}{\mathbf{w}_E} = \frac{D_E}{D_R} = \text{ideal mechanical advantage (IMA)}$$

$$\text{efficiency} = \frac{\text{work output}}{\text{work input}} = \frac{\text{MA}}{\text{IMA}} = \frac{\text{ideal effort}}{\text{actual effort}}$$

$$\frac{\text{length of plane}}{\text{height of plane}} = \text{IMA} = \frac{\text{weight of object}}{\text{ideal effort}}$$

$$\text{density} = \frac{\text{mass}}{\text{volume}}$$

$$P = \frac{\mathbf{F}}{A}$$

$$\text{IMA} = \frac{\mathbf{F}}{\mathbf{f}} = \frac{A}{a} = \frac{(\text{diameter of large piston})^2}{(\text{diameter of small piston})^2}$$

$$\frac{V_1}{V_2} = \frac{T_1}{T_2}$$

$$P_1 V_1 = P_2 V_2$$

$$\frac{P_1 V_1}{T_1} = \frac{P_2 V_2}{T_2}$$

heat lost by hot object = heat gained by cold object

heat lost (or gained) = mass × sp. heat × temp. change

heat required for melting = mass × H_F

heat required for vaporization = mass × H_v

heat gained (or lost) = mass × sp. heat × temp. change
+ mass melted × heat of fusion
+ mass vaporized × heat of vaporization

$$Q = \Delta U + W$$

$$T = \frac{1}{f}$$

$$v = f\lambda$$

number of beats = difference between two frequencies

$\lambda = 4l_a$

$\lambda = 2l_a$

$\lambda = 2l_s$

decibels = 100 log $\frac{P_1}{P_2}$

$f = \frac{R}{2}$

$$\text{index of refraction} = \frac{\text{sine of the angle in rarer medium}}{\text{sine of the angle in denser medium}}$$

$$= \frac{\text{speed of light in air (or vacuum)}}{\text{speed of light in substance}}$$

$$\frac{1}{\text{object distance}} + \frac{1}{\text{image distance}} = \frac{1}{\text{focal length}}; \quad \frac{1}{d_o} + \frac{1}{d_i} = \frac{1}{f}$$

$$\frac{\text{size of image}}{\text{size of object}} = \frac{\text{image distance}}{\text{object distance}} = \text{magnification } (m)$$

$$\text{telescope magnification} = \frac{\text{focal length of objective}}{\text{focal length of eyepiece}}$$

$$\text{illumination} = \frac{\text{intensity of source}}{(\text{distance})^2}$$

$$\frac{\lambda}{\mathbf{d}} = \frac{x}{L}$$

$$\mathbf{F} = \frac{kq_1q_2}{\mathbf{d}^2}$$

$$\mathbf{E} = \frac{\mathbf{F}}{q}$$

$$V = \frac{\text{work}}{q} \quad \text{or} \quad \text{volts} = \frac{\text{joules}}{\text{coulombs}}$$

$$\mathbf{E} = \frac{V}{\mathbf{d}}$$

$$R = \frac{kL}{A}$$

$V_T = \text{emf} - Ir$

$H = I^2Rt$ (joules)

$$P = VI = I^2R = V^2/R$$

$$\text{energy} = \text{power} \times \text{time}$$

$$\mathbf{F} = IL\mathbf{B}$$

$$\mathbf{F} \propto \frac{I_1 I_2}{d}$$

$$\mathbf{F} = q\mathbf{v}\mathbf{B}$$

$$\text{emf} = I\mathbf{v}\mathbf{B}$$

$$\frac{\text{secondary emf}}{\text{primary emf}} = \frac{\text{number of turns on secondary}}{\text{number of turns on primary}}$$

$$\text{power supplied by secondary} = \text{efficiency} \times \text{power supplied to primary}$$

$$V_s I_s = V_p I_p \times \text{efficiency}$$

$$Q = CV$$

$$\text{P.E.} = \tfrac{1}{2}CV^2 = \tfrac{1}{2}QV = \tfrac{1}{2}\frac{Q^2}{C}$$

$$E_k = hf - \phi$$

$$\phi = hf_0$$

$$p = \frac{h}{\lambda} \quad \text{or} \quad \lambda = \frac{h}{mv}$$

$$E = m\mathbf{c}^2$$

$$M = \frac{M_0}{\sqrt{1 - \left(\frac{\mathbf{v}^2}{\mathbf{c}^2}\right)}}$$

$$L = L_0\sqrt{1 - \left(\frac{\mathbf{v}^2}{\mathbf{c}^2}\right)}$$

$$t = \frac{t_0}{\sqrt{1 - \left(\frac{\mathbf{v}^2}{\mathbf{c}^2}\right)}}$$

APPENDIX III

Review of Mathematics

Algebra

Equations and Relationships

Physical relationships are often expressed as mathematical equations. The techniques of algebra are often used to solve these equations as a part of the process of analyzing the physical world. In general, the letters *u, v, w, x, y,* and *z* are used as variable unknown quantities (with *x, y,* and *z* the most popular choices). The letters *a, b, c, d, e, . . .* often represent constants or coefficients (with *a, b,* and *c* the most popular).

An equation of the form $y = ax + b$ is called a *linear equation,* and its graph is a straight line. The coefficient *a* is the *slope* of the line, and *b* is termed the *y-intercept,* the point at which the line crosses the *y*-axis (using standard Cartsian coordinates). An equation of the form $y = ax^2 + bx + c$ is a *quadratic* equation, and a graph of this relationship is a parabola. The *order* of the equation is the highest power of *x*. The variables *x* and *y* can represent any physical quantities being studied. For example, the following quadratic equation represents the displacement of a particle undergoing one-dimensional uniformly accelerated motion:

$$x = x_0 + v_0 t + \frac{1}{2}at^2$$

In this equation x_0, v_0, and *a* are all constants, while the letters *x* and *t* are the variables (*x* representing the displacement, and *t*, the time).

While there is one solution to a linear equation, a quadratic equation has two solutions. However, in physics, it is possible that only one solution is physically reasonable (e.g., there is no "negative time").

Solutions of Algebraic Equations

A set of point (x, y) is a solution of an algebraic equation if, for each *x,* there is one and only one *y*-value (this is also the definition of a *function*) when the value of *x* is substituted into the equation and the subsequent arithmetical operations are performed. In other words, the equation $3x + 2 = -4$ has only one solution since there is only one variable, *x* (the solution being $x = -2$). The equation $y = 3x + 2$ requires a pair of numbers (x, y) since there are two variables.

Quadratic equations can be solved using the *quadratic formula.* Given the quadratic equation $ax^2 + bx + c = 0$, we can write

$$x = \frac{-b \pm \sqrt{b^2 - 4ac}}{2a}$$

The quantity under the radical sign is called the *discriminant* and can be positive, negative, or zero. If the discriminant is negative, the square root is an *imaginary number* that usually has no physical meaning for our study.

For example, suppose we wish to solve $x^2 + 3x + 2 = 0$ using the quadratic formula. In this equation $a = 1$, $b = 3$, and $c = 2$. Direct substitution and taking the necessary square roots give $x = -1$ or $x = -2$.

Often equations are given in one form and need to be expressed in an alternative form. The rules of algebra allow us to manipulate the form of an equation. For example:

$$\text{Given } x = \left(\frac{1}{2}\right)at^2\text{, solve for } t.$$

First, we clear the fraction by multiplying by 2 to get $2x = at^2$. Now, we divide both sides by a: $2x/a = t^2$. Finally, to solve for t, we take the square root of both sides. Mathematically, there are two solutions. However, in physics, we must allow for the physical reality of a solution. Since this equation represents uniformly accelerated motion from rest, the concept of negative time is not realistic. Hence, we discard the negative square root solution and simply state that $t = \sqrt{2x/a}$.

Exponents and Scientific Notation

Any number, n, can be written in the form of some base number, B, raised to a power, a. The number a is called the exponent of the base number B. In other words, we can write $n = B^a$. One common base number is the number 10. The use of products of numbers with powers of 10 is called *scientific notation.* Some examples of powers of 10 follow: $10 = 10^1$, $100 = 10 \times 10 = 10^2$, $1{,}000 = 10 \times 10 \times 10 = 10^{-3}$. By definition, any number raised to the 0 power is equal to 1. That is, $10^0 = 1$ by definition.

Numbers less than 1 have *negative exponents* since they are fractions of the powers of 10 discussed above. Some examples follow: $0.1 = 10^{-1}$, $0.01 = 10^{-2}$, $0.001 = 10^{-3}$. By definition, any number raised to the 0 power is equal to 1. That is, $10^0 = 1$ by definition.

Numbers less than 1 have *negative exponents* since they are fractions of the powers of 10 discussed above. Some examples follow: $0.1 = 10^{-1}$, $0.01 = 10^{-2}$, $0.001 = 10^{-3}$.

The use of negative exponents for small numbers comes from the *law of division of exponents,* which is defined as follows:

$$\frac{10^a}{10^b} = 10^{a-b}$$

Since a reciprocal means "one over . . ." and 10 raised to the power of 0 is equal to 1, then, in the division example, if $a = 0$ and $b =$ any number, we have the negative exponents for fractions less than 1.

The law of multiplication of exponents is as follows:

$$10^a 10^b = 10^{a+b}$$

$$(10^a)^b = 10^{ab}$$

Scientific notation involves the use of numbers and power of 10 as products. For example, the number 200 can be expressed in scientific notation as 2.0×10^2. The number 3450 can be expressed in scientific notation as 3.45×10^3. Finally, the number 0.045 can be expressed in scientific notation as 4.5×10^{-2}.

On most scientific calculators, scientific notation can be activated using the "exp" button. It is not necessary to press "× 10" (or "times 10") on the calculator. When you press the "exp" button, it automatically implies the "times 10 to the . . ." in the notation.

The following prefixes often appear in physics:

10^{18}	exa-	10^{-1}	deci-
10^{15}	peta-	10^{-2}	centi-
10^{12}	tera-	10^{-3}	milli-
10^{9}	giga-	10^{-6}	micro-
10^{6}	mega-	10^{-9}	nano-
10^{3}	kilo-	10^{-12}	pico-
10^{2}	hecto-	10^{-15}	femto-
10^{1}	deca-	10^{-18}	atto-

Since any number, n, can be written as a base number, B, raised to a certain power, a, we have a process inverse to *exponentiation.* In the designation, $n = B^a$, we can also state that a is the *logarithm* of n with the *base* B: $a = \log_B n$. If we are using base 10, we usually do not write the base number but simply use the designation "log" for what we call the *common logarithm.* For example, the logarithm of 100 is 2 (since 2 is the power 10 is raised to in order to make 100); that is, log 100 = 2. Since logarithms are basically exponents, the rules of multiplication and division of exponents translate to the following:

$$\log (xy) = \log x + \log y$$

$$\log (x/y) = \log x - \log y$$

$$\log (x^n) = n \log x$$

Another common base number is the irrational number e, which is equal to 2.71828. . . . Logarithms to the base e are called *natural logarithms* and are designated by "ln." These logarithms follow the same rules as common logarithms. The use of scientific calculators makes the necessity for extensive logarithm tables obsolete. Consult your own calculator's instructions for using logarithms.

Geometry

Some common formulas from geometry are of use in physics. We first review some of the more common geometric shapes and their equations:

Straight line with slope a and y-intercept b: $y = ax + b$

Circle of radius R centered at the origin: $x^2 + y^2 = R^2$

Parabola whose vertex is at $y = b$: $y = ax^2 + b$

Hyperbola: xy = constant

Ellipse with semimajor axis a and semiminor axis b: $\frac{x^2}{a^2} + \frac{y^2}{b^2} = 1$

We now review some of the physical characteristics of certain shapes:

Circle of radius R: area = πR^2; circumference = $2\pi R$

Sphere of radius R: volume = $\frac{4}{3}\pi R^3$; surface area = $4\pi R^2$

Cylinder of radius R and length l: volume = $\pi R^2 l$

Right circular cone of base radius R and height h: volume = $\frac{1}{3}\pi R^2 h$

Ellipse with semimajor axis a and semiminor axis b: area = πab

Rectangle of length l and width w: area = lw

Triangle with base b and altitude h: area = $\frac{1}{2}bh$

Trigonometry

The branch of mathematics called *trigonometry* involves the algebraic relationships between angles in triangles. If we have a right triangle with hypotenuse c and sides a and b, the relationship between these sides is given by the Pythagorean theorem: $a^2 + b^2 = c^2$.

The ratios of the sides of the right triangle to its hypotenuse define the *trigonometric functions* of sine and cosine.

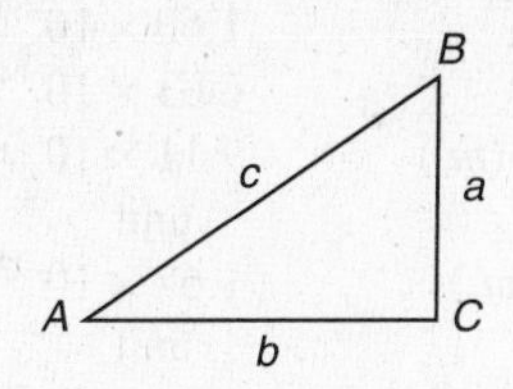

Right triangle

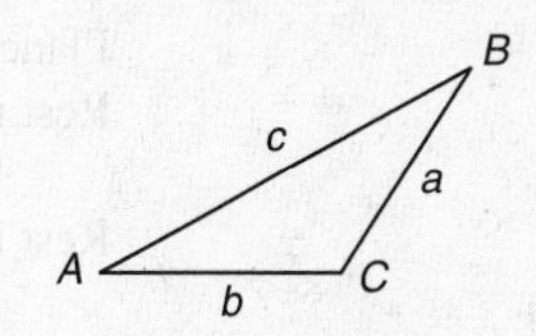

Arbitrary triangle

$$\sin A = \frac{a}{c} \qquad \cos A = \frac{b}{c} \qquad \tan A = \frac{a}{b} \qquad \tan A = \frac{\sin A}{\cos A}$$

$$\sec A = \frac{c}{a} \qquad \csc A = \frac{c}{b} \qquad \cot A = \frac{b}{a}$$

The second set of functions are the reciprocals of the first, main set. For any angle θ in the right triangle, $\sin\theta = \cos(90 - \theta)$ and $\cos\theta = \sin(90 - \theta)$.

In the arbitrary general triangle, the sides of the triangle are related through the law of cosines and the law of sines. These formulas are extremely useful in the study of vectors:

$$\text{law of cosines: } c^2 = a^2 + b^2 - 2ab\cos C$$

$$\text{law of sines: } \frac{a}{\sin A} = \frac{b}{\sin B} = \frac{c}{\sin C}$$

For angles between 0° and 90°, all the trigonometric functions have positive values. For angles between 90° and 180°, only the sine function is positive. For angles between 180° and 270°, only the tangent function is positive. Finally, between 270° and 360° (0°), only the cosine function is positive.

APPENDIX IV

Physics Reference Tables

LIST OF PHYSICAL CONSTANTS

Gravitational constant (G)	6.67×10^{-11} newton • meter^2/kilogram^2
Acceleration due to gravity (g) (near Earth's surface)	9.81 meters/second^2
Speed of light (c)	3.00×10^8 meters/second
Speed of sound at STP	3.31×10^2 meters/second
Mass-energy relationship	1 atomic mass unit = 9.31×10^2 million electron volts
Electrostatic constant	$k = 9.00 \times 10^9$ newtons • meters^2/coulomb^2
Charge of electron = 1 elementary charge	1.60×10^{-19} coulomb
One coulomb	6.25×10^{18} electrons
	6.25×10^{18} elementary charges
Electron volt (eV)	1.60×10^{-19} joule
Planck's constant (h)	6.63×10^{-34} joule • second
Rest mass of the electron (m_e)	9.11×10^{-31} kilogram = 0.0005486 atomic mass unit
Rest mass of the proton (m_p)	1.67×10^{-27} kilogram = 1.007277 atomic mass unit
Rest mass of the neutron (m_n)	1.67×10^{-27} kilogram = 1.008665 atomic mass unit

WAVELENGTHS OF LIGHT IN A VACUUM

Violet	$4.0–4.2 \times 10^{-7}$ meter
Blue	$4.2–4.9 \times 10^{-7}$ meter
Green	$4.9–5.7 \times 10^{-7}$ meter
Yellow	$5.7–5.9 \times 10^{-7}$ meter
Orange	$5.9–6.5 \times 10^{-7}$ meter
Red	$6.5–7.0 \times 10^{-7}$ meter

HEAT CONSTANTS

	Specific Heat (Average) (kJ/kg • C°)*	Melting Point (°C)	Boiling Point (°C)	Heat of Fusion (kJ/kg)*	Heat of Vaporization (kJ/kg)*
Alcohol (ethyl)	2.43 (liq.)	−117	79	109	855
Aluminum	0.90 (sol.)	660	2467	396	10500
Ammonia	4.71 (liq.)	−78	−33	332	1370
Copper	0.39 (sol.)	1083	2567	205	4790
Iron	0.45 (sol.)	1535	2750	267	6290
Lead	0.13 (sol.)	328	1740	25	866
Mercury	0.14 (liq.)	−39	357	11	295
Platinum	0.13 (sol.)	1772	3827	101	229
Silver	0.24 (sol.)	962	2212	105	2370
Tungsten	0.13 (sol.)	3410	5660	192	4350
Water					
Ice	2.05 (sol.)	0	—	334	—
Water	4.19 (liq.)	—	100	—	2260
Steam	2.01 (gas)	—	—	—	—
Zinc	0.39 (sol.)	420	907	113	1770

*Multiply by (1 kilocalorie/4.19 kilojoules) to obtain kilocalories.

ABSOLUTE INDICES OF REFRACTION

($\lambda = 5.9 \times 10^{-7}$ m)

Air	1.00	Carbon tetrachloride	1.46	Glycerol	1.47
Alcohol	1.36	Diamond	2.42	Lucite	1.50
Benzene	1.50	Glass, crown	1.52	Quartz, fused	1.46
Canada balsam	1.53	Glass, flint	1.61	Water	1.33

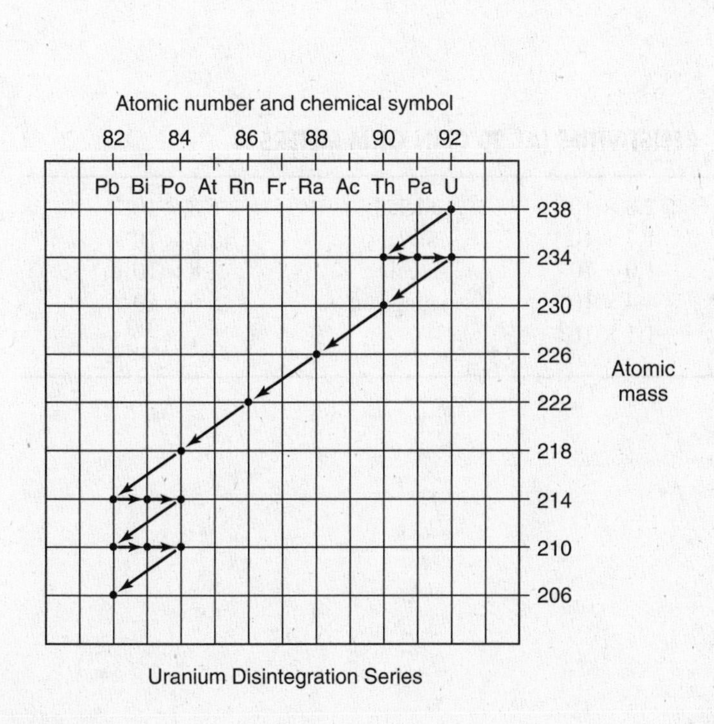

Uranium Disintegration Series

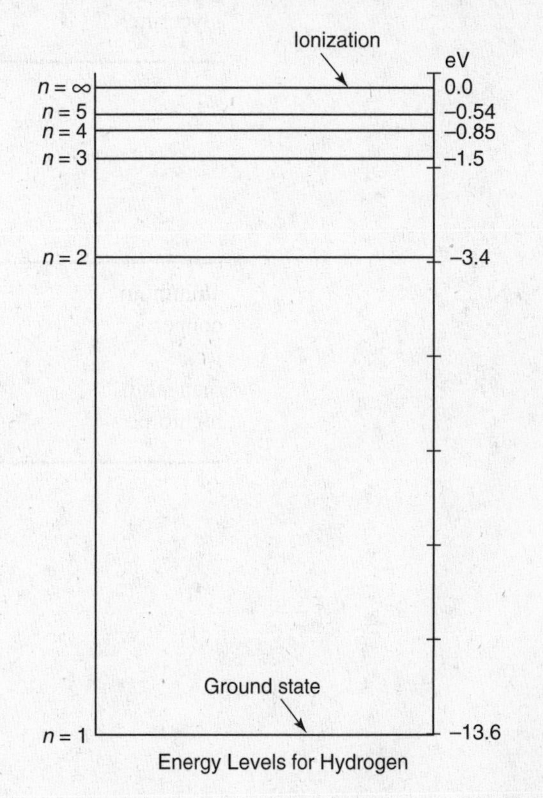

Energy Levels for Hydrogen

APPENDIX V

Densities of Some Substances

(grams/cubic centimeter)

SOLIDS

aluminum	2.7	lead	11.3
brass	8.5	marble	2.6
copper	8.9	paraffin	0.9
cork	0.24	platinum	21.5
glass	2.6	silver	10.5
gold	19.3	steel	7.7
ice	0.92	sulfur	2.0
iron	7.8		

LIQUIDS

alcohol	0.79	kerosene	0.8
carbon tetrachloride	1.6	mercury	13.6
gasoline	0.7	sulfuric acid	1.84
glycerine	1.3	water	1.0

RESISTIVITIES (AT 20°C) IN OHM-METERS

aluminum	2.8×10^{-8}	nickel	6.8×10^{-8}
copper	1.7×10^{-8}	silver	1.6×10^{-8}
iron	1.0×10^{-7}	steel	1.8×10^{-7}
manganin	4.4×10^{-7}	tungsten	5.6×10^{-8}
nichrome	1.1×10^{-6}		

APPENDIX VI

International Atomic Masses*

Based on the atomic mass of C-12 = 12 atomic mass units (amu or u)

Name	Symbol	Atomic Number*	Atomic Mass*	Name	Symbol	Atomic Number*	Atomic Mass*
Actinium	Ac	89	[227]	Holmium	Ho	67	164.93
Aluminum	Al	13	26.98	Hydrogen	H	1	1.01
Americium	Am	95	[243]	Indium	In	49	114.82
Antimony	Sb	51	121.75	Iodine	I	53	126.90
Argon	Ar	18	39.95	Iridium	Ir	77	192.2
Arsenic	As	33	74.92	Iron	Fe	26	55.85
Astatine	At	85	[210]	Krypton	Kr	36	83.80
Barium	Ba	56	137.34	Lanthanum	La	57	138.91
Berkelium	Bk	97	[247]	Lawrencium	Lr	103	[261]
Beryllium	Be	4	9.01	Lead	Pb	82	207.19
Bismuth	Bl	83	208.98	Lithium	Li	3	6.94
Boron	B	5	10.81	Lutetium	Lu	71	174.97
Bromine	Br	35	79.90	Magnesium	Mg	12	24.31
Cadmium	Cd	48	112.40	Manganese	Mn	25	54.94
Caesium	Cs	55	132.91	Meitnerium**	Mt	109	[266]
Calcium	Ca	20	40.08	Mendelevium	Md	101	[258]
Californium	Cf	98	[251]	Mercury	Hg	80	200.59
Carbon	C	6	12.01	Molybdenum	Mo	42	95.94
Cerium	Ce	58	140.12	Neilsbohrium**	Ns	107	[262]
Chlorine	Cl	17	35.45	Neodymium	Nd	60	144.24
Chromium	Cr	24	52.00	Neon	Ne	10	20.18
Cobalt	Co	27	58.93	Neptunium	Np	93	237.05
Copper	Cu	29	63.55	Nickel	Ni	28	58.71
Curium	Cm	96	[247]	Niobium	Nb	41	92.90
Dysprosium	Dy	66	162.50	Nitrogen	N	7	14.01
Einsteinium	Es	99	[254]	Nobelium	No	102	[259]
Erbium	Er	68	167.26	Osmium	Os	76	190.2
Europlum	Eu	63	151.96	Oxygen	O	8	16.00
Fermium	Fm	100	[257]	Palladium	Pd	46	106.4
Flourine	F	9	19.00	Phosphorus	P	15	31.97
Francium	Fr	87	[223]	Platinum	Pt	78	195.09
Gadolinium	Gd	64	157.25	Plutonium	Pu	94	[244]
Gallium	Ga	31	69.72	Polonium	Po	84	[210]
Germanium	Ge	32	72.59	Potassium	K	19	39.10
Gold	Au	79	196.97	Praseodymium	Pr	59	140.90
Hafnium	Hf	72	178.49	Promethium	Pm	61	[145]
Hahnium**	Ha	105	[262]	Protactinium	Pa	91	231.04
Hassia**	Hs	108	[265]	Radium	Ra	88	[226]
Helium	He	2	4.00	Radon	Rn	86	[222]

Name	Symbol	Atomic Number*	Atomic Mass*	Name	Symbol	Atomic Number*	Atomic Mass*
Rhenium	Re	75	186.2	Terbium	Tb	65	158.92
Rhodium	Rh	45	102.91	Thallium	Tl	81	204.37
Rubidium	Rb	37	85.47	Thorium	Th	90	232.03
Ruthenium	Ru	44	101.07	Thulium	Tm	69	168.93
Rutherfordium**	Rf	104	[261]	Tin	Sn	50	118.69
Samarium	Sm	62	150.35	Titanium	Ti	22	47.90
Scandium	Sc	21	44.95	Tungsten	W	74	183.85
Selenium	Se	34	78.96	Unnilhexium**	Unh	106	[263]
Silicon	Si	14	28.09	Uranium	U	92	238.03
Silver	Ag	47	107.87	Vanadium	V	23	50.94
Sodium	Na	11	22.99	Xenon	Xe	54	131.30
Strontium	Sr	38	87.62	Ytterbium	Yb	70	173.04
Sulfur	S	16	32.06	Yttrium	Y	39	88.91
Tantalum	Ta	73	180.95	Zinc	Zn	30	65.37
Technetium	Tc	43	[99]	Zirconium	Zr	40	91.22
Tellurium	Te	52	127.60				

*Most of the listed atomic masses are for naturally occurring mixtures of the isotopes, also known as atomic weights. A value given in brackets is the mass number of the most stable known isotope.

**The names of the elements with atomic numbers of 104 and above have not been officially adopted.

Glossary

Absolute index of refraction: In a transparent material, a number that represents the ratio of the speed of light in vacuum to the speed of light in the material.

Absolute temperature: A measure of the average kinetic energy of the molecules in an object; the Kelvin temperature is equal to the Celsius temperature of an object plus 273.16.

Absolute zero: A temperature of –273°C, designated as 0 K.

Absorption spectrum: A continuous spectrum crossed by dark lines representing absorption of particular wavelengths of radiation by a cooler medium.

Acceleration: A vector quantity representing the time rate of change in velocity.

Activity: The number of radioactive decays of a radioactive atom per unit of time.

Adhesion: The force holding the surfaces of unlike substances together.

Adiabatic: In thermodynamics, the process that occurs when no heat is added to or subtracted from a system.

Alpha decay: The spontaneous emission of a helium nucleus from certain radioactive atoms.

Alpha particle: A helium nucleus ejected from a radioactive atom.

Alternating current: An electric current that varies its direction and magnitude according to a regular frequency.

Ammeter: A device for measuring the current in an electric circuit when placed in series; a galvanometer with a low-resistance coil placed across it.

Ampere: The SI unit of electric current equal to 1 C/s.

Amplitude: The maximum displacement of an oscillating particle, medium, or field relative to its rest position.

Angle of incidence: The angle between a ray of light and the normal to a surface that the ray intercepts.

Angle of reflection: The angle between a reflected light ray and the normal to a reflecting surface.

Angle of refraction: The angle between the emerging light ray in a transparent material and the normal to the surface.

Angström: A unit of length measurement equal to 1×10^{-10} m.

Antinodal lines: A region of maximum displacement in a medium where waves interact with each other.

Aphelion: In the solar system, the point in a planet's orbit when it is farthest from the Sun.

Armature: The rotating coil of wire in an electric motor.

Atomic mass unit: A unit of mass equal to one-twelfth the mass of a carbon-12 atom.

Atomic number: The number of protons in an atom's nucleus.

Atmosphere: A unit of pressure defined at sea level to be equal to 101 kPa.

Average speed: A scalar quantity equal to the ratio of the total distance to the total elapsed time.

Avogadro's number: The number of molecules in 1 mol of an ideal gas, equal to 6.02×10^{23}; named for Italian scientist Amadeo Avogadro.

Back emf: The induced potential difference in the armature of an electric motor that opposes the applied potential difference.

Balmer series: In a hydrogen atom, the visible spectral emission lines that correspond to electron transitions from higher excited states to lower level 2.

Battery: A combination of two or more electric cells.

Beats: The interference caused by two sets of sound waves with only a slight difference in frequency.

Beta decay: The spontaneous ejection of an electron from the nuclei of certain radioactive atoms.

Beta particle: An electron that is spontaneously ejected by a radioactive nucleus.

Binding energy: The energy required to break apart an atomic nucleus; the energy equivalent of the mass defect.

Boyle's law: A principle stating that, for an ideal gas at constant temperature, the pressure of the gas varies inversely with its volume; named for English scientist Robert Boyle.

Bright-line spectrum: The display of brightly colored lines on a screen or in a photograph indicating the discrete emission of radiation by a heated gas at low pressure.

Capacitance: The ratio of the total charge to the potential difference in a capacitor.

Capacitor: A pair of conducting plates, with either a vacuum or an insulator between them, used in an electric circuit to store charge.

Capillarity: The action of liquid rising through small tube openings.

Carnot cycle: In thermodynamics, a sequence of four phases experienced by an ideal gas confined in a cylinder with a movable piston (a Carnot engine); includes an isothermal expansion, an adiabatic expansion, an isothermal compression, and an adiabatic compression; named for French physicist Sadi Carnot.

Cartesian coordinate system: A set of two or three mutually perpendicular reference lines called *axes*, usually designated as *x, y,* and *z,* used to define the location of an object in a frame of reference; named for French scientist Rene Descartes.

Cathode ray tube: An evacuated gas tube in which a beam of electrons is projected and their energy produces an image on a fluorescent screen; they are deflected by external electric or magnetic fields.

Celsius temperature scale: A metric temperature scale in which, at sea level, water freezes at 0°C and boils at 100°C.

Center of curvature: A point that is equidistant from all other points on the surface of a spherical mirror; a point equal to twice the focal length of a spherical mirror.

Center of mass: The weighted mean point where all the mass of an object can be considered to be located; the point at which, if a single force is applied, purely translational motion will result.

Centripetal acceleration: The acceleration of mass moving uniformly in a circle at a constant speed directed radially inward toward the center of the circular path.

Centripetal force: The deflecting force, directed radially inward toward a given point, that causes an object to follow a circular path.

Chain reaction: In nuclear fission, the reaction of neutrons splitting uranium nuclei and creating more neutrons that continue the self-sustained process.

Charles' law: In thermodynamics, a principle stating that at constant pressure the volume of an ideal gas varies directly with its absolute temperature; named for Jacques Charles, French physicist.

Chromatic aberration: In optics, the defect in a converging lens that causes the dispersion of white light into a continuous spectrum and results in the lens refracting the colors to different focal points.

Coefficient of friction: The ratio of the force of friction to the normal force when one surface is sliding (or attempting to slide) over another surface.

Coherent: Describing a set of waves that have the same wavelength, frequency, and phase.

Cohesion: The ability of liquid molecules to stick together.

Commutator: A split ring in a dc motor whose segments are connected to opposite terminals of a battery; as the armature turns, the split in the rings acts to reverse the direction of the current every half-revolution.

Component: One of two mutually perpendicular vectors that lie along the principal axes in a coordinate system and can be combined to form a given resultant vector.

Concave lens: A lens that causes parallel rays of light to emerge in such a way that they appear to diverge away from a focal point behind the lens.

Concave mirror: A spherical mirror that causes parallel rays of light to converge to a focal point in front of the mirror.

Concurrent forces: Two or more forces that act at the same point at the same time.

Conductor: A substance, usually metallic, that allows the relatively easy flow of electric charges.

Conservation of energy: A principle stating that, in an isolated system, the total energy remains the same during all interactions within the system.

Conservation of mass-energy: A principle stating that, in the conversion of mass into energy or energy into mass, the total mass-energy remains the same.

Conservation of momentum: A principle stating that, in the absence of any external forces, the total momentum of an isolated system remains the same.

Conservative force: A force whose work is independent of the path taken; when work is done by this force, the work can be recovered without any loss.

Constructive interference: The additive result of two or more waves interacting with the same phase relationship as they move through a medium.

Continuous spectrum: A continuous band of colors, consisting of red, orange, yellow, green, blue, and violet, formed by the dispersion or diffraction of white light.

Control rod: A device (usually made of cadmium) used in a nuclear reactor to control the rate of fission.

Converging lens: A lens that causes parallel rays of light incident on its surface to be refracted and converge to a focal point; a convex lens.

Converging mirror: A spherical mirror that causes parallel rays of light incident on its surface to reflect and converge to a focal point; a concave mirror.

Convex lens: A converging lens.

Convex mirror: A diverging mirror.

Coordinate system: A set of reference lines, not necessarily perpendicular, used to locate the position of an object within a frame of reference using the rules of analytic geometry.

Core: The interior of a solenoid electromagnet usually made of a ferromagnetic material; the part of a nuclear reactor where the fission reaction occurs.

Coulomb: The SI unit of electric charge, defined to be the amount of charge 1 A of current delivers each second while passing a given point; named for French physicist Charles de Coulomb.

Coulomb's law: A principle stating that the electrostatic force between two point charges is directly proportional to the product of the charges and inversely proportional to the square of the distance separating them.

Critical angle of incidence: The angle of incidence with a transparent substance such that the angle of refraction is 90° relative to the normal drawn to the surface.

Current: A scalar quantity that measures the amount of charge passing a given point in an electric circuit each second.

Current length: A relative measure of the magnetic field strength produced by a length of wire carrying current and equal to the product of the current and the length.

Cycle: One complete sequence of periodic events or oscillations.

Damping: The continuous decrease in the amplitude of mechanical oscillations due to a dissipative force.

Decibel: A logarithmic unit of sound intensity defined so that 0 dB = 1.0×10^{-12} W/m^2.

Deflecting force: Any force that acts to change the direction of motion of an object.

Derived unit: Any combination of fundamental physical units.

Destructive interference: The result produced by the interaction of two or more waves with opposite phase relationships as they move through a medium.

Deuterium: An isotope of hydrogen containing one proton and one neutron in its nucleus; heavy hydrogen (a component of heavy water).

Dielectric: An electric insulator placed between the plates of a capacitor to alter its capacitance.

Diffraction: The spreading of waves around or past obstacles.

Diffraction grating: A reflecting or transparent surface with many thousands of lines ruled on it that is used to diffract light into a spectrum.

Direct current: Electric current that is moving in only one direction around an electric circuit.

Dispersion: The separation of light into its component colors or spectrum.

Dispersive medium: Any medium that produces a dispersion of light; any medium in which the velocity of a wave depends on its frequency.

Displacement: A vector quantity that determines the change in position of an object by measuring the straight-line distance and direction from the starting point to the ending point.

Dissipative force: Any force, such as friction, that removes kinetic energy from a moving object; a nonconservative force.

Distance: A scalar quantity that measures the total length of a path taken by a moving object.

Diverging lens: A lens that causes parallel rays of light incident on its surface to be refracted and to diverge away from a focal point on the incoming light side of the lens; a concave lens.

Diverging mirror: A spherical mirror that causes parallel rays of light incident on its surface to be reflected and diverge away from a focal point on the center-of-curvature side of the mirror.

Doppler effect: The apparent change in the wavelength or frequency of a wave as the source of the wave moves relative to an observer.

Dynamics: The branch of mechanics that studies the effect of forces on objects.

Effective current: For an ac circuit, a measure of the root mean square (i.e., the square root of the average of the square) of the time-varying current; equal to 0.707 times the maximum current.

Effective voltage: For an ac circuit, a measure of the root mean square of the time-varying voltage; equal to 0.707 times the maximum voltage.

Efficiency: In a machine, the ratio of work output to work input; in a transformer, the ratio of the power from the secondary coil to the power into the primary coil.

Elastic collision: A collision between two objects in which there is a rebounding with no loss of kinetic energy.

Elastic potential energy: The energy stored in a spring when work is done to stretch or compress it.

Electric cell: A chemical device for generating electricity.

Electric circuit: A closed conducting loop consisting of a source of potential difference, conducting wires, and other devices that operate on electricity.

Electric field: The region where an electric force is exerted on a charged object.

Electric field intensity: A vector quantity that measures the ratio of the magnitude of the force to the magnitude of the charge on an object.

Electric motor: A coil of wire carrying current that is caused to rotate in an external magnetic field because of the alternation of electric current.

Electrical ground: The passing of charges to or from Earth to establish a potential difference between two points.

Electromagnet: A coil of wire wrapped around a ferromagnetic core (usually made of iron) that generates a magnetic field when current is passed through it.

Electromagnetic field: The field produced by an electromagnet or moving electric charges.

Electromagnetic induction: The production of a potential difference in a conductor as a result of the relative motion between the conductor and an external magnetic field; the production of a potential difference in a conductor as a result of the change in an external magnetic flux near the conductor.

Electromagnetic spectrum: The range of frequencies covering the discrete emission of energy from oscillating electromagnetic fields; included are radio waves, microwaves, infrared waves, visible light, ultraviolet light, X rays, and gamma rays.

Electromagnetic wave: A wave generated by the oscillation of electric charges, producing interacting electric and magnetic fields that oscillate in space and travel at the speed of light in a vacuum.

Electromotive force (emf): the potential difference caused by the conversion of different forms of energy into electric energy; the energy per unit charge.

Electron: A negatively charged particle that orbits a nucleus in an atom; the fundamental carrier of negative electric charge.

Electron capture: The process in which an electron is captured by a nucleus possessing too many neutrons with respect to protons; K capture.

Electron cloud: A theoretical probability distribution of electrons around a nucleus due to the uncertainty principle, the most probable location for an electron is the densest regions of the cloud.

Electronvolt: The energy associated with moving a one-electron charge through a potential difference of 1 V; equal to 1.6×10^{-19} J.

Electroscope: A device for detecting the presence of static charges on an object.

Elementary charge: The fundamental amount of charge of an electron.

Emission spectrum: The discrete set of colored lines representing electromagnetic energy produced when atomic compounds are excited into emitting light as a result of heat, sparks, or atomic collisions.

Energy: A scalar quantity representing the capacity to work.

Energy level: One of several regions around a nucleus where electrons are considered to reside.

Entropy: The degree of randomness or disorder in a thermodynamic system.

Equilibrant: The force equal in magnitude and opposite in direction to the resultant of two or more forces that bring a system into equilibrium.

Equilibrium: The balancing of all external forces acting on a mass; the result of a zero vector sum of all forces acting on an object.

Escape velocity: The minimum initial velocity of an object away from a planet at which the object will not be pulled back toward the planet.

Excitation: The process by which an atom absorbs energy and causes its orbiting electrons to move to higher energy levels.

Excited state: The situation in an atom in which its orbiting electrons are residing in higher energy levels.

Farad: A unit of capacitance equal to 1 C/V.

Faraday's law of electromagnetic induction: A principle stating that the magnitude of the induced emf in a conductor is equal to the rate of change in the magnetic flux; named for English scientist Michael Faraday.

Ferromagnetic substances: Metals or compounds made of iron, cobalt, or nickel that produce very strong magnetic fields.

Field: A region characterized by the presence of a force on a test body, such as a unit mass in a gravitational field or a unit charge in an electric field.

Field intensity: A measure of the force exerted on a unit test body; force per unit mass; force per unit charge.

First law of thermodynamics: The law of conservation of energy as applied to thermodynamic systems.

Fission: The splitting of a nucleus into two smaller and more stable nuclei by means of a slow-moving neutron, which releases a large amount of energy.

Fluid: A form of matter that assumes the shape of its container; includes liquids and gases.

Flux: A measure of the product of the perpendicular component of a magnetic field crossing an area and the magnitude of the area (in webers).

Flux density (magnetic): A measure of the field intensity per unit area (in webers per square meter).

Focal length: The distance along the principal axis from a lens or spherical mirror to the principal focus.

Focus: The point of convergence of light rays caused by a converging mirror or lens; either of two fixed points on an ellipse that determine its shape.

Force: A vector quantity that corresponds to any push or pull due to an interaction that changes the motion of an object.

Forced vibration: Vibrations caused by the application of an external force.

Frame of reference: A point of view consisting of a coordinate system in which observations are made.

Frequency: The number of completed periodic cycles per second in an oscillation or wave motion.

Friction: A force that opposes the motion of an object as it slides over another surface.

Fuel rods: Rods packed with fissionable material that are inserted into the core of a nuclear reactor.

Fundamental unit: An arbitrary scale of measurement assigned to any of certain physical quantities such as length, time, mass, and charge, that are considered to be the basis for all other measurements.

Fusion: The combination of two or more light nuclei that produces a more stable heavier nucleus with the release of energy.

Galvanometer: A device used to detect the presence of small electric currents when connected in series in a circuit.

Gamma radiation: High-energy photons emitted by certain radioactive substances.

Gravitation: The mutual force of attraction between two uncharged masses.

Gravitational field strength: A measure of the gravitational force per unit mass in a gravitational field.

Gravity: Another name given to gravitation or gravitational force; the tendency of objects to fall to Earth.

Ground state: The lowest energy level in an atom.

Half-life: The time required for a radioactive material to decay to one half its original amount.

Heat: The net disordered kinetic energy associated with molecular motion in matter.

Heat of fusion: The amount of energy per kilogram needed to change the state of matter from a solid to a liquid or from a liquid to a solid.

Heat of vaporization: The amount of energy per kilogram needed to change the state of matter from a liquid to a gas or from a gas to a liquid.

Hertz: The SI unit of frequency equal to 1 s^{-1}; the number of cycles/s in a vibration.

Hooke's law: A principle stating that the stress applied to an elastic material is directly proportional to the strain produced; named for English scientist Robert Hooke.

Ideal gas: A gas for which the assumptions of the kinetic theory are valid.

Image: An optical reproduction of an object by means of a lens or mirror.

Impulse: A vector quantity equal to the product of the average force applied to a mass and the time interval over which the force acts; the area under a force versus time curve.

Induced potential difference: A potential difference created in a conductor as a result of its motion relative to an external magnetic field.

Inductance: In an electric circuit containing a solenoid, the ratio of the induced emf in the solenoid to the rate of change in current; designated as L and measured in henrys.

Induction coil: A transformer in which a variable potential difference is produced in a secondary coil when a direct current applied to the primary is turned on and off.

Inelastic collision: A collision in which two masses interact with a loss of kinetic energy.

Inertia: The property of matter than resists the action of an applied force trying to change the motion of an object.

Inertial frame of reference: A frame of reference in which the law of inertia holds; a frame of reference moving with constant velocity relative to Earth.

Instantaneous velocity: The slope of a tangent line to a point on a displacement versus time graph; the velocity of a particle at any instant in time.

Insulator: A substance that is a nonconductor of electricity because of the absence of free electrons.

Interference: The interaction of two or more waves producing an enhanced or diminished amplitude at the point of interaction; the superposition of one wave on another.

Interference pattern: The pattern produced by the constructive and destructive interference of waves generated by two point sources.

Isobaric: In thermodynamics, a process in which the pressure of a gas remains the same.

Isochoric: In thermodynamics, a process in which the volume of a gas remains the same.

Isolated system: Two or more interacting objects that are not being acted on by an external force.

Isotopes: Atoms with the same number of protons but different numbers of neutrons.

Joule: The SI unit of work equal to 1 N-m; the SI unit of mechanical energy equal to 1 kg • m^2/s^2.

Junction: The point in an electric circuit where a parallel connection branches off.

K capture: Electron capture.

Kelvin: The unit of absolute temperature, defined so that 0 K = –273°C.

Kepler's first law: A principle stating that the orbital paths of all planets are elliptical with the Sun at one focus; named for German astronomer Johannes Kepler.

Kepler's second law: A principle stating that a line from the Sun to a planet sweeps out equal areas in equal times.

Kepler's third law: A principle stating that the ratio of the cube of the mean radius to the Sun to the square of the period is a constant for all planets orbiting the Sun.

Kilogram: The SI unit of mass.

Kilojoule: A number representing 1,000 J.

Kilopascal: A number representing 1,000 Pa.

Kinematics: In mechanics, the study of how objects move.

Kinetic energy: The energy possessed by a mass as a result of its motion relative to a frame of reference.

Kinetic friction: The friction induced by sliding one surface over another.

Kinetic theory of gases: A theory stating that all matter is made of molecules that are in a constant state of motion.

Kirchhoff's first law: A principle stating that the algebraic sum of all currents at a junction in a circuit equals zero; named for German physicist Gustav Kirchhoff.

Kirchhoff's second law: A principle stating that the algebraic sum of all potential drops around any closed loop in a circuit equals zero.

Laser: A device for producing an intense coherent beam of monochromatic light; an acronym for **l**ight **a**mplification by the **s**timulated **e**mission of **r**adiation.

Lens: A transparent substance with one or two curved surfaces used to direct rays of light by refraction.

Lenz's law: A principle stating that an induced current in a conductor is always in a direction such that its magnetic field opposes the change in the magnetic field that induced it; named for German physicist Heinrich Lenz.

Line of force: An imaginary line drawn in a gravitational electric, or magnetic field that indicates the direction a test particle takes when experiencing a force in that field.

Longitudinal wave: A wave in which the oscillating particles vibrate in a direction parallel to the direction of propagation.

Magnet: An object that exerts a force on ferrous materials or a current.

Magnetic field: A region surrounding a magnet in which ferrous materials or moving charged particles experience a force.

Magnetic field strength: The force on a unit current length in a magnetic field.

Magnetic flux density: The total magnetic flux per unit area in a magnetic field.

Magnetic pole: The region on a magnet where magnetic lines of force are the most concentrated.

Mass: The property of matter used to represent the inertia of an object; the ratio of the net force applied to an object to its subsequent acceleration as specified by Newton's second law of motion.

Mass defect: The difference between the actual mass of a nucleus and the sum of the masses of the protons and neutrons it contains.

Mass number: The total number of protons and neutrons in an atomic nucleus.

Mass spectrometer: A device that uses magnetic fields to cause nuclear ions to assume circular trajectories and thereby determines their masses based on their charge, the radius of the path, and the external field strength.

Mean radius: For a planet, the average distance to the Sun.

Meter: The standard SI unit for length: equal to the distance light travels in 1/299,792,458 s.

Moderator: A material, such as water or graphite, used to slow neutrons in fission reactions.

Mole: The amount of a substance that contains Avogadro's number of molecules.

Moment arm distance: A line drawn from a pivot point.

Momentum: A vector quantity equal to the product of the mass and velocity of a moving object.

Monochromatic light: Light consisting of only one frequency.

Natural frequency: The frequency with which an elastic body will vibrate if disturbed.

Net force: The vector sum of all forces acting on a mass.

Neutron: A subatomic particle, residing in the nucleus, that has a mass comparable to that of a proton but is electrically neutral.

Newton: The SI unit of force equal to 1 kg • m/s^2; named for English mathematician/physicist Sir Isaac Newton.

Newton's first law of motion: A principle stating that an object at rest tends to stay at rest, and that an object in motion tends to stay in motion at a constant velocity, unless acted on by an external force; the law of inertia.

Newton's second law of motion: A principle stating that the acceleration of a body is directly proportional to the applied net force; $\mathbf{F} = ma$.

Newton's third law of motion: A principle stating that for every action force there is an equal but opposite reaction force.

Nodal line: A line of minimum displacement occurring when two or more waves interfere.

Nonconservative force: A force, such as friction, that decreases the amount of kinetic energy after work is done by the force.

Normal: A line perpendicular to a surface.

Normal force: A force that is directed perpendicularly to a surface when two objects are in contact.

Nuclear force: The force between nucleons in a nucleus that opposes the Coulomb force repulsion of the protons; the strong force.

Nucleon: Either a proton or a neutron as it exists in a nucleus.

Ohm: The SI unit of electric resistance equal to 1 V/A; named for German physicist Georg Ohm.

Ohm's law: A principle stating that, in a circuit at constant temperature, the ratio of the potential difference to the current is a constant (called the *resistance*).

Optical center: The geometric center of a converging or diverging lens.

Parallel circuit: A circuit in which two or more devices are connected across the same potential difference, providing two parallel paths for current flow.

Pascal: The SI unit of pressure equal to 1 N/m^2; named for French mathematician Blaise Pascal.

Pascal's principle: A law stating that, in a confined static fluid, any external pressure applied is distributed uniformly throughout the fluid.

Perihelion: The point of closest approach of a planet to the Sun.

Period: The time (in seconds) required to complete one cycle of a repetitive oscillation or uniform circular motion.

Permeability: The property of matter that affects the external magnetic field the material is placed in, causing it to align free electrons within the substance and hence magnetize it; a measure of a material's ability to become magnetized.

Phase: The fraction of a period or wavelength from a chosen reference point on the wave.

Photoelectric effect: The process by which surface electrons in a metal are freed through the incidence of electromagnetic radiation above a certain minimum frequency.

Photon: The quantum of electromagnetic energy.

Planck's constant: A universal constant representing the ratio of the energy of a photon to its frequency; the fundamental quantum of action having a value of 6.63×10^{-34} kg • m^2/s; named for German physicist Max Planck.

Polar coordinate system: A coordinate system in which the location of a point is determined by a vector from the origin of a Cartesian coordinate system and the angle it makes with the positive horizontal axis.

Polarization: The process by which the vibrations of a transverse wave are selected to lie in a preferred plane.

Polarized light: Light that has been polarized by passing it through a suitable polarizing filter.

Positron: A subatomic particle having the same mass as an electron except for a positive electric charge; an antielectron.

Potential difference: The work necessary to move a positive unit charge between any two points in an electric field.

Potential energy (gravitational): The energy possessed by a body as a result of its vertical position relative to the surface of Earth or any other arbitrarily chosen base level.

Power: The ratio of the work done to the time needed to complete the work.

Pressure: The force per unit area.

Primary coil: In a transformer, the coil connected to an alternating potential difference source.

Principal axis: An imaginary line passing through the center of curvature of a spherical mirror or the optical center of a lens.

Prism: A transparent triangular shape that disperses white light into a continuous spectrum.

Pulse: A localized disturbance in an elastic medium.

Quantum: A discrete packet of electromagnetic energy; a photon.

Radian: The angle that intercepts an arc length; a unit of angle measure based on the unit cycle such that $360° = 2\pi$ rad.

Radioactive decay: The spontaneous disintegration of a nucleus as a result of its instability.

Radius of curvature: The distance from the center of curvature to the surface of a spherical shape.

Ray: A straight line used to indicate the direction of travel of a light wave.

Real image: An image formed by a converging mirror or lens that can be focused onto a screen.

Refraction: The bending of a light ray when it passes obliquely from one transparent substance to another in which its velocity is different.

Resistance: The ratio of the potential difference across a conductor to its current.

Resistivity: A property of matter that measures the ability of a substance to act as a resistor.

Resolution of forces: The process by which a given force is decomposed into a pair of perpendicular forces.

Resonance: Vibrations in a body at its natural vibrating frequency, caused by the transfer of energy.

Rest mass: The mass of an object when the object is at rest with respect to the measuring instrument.

Rotational equilibrium: The point at which the vector sum of all torques acting on a rotating mass equals zero.

Scalar: A physical quantity, such as mass or speed, that is characterized by only magnitude or size.

Second: The SI unit of time.

Second law of thermodynamics: A principle stating that heat flows naturally only from a hot body to a cold one; according to this law, the entropy of the universe is always increasing and no process is possible from which the sole result is the complete conversion of heat from a source into work.

Secondary coil: In a transformer, a coil in which an alternating potential difference is induced.

Series circuit: A circuit in which electric devices are connected sequentially in a single conducting loop, allowing only one path for charge flow.

Shunt: A low-resistance resistor connected parallel across a galvanometer, turning it into an ammeter.

Sliding friction: A force that resists the sliding motion of one surface over another; kinetic friction.

Snell's law: A principle stating that, for light passing obliquely from one transparent substance to another, the ratio of the sine of the angle of incidence to the sine of the angle of refraction is equal to the relative index of refraction for the two media; named for Dutch mathematician Willebrord Snellius.

Solenoid: A coil of wire used for electromagnetic induction.

Specific heat: The amount of energy (in kilojoules) needed to change the temperature of 1 kg of a substance by 1°C.

Speed: A scalar quantity measuring the time rate of change in distance.

Spherical aberration: In a spherical mirror or a lens, the inability to properly focus parallel rays of light because of the shape of the mirror or lens.

Spring constant: The force needed to stretch or compress a spring by a unit length; force constant; Young's modulus; modulus of elasticity.

Standard pressure: At sea level, 1 atm, which is equal to 101.3 kPa.

Standing wave: A stationary wave pattern formed in a medium when two sets of waves with equal wavelengths and amplitudes pass through each other, usually after a reflection.

Starting friction: The force of friction overcome when two objects initially begin to slide over one another.

Static electricity: Stationary electric charges.

Static friction: The force that prevents one object from beginning to slide over another.

Statics: The study of the forces acting on an object that is at rest and unaccelerated relative to a frame of reference.

Strain: The deformation in a solid due to the application of a stress.

Stress: A force applied to a solid per unit area.

Superposition: When waves overlap, a condition in which the resultant wave is the algebraic sum of the individual waves.

Surface tension: In a liquid, the tendency of the top layers to behave like a membrane under tension because of adhesive molecular forces.

Temperature: The relative measure of warmth or cold based on a standard; a measure of the average kinetic energy of molecules in matter.

Tesla: The SI unit of magnetic field strength equal to 1 W/m^2; named for American engineer Nikola Tesla.

Test charge: A small, positively charged mass used to detect the presence of a force in an electric field.

Thermionic emission: The emission of electrons from a hot filament.

Thermometer: An instrument that makes a quantitative measurement of temperature based on an accepted scale.

Third law of thermodynamics: A principle stating that the temperature of any system can never be reduced to absolute zero.

Threshold frequency: The minimum frequency of electromagnetic radiation necessary to induce the photoelectric effect in a metal.

Torque: The application of a force at right angles to a designated line, from a pivot, that tends to produce circular motion; the product of a force and a perpendicular moment arm distance.

Total internal reflection: The process by which light is incident on an interface from a higher to a lower index-of-refraction medium in which the velocity of the light increases so as to imply an angle of refraction greater than 90°, reflecting the light back into the original medium.

Total mechanical energy: In a mechanical system, the sum of the kinetic and potential energies.

Transformer: Two coils used to change the alternating potential difference across one of them (the primary) to either a larger or smaller value in the secondary.

Transmutation: The process by which one atomic nucleus changes into another by radioactive decay or nuclear bombardment.

Transverse wave: A wave in which the vibrations of a medium or field are at right angles to the direction of propagation.

Uniform circular motion: Motion around a circle at a constant speed.

Uniform motion: Motion at a constant speed in a straight line.

Unit: An arbitrary scale assigned to a physical quantity for measurement and comparison.

Universal law of gravitation: A principle stating that the gravitational force between any two masses is directly proportional to the product of their masses and inversely proportional to the square of the distance between them.

Vector: A physical quantity characterized by both magnitude and direction; a directed arrow drawn in a Cartesian coordinate system used to represent a quantity such as force, velocity, or displacement.

Velocity: A vector quantity representing the time rate of change in displacement.

Virtual focus: The point at which the rays from a diverging lens would meet if they were traced back with straight lines.

Virtual image: An image formed by a mirror or lens that cannot be focused onto a screen.

Viscosity: In a fluid, the measure of the resistance that adjacent vertical layers have sliding over each other; the internal resistance of a fluid to flowing.

Visible light: The portion of the electromagnetic spectrum (ROYGBIV) that can be detected by the human eye.

Volt: The SI unit of potential difference equal to 1 J/C; named for Italian physicist Alessandro Volta.

Volt per meter: The SI unit of electric field intensity equal to 1 N/C.

Voltmeter: A device used to measure potential difference between two points in a circuit when connected in parallel; a galvanometer with a large resistor placed in series with it.

Watt: The SI unit of power equal to 1 J/s; named for Scottish inventor James Watt.

Wave: A series of periodic disturbances in an elastic medium or field.

Wavelength: The distance between any two successive points in phase on a wave.

Weight: The force of gravity exerted on a mass at the surface of a planet.

Work: A scalar measure of the amount of change in mechanical energy; the product of the magnitude of the displacement of an object and the component of applied force in the same direction as the displacement; the area under a plot of force versus displacement.

Work function: The minimum amount of energy needed to free an electron from the surface of a metal using the photoelectric effect.

Young's modulus: The ratio of the applied stress to the linear strain in an elastic solid; named for English physicist Thomas Young.

Index